Best of AMT Magazine
Airframe Technology
Accessory Technology

Aviation Supplies & Academics, Inc.
Newcastle, Washington

Airframe & Accessory Technology

The Best of Aircraft Maintenance Technology Magazine:
Airframe Technology/Accessory Technology
by Aircraft Maintenance Technology Magazine
(ISSN 1072-3145)

Kathy Marr, Publisher
Greg Napert, Editor

© 1996 *Aircraft Maintenance Technology Magazine*, Johnson Hill Press Inc., a subsidiary of PTN Publishing Company; Stanley S. Sills, chairman and CEO; Richard A. Reiff, president and COO.

The material presented in *Aircraft Maintenance Technology* is intended to complement technical information that is currently available from supplier and regulatory sources such as aircraft manufacturers and the Federal Aviation Administration. Every effort is made to ensure that the information provided is accurate. However, if information presented by *Aircraft Maintenance Technology* is in conflict with supplier and regulatory sources, the latter shall take precedence.

All rights reserved. No part of this publication may be reproduced or transmitted in any form or by any means, electronic or mechanical, including photocopy, recording or any information storage or retrieval system, without written permission from the publisher.

Published 1996 by Aviation Supplies & Academics, Inc.

Printed in Canada

99 98 97 96   9 8 7 6 5 4 3 2 1

ASA-BAMT-A
ISBN 1-56027-270-8

Aviation Supplies & Academics, Inc. (ASA)
7005 132nd Place SE
Newcastle, WA 98059-3153

A special thanks to all contributing columnists:

Airframe Technology
Jim Benson
John Boyce
Gary Eiff
Richard Floyd
Cynthia Foreman
Joseph Hahn
B.J. High
Douglas Latia
Peter S. Lert
Nick Levy
Greg Napert
Jim Sparks
Gerald R. Stoehr

Accessory Technology
John Bakos
Dennis Dryden
Andrew E. Geist
Eric Kornaw
Scott Marvel
Joseph F. Mibelli
Greg Napert
Jeff Rogers
Don Ross
Jim Sparks
Rob Starr
Rudy Swider
Ted Wilmot

Contents

Preface .... *vii*

Airframe Technology

Composite machining *3*
Interior refurbishing *6*
    AMT's interior services and supplies directory *9*
Corrosion detection methods *13*
    Forms of corrosion *15*
Window care and repair *17*
    A short list of certificated window repair companies *20*
Finding solutions for environmental concerns at paint shops is not easy *21*
Pitot-static system testing *23*
Engine-driven fuel pumps *26*
    More fuel pump tips *28*
    What every technician should know about turbine fuel contamination *29*
The importance of cleanliness in hydraulic maintenance *30*
Auto pilot INOP: Oh no! *33*
Large aircraft air conditioning *36*
Keeping corrosion at bay *40*
Going beyond the requirements in a helicopter airframe inspection *43*
Getting a grasp on hydraulic systems *45*
Taking command of composites *49*
Pneumatic system maintenance *51*
    Air pump pointers *54*
This is not your father's fabric *55*
Aircraft fluorescent lighting systems *58*
Lead-acid battery servicing tips *61*
    Technicians can't jump *63*
Helpful tips for handling flexible hose *64*
King Air five-year landing gear inspection tips *68*
Touch-up painting pointers *71*
    Paint compatibility rules *72*
Fire detection/extinguishing systems *73*
Magnetic particle inspection *76*
The VOM and electrical measurement errors *79*
Fuel Quantity Indicating Systems *82*
Non-destructive testing *86*
    Penetrant waste water disposal *89*
The Art of Welding *90*
Autopilots *92*

Continued

Airframe Technology *continued*

Deicer systems *95*
Don't fuel with it *99*
Electrical troubleshooting basics *101*
The pressurized window *104*
    Who's qualified to repair your windows? *107*
Corrosion *108*
    ACF-50 .... *110*
    Types of corrosion *111*
Sheet metal repairs *112*
    Blind riveting guidelines *114*
Oxygen systems *115*
Composite rotor blade inspection and repair *118*
Avionics removal and replacement *121*

Accessory Technology

Troubleshooting aircraft alternator systems *125*
Battery care *129*
    A new approach to battery maintenance *131*
    A closer look at inspecting aircraft windows *134*
Certified vs. qualified welders *136*
    Causes and cures of common welding troubles *138*
Starter-generator overhaul *140*
Phosphate ester-based hydraulic fluids *143*
    Skydrol fluid safe handling summary *148*
Helicopter rotor track and balance *149*
Hydromatic propeller governors *152*
Tire care and maintenance *157*
    A few tire definitions to keep you straight *162*
    Tire inspection criteria *163*
Aviation ignition exciters *165*
    General maintenance tips for turbine ignition systems *167*
Lubricants as tools *169*
Slip sliding away *172*
    Typical escape slide folding sequence *174*
The composite propeller *175*
Troubleshooting the Bendix DP fuel control system *179*
Carbon brake repair *182*
A few D.C. generator basics *185*
    Upgrading to an alternator *187*
Raw data *188*
Ultrasonic testing basics *191*

Airframe & Accessory Technology

A look inside the Bell 206L transmission *193*
    Gulf Coast corrosion protection is part of PHI's maintenance program *195*
Safety wiring basics *196*
    New technology in safety wiring *197*
Do you know Eddy? Current, that is *198*
More than meets the eye *201*
Precision bearing inspection *204*
    Typical bearing defects *206*
The neglected ELT *207*
Rotor track and balance *210*
    Technician's input alters magneto *212*
Nickel-cadmium battery maintenance *213*
Recip engine synchronization *216*
Turbine engine synchronization *219*
Braking tradition *221*
Rotor head inspection tips *224*
Recip engine oil *227*
Hot cooler tips *230*
    The basic overhaul process *231*
The combustion heater *232*
    Run your heater on a Hobbs™ *235*
    Check Mate® expedites troubleshooting *235*
Maintenance of the Precision Airmotive RSA fuel system *236*
Starter-generators *240*
Wheel and brake servicing *243*
Aging props prompt closer attention to maintenance *246*
    McCauley meeting produces common questions and good answers *248*
Starter systems *249*
    Cold weather operation *251*
Vacuum pumps *252*
    Slick's new magneto timing procedure is truly "slick" *253*
    Hot brakes or cool? *254*

Preface

Technical reading material is, by its very nature, "dry." No laughs or chuckles can be heard as a technician reads through the typical A&P training manual, nor can you find the average technician sitting down on the weekend entertaining himself with a good A&P regulations book.

When we started publishing *Aircraft Maintenance Technology Magazine* in 1989, we wanted to provide a technical training curriculum that technicians could use to advance themselves in their profession; yet, we realized that if it were to work, we would have to provide material that was not only technically accurate and relevant to the profession, but fun to read. We feel that we've accomplished this over the years, as our subscribers relate stories of actually reading the material during their leisure hours—and learning something from it.

The fact that the magazine is required reading material in many A&P schools and that the FAA and Transport Canada have approved the material for recurrency training purposes is testament to the fact that the material is also technically relevant.

After being hounded repeatedly for permission to reprint articles and to provide back issues to individuals, and after realizing that much of the material in the magazine was actually timeless, we finally decided to make the material available in book format.

The Best of Aircraft Maintenance Technology Magazine series represents the best articles published during the years 1989 through 1995. It is divided into a three-volume set, which includes the Recip & Turbine Technology Series, Airframe & Accessory Technology Series, and Professional & Legal Series. The articles appear only in the order they appeared in AMT Magazine, not according to subject matter. So read each volume from back to front, front to back, or select only the subject matter you are interested in.

But most importantly, find a nice quite place, and sit back and enjoy.

Happy reading!

Greg Napert
Editor, AMT Magazine

Airframe Technology

Airframe Technology

Composite machining
Basic techniques for working composites

By Cynthia Foreman

Composite materials are quickly becoming recognized as the most advanced substance for fabrication. Composite structures are made from a combination of materials in the form of fabrics, fibers, foams, and honeycomb materials bonded by a matrix or resin system.

The term composite is used to describe two or more materials that are combined to form a much stronger structure than either material by itself. The most simple composite is composed of two elements: a matrix which serves as a bonding substance, and a reinforcing material.

One of the most common problems associated with the use of composites is that there are too few technicians who are trained in the techniques and methods of composite repair. Composites represent new materials and techniques which must be mastered by those persons who want to stay in tune with the repair industry.

It's not difficult to complete an airworthy repair to a composite structure; however, the techniques, materials, and tools which are used are different than those which are used on conventional repairs. If care is not taken to do a composite repair correctly, the repair will not develop the full-strength characteristics that are desirable in a composite structure.

Besides having a good understanding of resins and the bonding process, the drilling, sanding, and cutting of the materials or "machining" is much more critical than most realize. Composite materials act differently than traditional aluminum or other common metals when machined. Each different type of fabric will machine differently, and understanding the interaction between the machining tools and the different fabrics can make a difference in the success of the repair.

Cutting uncured fabrics

Before a fiberglass or carbon/graphite fabric is combined with a matrix and cured, it can be cut with conventional fabric scissors. Aramid fabric or Kevlar®, which is a trade name of DuPont, in its raw state is more difficult to cut, however. Scissors with special steel blades containing serrated or diamond edges are used to cut through aramid. Also desirable for use on aramid fabric are ceramic blades with serrated edges. These scissors will cut through aramid with ease and last many times longer. The serrated edges will hold the fabric and prevent it from sliding, while it cuts without fraying the edges.

Scissors that are used to cut aramid should only be used to cut aramid, never fiberglass or carbon/graphite. The reason for not interchanging scissors is that the different fabrics tend to dull the cutting surface in different ways. Keep your scissors and tools reserved for specific materials and the life of the tool will be dramatically extended.

Similarly, conventional fabric scissors can be used to cut fiberglass or carbon/graphite. However, scissors which are intended to cut fiberglass should never be used to cut carbon/graphite, and visa versa. Although fiberglass and carbon/graphite can be cut with the same type of scissors, they are not interchangeable.

Preimpregnated materials can be cut with a razor blade/utility knife, and a template or straight edge. The resin tends to hold the preimpregnated fibers in place while the razor edge cuts through the fiber. Very sharp, defect-free cutting edges are necessary to work with composite fabrics.

Machining cured composites

Because of the high strength of cured composites, different machining tools and techniques are used as compared with metal structures. Machining characteristics of composites vary with the type of reinforcement fiber being used.

A note of warning however: Machining of cured composite structures will produce dust that may cause skin irritations. Breathing excessive amounts of this dust may irritate your lungs. Also, some composites decompose when being trimmed or drilled at high speeds. Because of the friction generated, you may be burning away various materials, creating toxic fumes. Composites vary in their toxicity, so you should consider all composites equally hazardous and should observe appropriate safety precautions while working with any of them.

Drilling and countersinking

The production of holes in composite materials presents different problems from those encountered in drilling metal. Composites are more susceptible than metal to material failures when machined. The proper selection and application of cutting tools can produce structurally sound holes.

Airframe Technology

Some problems that may occur when drilling composites are delamination, fracture, breakout, and separation:

Delamination... most often occurs as a peeling way of the bottom layer when the force of the drill pushes the layers apart, rather than cutting through the last piece.

Fracture... occurs when a crack forms along one of the layers due to the force of the drill.

Breakout... occurs when the bottom layer splinters as the drill completes the hole.

Separation... occurs when a gap opens between layers as the drill passes through the successive layers—usually from using too much pressure.

To combat these problems in drilling, the material being drilled should be backed with wood whenever possible. When the backside is inaccessible, a wood backup may not be possible. A drill stop is useful to limit how deep the drill will go through the composite structure. By limiting the depth of the drill passage, breaking the fibers on the backside can be eliminated. When exiting the backside of a hole with a drill, very light or no pressure should be used. A very sharp drill should be used to cut through the laminate, not push through. This will prevent the delamination of the last ply.

Carbide drill bits will work on all types of composites and have a longer life than a standard steel drill. Diamond dust charged cutters perform well on fiberglass and carbon; however, they will produce excessive fuzzing around the cut if used on an aramid or Kevlar component. Drill motor speed is important. A high speed will work best for most types of materials being drilled. However, do not use excessive pressure.

Drilling aramid

Machining and drilling materials reinforced with aramid or Kevlar fibers require different tools from those made of fiberglass or carbon/graphite fibers. The physical properties of aramid fibers are unique.

Because of the flexibility of the aramid fiber, the drill will pull a fiber to the point of breaking instead of cutting it. As each fiber is pulled before it is cut, a fuzzing appearance is produced around the edge of the drilled hole. Drilled holes in aramid often measure a smaller diameter than the drill which is used, because of the fuzzing of the fibers. The fuzzing around the hole may not produce a problem in itself; however, if a fastener is to be installed, it may not seat properly in the hole. Consequently, if the fastener doesn't seat properly, mechanical failure may occur when stresses are not properly distributed.

There are drill bits which are made specifically for aramid. The special bits cut through the fibers without fraying the material. The bits will last longer than conventional drill bits, and will usually produce a cleaner hole. If possible, use a drill made specifically for aramid, and use it only on aramid.

The brad-type bit used for Kevlar fabrics. *The dagger or spade bit for carbon/graphite or fiberglass.* *Diamond cut bit for routing out honeycomb cores.*

One such bit is a brad point which is designed specifically for aramid fabric. It is produced with a C-shape cutting edge to pull the fiber out, then cut through the fiber without stretching. Although they were specifically designed for aramid composites, they will also produce good holes in fiberglass and carbon/graphite.

Aramid should be drilled at a high speed. A better cut will be produced if the drill is very sharp. The pressure on the drill should be light; the weight of the drill motor alone is usually sufficient. When exiting out the backside of the hole with the drill, less pressure should be used in order to prevent breakout. This problem can be eliminated if a drill stop is used that is set so that just the tip of the drill will have clearance past the backside of the material.

Drilling fiberglass or carbon/graphite

Drilling fiberglass or carbon/graphite can be accomplished with most conventional tools; however, the abrasiveness of these composite materials will reduce the quality of the cutting edge and shorten the life of the drill drastically.

Carbide, diamond-charged, or carbide-coated tools are used to obtain better results and longer tool life. Diamond-charged tools are usually steel drills which have a coat of diamond dust to cut through the material. This type of drill works well in carbon/graphite and fiberglass components. When cutting fiberglass, the fibers in most cases fracture at the cutting edge of the tool.

Carbon/graphite fibers are stiffer and stronger and resist the cutting action of the tool. If a dull or improper drill starts to cut the individual fiber, it may break inside the composite structure, causing the hole size to be larger than that of the drill. Drilled holes in carbon/graphite will often show larger diameters than the drill

Airframe Technology

Sanding composite materials can be difficult but mastered with practice.

which is used. Dust chips which are allowed to remain in the holes during the drilling process can also cut, thus enlarging the hole diameter. This creates a problem in that the excessive hole size will cause the fastener to wear in the hole, and will not offer the required strength.

For fiberglass or carbon/graphite drilling, a dagger or spade bit can be used. The use of these bits will reduce the tendency of the fibers to break rather than be cut. This type of bit has a single cutting edge.

Mechanical sanding

Mechanical sanding is the fastest and easiest method, but it's also more likely to cause additional damage by sanding away too much material. One of the best tools for mechanically sanding of composites is a small pneumatic right-angle sander. Adequate control of the sanding operation can best be achieved with patience and experience.

Finer grit sandpaper will usually keep the fuzzing down when sanding aramid fabric. The finer grit also removes the material slowly, allowing more time to find the individual plies. The sanding operation may be accomplished by step-cutting or by scarfing.

When sanding laminates during a repair operation, a right-angle sander or drill motor should be used. The tool should be capable of 20,000 rpm and equipped with either 1-, 2- or 3-inch sanding disc. The sanding disc can be used in combination with a drill motor or with a sander. These come in many different diameters, but a smaller 1- or 2-inch disc will give you more control when step-sanding or scarfing the composite structure.

The drill motor is widely used in the repair industry. It is used primarily for drilling, but can also be used with a disc as a sander. A right-angle sander is used for scarfing and step-cutting the repair. There is much more control, however, with a right-angle sander than with a drill motor because your hand is closer to the work.

Each material sands differently, and various techniques should be used with each material. When sanding aramid, expect the material to fuzz. When the sanding is almost through the layer, a lighter color of fuzz will be seen and spots of "gloss area" may appear. During the sanding process it is important to look carefully for a gloss area. When an area begins to gloss, it is indicating that one layer of laminate has been removed and the sander is just above the following layer.

Carbon/graphite material will produce a very fine powder when it is sanded. It is usually easier to see the layers of carbon/graphite than with aramid.

Another way to tell if sanding through one layer has been completed is to look at the weave. Since most composites are made with each layer's weave in different directions, seeing a change in weave direction may be an indication of a new layer. As the top ply is sanded, the next layer will produce the weave in a different orientation, signaling that one layer has been removed.

Use extreme care when sanding composites. The layers of a composite laminate are very thin, and it's not uncom-mon to sand with too much pressure, moving too quickly through the layers. Because of this, two layers can be mistaken for one. This may present a problem if there are only three layers in the laminate over a core structure and the repair calls for sanding down to the core. If the first two layers are sanded down and counted as one layer, then when the next layer is sanded down, the honeycomb core is exposed, and there will not be enough surface area to laminate a new patch over the plies.

Trimming cured laminates

Standard machining equipment can be used to trim composites; however, some modifications to the tooling may be necessary. All cutting surfaces should be carbide-coated whenever possible. Diamond-edged blades work well on carbon/graphite and fiberglass.

Routers

The most common types of routers operate at 25,000 to 30,000 rpm. They are used to trim composite laminates and to route out damaged core material.

For routing honeycomb, carbon/graphite or fiberglass laminates, a carbide blade diamond cut router bit works best. A diamond cut router bit does not refer to diamond chips or dust on the cutting surface, but rather to the shape of the cut on the flutes. These bits can be used for routing fiberglass, carbon/graphite, and for Nomex® honey-comb.

This brief overview of composite machining techniques reflects some of the more common problems and techniques involved with the repair of commercial aircraft. The techniques are not difficult to master, but may require some practice. *November/December 1995*

Cynthia Foreman is the chief executive officer of Composite Educational Services and CES Composites. She is also the author of Advanced Composites, *published by International Aviation Publishers.*

Airframe Technology

Interior refurbishing
An inside look on the inside jobs

By Greg Napert

Although interior refurbishing may seem quite simple, at first glance, when viewed more closely, it involves a host of skills, talents, and FAA approvals. For this reason, many maintenance facilities choose to leave interior refurbishment up to specialty shops and/or facilities who offer the service.

Returning an aircraft to its original condition is about the simplest form of refurbishment. But it still takes the skills and talents required to upholster, carpet, and finish surfaces—skills and talents that are not within the scope of the average maintenance facility. And add one simple modification, upgrade, or accessory, and you're staring down the face of an FAA approval procedure that may require testing of materials, proof of performance, and a recalculation of weight and balance, at a minimum.

In the corporate market, there are also many aircraft operators requesting out-of-the-ordinary interior refurbishment options and installations. And designing, building, and approving these requests is time-consuming to say the least.

Fortunately, there is an array of facilities available which are designed to either help the average maintenance department, or to provide full-service interior refurbishment to aircraft operators. Upholstery shops, cabinet shops and entertainment specialists are among some of the specialty shops available.

Then there are the full-service refurbishers and completion centers. These facilities tend to offer everything from maintenance to cabinet work, to upholstery services all under one roof. *AMT* magazine traveled to one of those facilities, K-C Aviation in Appleton, WI, to get a taste of what they go through to refurbish interiors. Although K-C focuses on the high-end corporate market, the experience it has gained in the interior refurb and completion business can give anyone in the industry a feel for what interior refurbishment is all about.

Materials selection

Because K-C's clientele is primarily the corporate market, much of its work involves customization.

Customization of interiors means that rarely is the same material used from aircraft to aircraft. Customers often choose leathers, suedes, carpeting, and other materials that must be tested to ensure they pass minimum FAA regulations. These regulations can include such things as "vertical burn" and "fire blocking" tests.

Technicians at K-C Aviation finish interior panels.

These regulations not only apply to fabric, they also apply to new foam, cushions, etc.

Dawn Jensen, design coordinator for K-C in Appleton, says, "The vertical burn test is something that is very important when we are talking about the type of material that can be used. Generally, we try very hard not to present anything to the customer that won't pass the fire block or vertical burn test.

"The materials that can be used, however, are different depending if you're talking about a fire-blocked airplane or not?

"If the aircraft is a Part 135 aircraft, it will need to be certified for fire block. Fire blocking is a much more stringent certification than the standard vertical burn requirement. Part 91 operators, which are usually corporate customers, don't typically choose to comply

Airframe Technology

with fire-blocking requirements. This gives them a larger selection of material to choose from.

"It's good to build in added safety, however. Part 25.853B addresses burn requirements that apply to every single item that goes into an airplane regardless if the airplane is operating under 91 or 135. Part 135 takes it one step further and requires fire blocking for materials such as upholstery and carpeting in the cabin," she says.

Vertical burn vs. fire blocking

Jensen explains, "The vertical burn test is a very specific test where, the fabric is cut to a special length, placed into a controlled environment where a controlled flame is applied to the fabric for a specific amount of time. Then the flame is removed and the burning fabric has to self-extinguish within so many seconds. At K-C, we have the capability of performing this test in-house.

"We have a policy here that we test all fabrics we use for vertical burn requirements in-house, regardless if the manufacturer says the fabrics meet the requirements. There are companies that say their materials pass, but we just don't take their word for it. Since we're ultimately held responsible for the certification, we want to make sure the fabrics we use pass the FAA's requirements.

"Fire blocking, on the other hand," says Jensen, "involves treating the fabric so that it meets certain burn requirements. Because the treatment and the test are more complex for fire blocking, we don't perform the test at our facility; instead we source it out.

"What we do is select and purchase the material and make up three seat cushions and three backs for testing. These items require up to 90 square feet of fabric or leather plus the foam materials which are ultimately destroyed for the tests.

"Generally, if a material passes a vertical burn test, it can pass the fire-blocking test. The fire-blocking test, however, looks for such things as material weight loss after the 'blow torch' test. Factors that can affect whether or not the seat cushions pass these tests include such things as how tightly the fabric is stretched over the cushion material. This is something we can easily control, and we have quite a bit of experience. We have occasionally had to send something back to retest," says Jensen.

"One thing that we do to ensure the materials are going to pass these tests is to always use natural fibers such as wool. Natural fibers are much more fire-resistant than synthetic materials," she says.

Typically, in order for many of these products to meet these fire-blocking tests, some sort of chemical treating is necessary.

Jensen says that "flame treating a fabric is a big operation, and when you're looking at 40 yards of fabric, flame treating is not a feasible operation for most FBOs. Also, treating leather can present many problems. If you're not careful, you can ruin the material."

Besides fabrics, all other materials need to be tested as well — counter tops for example.

STC vs. field approvals

Norm Kopesky, avionics systems coordinator for K-C works very closely with the FAA on many of the company's completions. "Coordinating FAA," he says, "is pretty much the same whether you're doing an avionics installation, an interior redesign, or a green completion."

There are several types of FAA approvals that K-C seeks, including: FAA Field Approved Form 337s, 81103/ FAA Form 337 combination, or STCs. The type of approval they seek depends on a couple of factors such as: what type of installation is being performed and how much time they have to get the approval. He says, "Because we have several FAA designated engineering representatives in our facility, the quickest form of approval is typically an 8110-3/FAA form 337 combination.

The 8110-3 is a statement of compliance that the DER fills out to certify the data (drawings or reports) conform to FAA specifications and regulations. This is essentially done on behalf of the engineering office of the FAA. A 337 then has to be filled out stating the installation of the system or component is done per the approved data; the aircraft can then be returned to service.

"In the case of a microwave oven installation, for instance," Kopesky says, "we would typically do an 8110-3 approving the data and then return the aircraft to service on an FAA Form 337. In order to determine if it meets basic FAA design requirements, we will look at things like wire type, wire gauge, and all of the structural aspects related to how it's installed. But it doesn't make any difference if it is a microwave, a black box, or a brick; all aspects of the installation have to be thoroughly reviewed and approved.

A decision to obtain a Supplemental Type Certificate (STC) depends on a number of factors including: whether or not we have time to apply for and complete the STC, whether or not we think the modification or installation will have application on other aircraft of the same type, and whether or not we think there will be a demand for the STC.

"We often take one of our existing STCs and make minor modifications to them. In these cases, we simply deviate from the STC and have our DERs approve the data. Then if we have future application, we will incorporate that deviation into a revision or amendment to the original STC. If the deviations are significant enough, they may become separate STCs unto themselves," he says.

Kopesky explains, "A complete refurbishment can include a variety of STCs and field approvals. We try to keep certain 'packages' under one STC, but it's often impossible to include everything. Items such as

Airframe Technology

emergency lighting or oxygen systems will be a separate STC. Avionics items are usually going to be separate approvals, STC'd or otherwise. Antenna installations, phones, and entertainment systems are so customized that they are usually field approvals.

"One challenge today is all of the personal electronics items that customers want installed such as personal computers, faxes, displays, etc.

"A lot of customers are also requesting outlets and computer jacks to be installed and that way they don't have to approve the computer installation, only the installation of the jack. We have many customers who want cellular phones installed with external antennas. The only problem here is that the FCC has ruled them to be illegal for use when in flight. However, they can be used anytime you are on the ground or when you are taxiing around the airport. If we have installed a cellular phone system, we typically wire the phone to the squat switch so that it is inoperative when once airborne, and we always placard the airplane stating the regulations related to cellular phone usage.

"Unfortunately, many customers now have the phones in their pockets and they do what they want to do," he says.

Soundproofing

Joe Thurman, supervisor for fabrication and installation at K-C, says that "soundproofing is something that we at K-C work very hard at. We try to reduce noise as much as possible when we design an interior. However, ultimately, any soundproofing you install adds weight to the aircraft. And that creates a penalty in terms of how many passengers or luggage you can carry.

"Our customers also sometimes specify cabin noise limits. We have to emphasize, however, that this may affect their selection of materials and/or design of the interior.

"To reduce noise, we can do things such as install fabric bulkheads instead of wood, or fabric on the seats instead of leather, or insulated window shades. Additional soundproofing beyond manufacturers' designs, will usually result in some sort of tradeoff," he says.

"Most aircraft manufacturers have a completion center manual with recommended sound packages. But these are more or less guidelines to go by.

"As far as soundproofing materials go, the primary type of material used in the past was lead vinyl. However, lead vinyl is very heavy and cumbersome, and the adhesives that you need to use are very caustic. The technicians have to completely suit up just to install it. Also, lead vinyl is only effective if completely sealed. It acts as a barrier, sealing out the sound rather than dampening it. If you have an opening anywhere in the lead vinyl, the sound gets through. So you have to make sure everything is sealed tightly, which is a tedious task.

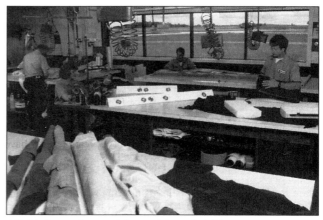

All fabrics and materials used in refinishing must pass rigid FAA burn standards.

"We have found that we have more success with sound dampening instead. This starts at the aircraft skin using a foam material with a foil backing. Skin dampening materials have improved considerably over the years. The problem in the past was that in extreme cold temperatures, at altitude, the foam would get hard and would no longer serve its purpose. Newer dampening foam can withstand extreme temperatures and still be effective."

Probably the most success that K-C has had with noise reduction is through the use of vibration isolation. Thurman explains, "We have found through experience that it is usually better to dampen or isolate vibrations rather than to block out the noise.

"For starters, we design our cabinetry to 'float' rather than hard-mount it. We shock-mount as many components as possible so the noise doesn't transmit from the aircraft frame into these units. If hard-mounted, the furnishings can become, essentially, big 'speakers' that tend to amplify these vibrations.

"As more people are getting involved in vibration isolation, it's becoming more of an exact science. It's quite a change from the basic concept of soaking up or blocking noise. The move to vibration isolation allows us to implement noise controls without adding additional weight to the aircraft," he says.

"As promising as vibration isolation is, however, there are some items that you just can't use it on. For example, you would not want to mount a table on vibration isolators because you don't want the table to flex; you want it to feel solid.

"But we're getting smarter. What we do in the case of the table now is to mount the table solidly onto a panel and then dampen the panel.

"We're also using headliner panels and side wall panels that are vibration isolated as well.

"In practice, what we do is incorporate many of these changes into a noise reduction package that we provide as part of a completion package," Thurman concludes.

Dealing with a variety of requests

Peg Docter, sales manager for K-C, says that they often deal with a variety of requests for installation of items in the aircraft. "Some of the most popular are new entertainment systems, phone systems, flat screen TVs and avionics requests for GPS installations. Most requests involve creature comforts.

"We also have many requests from pilots and from maintenance people. Maintenance folks want interiors that are easy to repair or clean, for instance. We take all of these requests under consideration in our initial discussion. We also take into consideration whether the aircraft is going to be for charter or if it's going to have a lot of different people in it. For heavy traffic we would steer the customer toward an ultra leather rather than suede, or a dark carpet rather than a light carpet," says Docter.

Thurman says that "we also try to make things as accessible as possible for maintenance. When we put in a galley, we try to make it easy to take that galley out. We use as many quick disconnects and access panels as possible."

As a final note, Kopesky says, "Any aircraft modification or installation requires a look at the aircraft weight and balance. If it's a minor installation and the aircraft has been weighed recently, we will use existing numbers and recalculate the weight and balance based on the existing numbers. If it's a major installation or modification, we will reweigh the entire aircraft for a fresh weight and balance report."

AMT's interior services and supplies directory

Whether you're in the business of working on aircraft interiors or not, you'll want to keep this handy list of interior maintenance facilities and service providers nearby. The following are companies that responded to a mailing conducted by *AMT* magazine:

Accent Interiors
Offers fabrication and modification of interiors, design, upholstery, and cabinet work.
Opa Locka Airport, Bldg. 147, Opa Loka, FL 33054, Ruth Cannon, (305) 681-4010.

ADI Interiors
The company offers full interior design capabilities in-house, or will work with a designer of your choice. Also offered are certified fire blocking services, an on-staff DER, and a broad selection of fabrics, leathers, and carpets from around the world
Oakland/Pontiac Airport, 6544 Highland Rd., P.O. Box 270100, Waterford, MI 48327-0100, Brian Wells (810) 666-3500.

Aero Air Inc.
A full service FBO specializing in extensive corporate jet and turboprop interior refurbishing and exterior paint. Complementing these services are aircraft maintenance, sales, charter, and entertainment systems installations.
2050 N.E. 25th Ave, Portland-Hillsboro Airport, Hillsboro, OR 97124, Tom Krueger, (503) 640-3711.

Airo Industries Inc.
Aircraft interior refurbishments include galleys, seats, and sidewall panels.
14675 Titus St., Panorama City, CA 91402, David Nejad, VP Sales/Mktg., (818) 780-8733.

Arizona Aircraft Interior Designs, Inc.
Offers complete interior refurbishing of aircraft including: cabinets, fire blocking, interior painting, and soundproofing. Specializes in turboprops and small jets for corporate customers.
5047 E. Roadrunner Dr., Mesa, AZ 85215,
Wayne Bryant, 602) 832-1330.

Brice Manufacturing
Manufactures aircraft seat components and offers refurbishment services for the integration of all telephonics and personal video system products. Also, turnkey service for the design, manufacture, and certification of seat kits.
10262 Norris Ave., Pacoima, CA 91331,
Sean Prendergast, (918) 665-2321.

Byerly Aviation Inc.
Complete business jet and turboprop refurbishing capability. Exterior paint and cabinet refinishing with complementary avionics support and GPS installations. Commander, Citation, and King Air maintenance and modifications.
Greater Peoria Airport, 1900 S. Maxwell Rd., Peoria, IL, 61607, R. Bruce Byerly, (309) 697-6300, Ext. 360.

Airframe Technology

Cameron Aircraft Interiors Inc.
Offers interior refurbishment of all categories of aircraft.
31W651 North Ave., Dupage Airport, West Chicago, 60185, Harry Cameron, (800) 866-4886, (708) 584-9359.

Cleveland Jet Center Inc.
Complete aircraft refurbishment, from design to paint, and wood shop. DAR, DER on staff. Aircraft worked include BAC 1-11 to piston twin. Full service maintenance facility.
38630 Jet Center Pl., Lost Nation Municipal Airport, Willoughby, OH 44094, Jack Barnett, (216) 942-0087.

Commodore Aviation Inc.
Perform interior refurbishment of seats, galleys, lavs, interior mods, and painting for Boeing, Douglas and other commuter aircraft. Also accomplish heavy maintenance services.
5300 N.W. 36 St., Box 661078, Miami, FL 33266-1078, Rick Weltmann, Dir.-Mktg./Contracts, (305) 871-1169, Ext. 201.

Custom Aircraft Interiors Inc.
Manufactures complete interior components for MDHC 500 Series aircraft. Prototype and design capabilities for all aircraft interiors.
3701 Industry Ave., Lakewood, CA 90712, Patricia Erwin, (310) 426-5098.

Downtown Airpark Inc.
Specializes in all types of interior work—from minor repairs to complete custom design.
P.O. Box 26027, 1701 South Western, Oklahoma City, OK 73121, Greg Groves, (800) 253-1456, (405) 634-1456.

Eagle Aviation Inc.
Full-service FBO offering such services as jet and turbo prop maintenance, avionics, completions, with 250 trained personnel.
2861 Aviation Way, Columbia Metro Airport, W. Columbia, SC 28170, Jim Neece, (803) 822-5586.

Elite Interior Designs
Offers aircraft interior refurbishment including: seats, headliners, sidepanels, carpets, and cabinet work. Also T.S.O. restraint belt refurbishment.
101 E. Reserve St., Pearson Airpark, Vancouver, WA 98661, Paul and Alana Sanchez, (206) 693-0051.

Gulfstream Aircraft Inc.
Now offers full-service interior refurbishment for Gulfstream aircraft in its Brunswick, Georgia facility.
500 Gulfstream Rd., Savannah, GA 31408, David F. Fulcher, (912) 965 3472.

Gulfstream Service Center
Complete interior refurbishment for Turbo Commander, Westwind, Sabreliner, and Citation aircraft. Complete woodwork returbishment and cabinet design capabilities.
7301 N.W. 50th, Bethany, OK 73008, Mark A. Fulton, (405) 789-5000, Ext. 486.

Heli-Dyne Systems Inc.
FAA-certified full-service helicopter support facility. Offers complete in-house capabilities to include custom design and engineering, avionics and communications.
9000 Trinity Blvd., P.O. Box 966, Hurst, TX 76053, David Likes, (817) 282-9804.

Hill Aircraft Interiors
Hill Aircraft leads the way in the provision of master quality interior design, repair, and installation on Beechcraft KingAir, Learjets, and twin-engine aircraft such as Cessna, Piper, and others.
3948 Aviation Circle, Fulton County Airport Brown Field, Atlanta, GA 30336, Jacques Escalere, (404) 691-3330.

J.A. Air Center
J. A. Air Center interior department offers the finest quality fabrics and leathers. Services range from simple repairs to full refurbishing of interior.
3N060 Powis Rd., DuPage Airport, West Chicago, IL 60185, Rick Milburn, (708) 594-3200.

JetCorp
Technical capabilities include: major interior refurbishment, non-destructive testing, engine repairs, hot sections, parts support, accessory repair and overhaul, airframe inspections, and avionics repair and installation.
18152 Edison Ave., Spirit of St. Louis Airport, Chesterfield, MO 63005, Jerry Moore, (314) 530-7000.

Jet East Inc.
Full service FBO—Citation and Learjet authorized service center providing minor to extensive refurbishments. Avionics total on-site support. Also offers parts, sales and charter services.
7363 Cedar Springs, Dallas, TX 75235, Leticia Chacon, (214) 350-8523.

KaiserAir Inc.
Provides custom interior refurbishment for both 91 and 135 operations. Custom cabinetry, fabrication, and custom interior design consultant available.
Airport Station, Box 2626, Oakland, CA 94614, Andrew F. Fitzgerald, (510) 569-9622.

Airframe Technology

Kal-Aero Inc.
Is a full "one stop shop" fixed based operator. Capabilities include state-of-the-art paint, refurbishment, and maintenance.
15745 S Airport Rd., Battle Creek, Ml 49017, John Hooskins, (616) 969-8400.

KC Aviation
Provides refurbishment, completion, maintenance services, avionics installation and repair, overhaul and repair services, and personnel and consulting services. DER and DAR on staff.
7440 Aviation Place, Dallas, TX 75235, Dallas Love Field, John Rahilly, (214) 618-7719.

King Aerospace Commercial Corp.
Aircraft brokers, interior design, installation and modification, maintenance and repair station, painting, coating and cleaners.
4444 Westgrove, #250, Dallas, TX 75348, Jerry King-Echevarria, (214) 248-4886.

Mechanical Enterprises Inc.
MEI is a manufacturer of plastic seat parts, tray tables, arm shrouds, etc. FAA-approved repair station.
2961 A Olympic Industrial Dr., Smyrna, GA 30080, Ron Kirschner, Pres., (404) 350-8489.

Midcoast Aviation
Offers modification, refurbishment, and maintenance services, combining state-of-the-art technology with old-world craftsmanship.
8 Archview Dr., Cahokia, IL 62207, Rodger Renaud, (618) 337-2100.

Million Air Inc.
Provides the highest quality craftsmanship on custom interior refurbishments, maintenance work, and avionics upgrades and installations. Specializes in corporate aircraft.
4300 Westgrove, Dallas, TX 75248, Bob Tharp, (214) 733-5821.

Omniflight
Offers completion/refurb. capabilities. Omniflight provides custom interiors for EMS, executive, law enforcement and new completions or refurbishments.
4650 Airport Parkway, Dallas, TX 75248, Allen Dales, (214) 233-6464.

The Oxford Aircraft Refurbishing Centre
Offers custom interior design service with owner participation. Standard refurbishing package includes: restyled seat cushions, carpeting, plastic paneling—ultrasuede covered, new headlining, cabinetry, and repolishing.
Oxford Airport, Kidlington, Oxford, OX5 1RA, Don Tempest, 01865-370848.

Premier Aviation Inc.
Full service helicopter completion/modification center providing executive transport air medical, public service, and special missions interiors. Cockpit design and instrumentation available.
2621 Aviation Pkwy, Grand Prarie, TX 75050, Tamera Bidelspach, (214) 988-6181.

Ranger Aviation Enterprises Inc.
Provides complete paint and interior on all aircraft from turboprop through jet. Have contract to paint all U.S. Airforce Lear 35s.
P.O. Box 61010, San Angelo, TX 76906-1010, John or Sandy Fields, (915) 949-3773.

Rocky Mountain Helicopters
Operates an in-house completion center for the design and installation of medical interiors. The company's specialists design and install each aircraft to customer specifications.
P.O. Box 1337, Provo, UT 84603, Jennifer Hunter-Jones, (801) 375-1124.

Sky Harbor Aircraft Refinishing Ltd.
Provides aircraft refurbishment services to include paint, interior, and maintenance. Over 40 years experience in complete interior and exterior aircraft refurbishment.
R.R. #5, Box 536, Goderich Airport, Goderich, Ontario N7A 4G7, D.E. (Sandy) Wellman, (519) 524-2165.

Skyworthy Interiors
Specializes in refurbishing and updating complete interiors in design and production. FAA-certified upholstery shop, meeting all specifications.
3112 N. 74th E. Ave, Hangar 23, Tulsa, OK 74115, D.A. Williams, (918) 835 4770.

South Coast Aircraft Interiors Inc.
Complete interior restoration of executive aircraft since 1982. Fire blocking, design and engineering services, window replacement, sound proofing, FAA repair station.
2898 Montecito Rd., Hangar A-1, Ramona, CA 92065, Andy Mirabelli (619) 788-9276, (800) 550-9276.

Airframe Technology

Special Products Aviation Inc.
Builds and repairs fabric-covered airplanes. Also offers painting and composite services.
850 9th Ave., Conway, AR 72032, Kenny Blalock, (501) 327-4339.

Tri-State Airmotive LLC
Accessory overhaul, paint, interior, sheetmetal and airframe inspection.
20 Tri-State Rd., Berryville, AR 72616, Chuck Bennett, (501) 423-4911.

United Beechcraft Tampa
Full service interior, cabinets, distinctive designs, S.T.C. approvals. Cabin displays, video and audio systems, and configuration changes.
2450 N. Westshore, Tampa, FL 33607, Skip Davies, (813) 878-4500.

West Star Aviation Inc.
Provides custom paint and interior refurbishments. Also log reproduction, super soundproofing, entertainment systems, commuter, medical, and custom business retrofits, and more.
796 Heritage Way, Grand Junction, CO 81506, J. Gregory Heaton, (970) 243-7500.

West Virginia Air Center
Offers aircraft maintenance, sheetmetal, interior, paint and composite on regional and corporate aircraft.
P.O. Box 908, 2400 Aviation Way, Bridgeport, WV 26330, Gary Palmer, (304) 842-6300.

September/October 1995

Airframe Technology

Corrosion detection methods
New technologies emerge as aircraft age

By Greg Napert

As the average age of the general aviation fleet climbs over 30 years, it has become evident to many in the industry, including the FAA, that new tools and nondestructive inspection (NDI) techniques are needed to locate corrosion, as well as other problems associated with aging. Corrosion prevention can be quite successful simply by keeping an aircraft clean, yet, even the most meticulous shop can't avoid the onset of corrosion on an aircraft that's exposed to the elements.

As a matter of economics, if a piece of NDI equipment can be employed to scan an aircraft for corrosion, the need to pull rivets and remove skins to perform the inspections visually—a process that is both time-consuming and expensive—can be eliminated.

But many questions arise as this new "high-tech" inspection equipment is introduced to the aviation marketplace: Which equipment and techniques for detecting corrosion are acceptable? What level of corrosion is acceptable before removing an aircraft from service? Who's going to give approval for these new inspection techniques?

Because of the flood of new technologies and equipment, combined with an urgent need to apply NDI equipment to the aging fleet, the FAA began a project in 1991 called the Aging Aircraft NDI Validation Center (AANC). The center, located in Albuquerque, NM, at Sandia National Laboratories, has been charged with testing and validating several emerging forms of NDI inspection for use on aircraft.

Part of the validation process developed by the AANC involves inviting actual industry technicians into the facility to evaluate each one of the NDI processes.

As a result, during a recent meeting in Albuquerque, the International Association of Machinists and Aerospace Workers Flight Safety Committee were invited to the facility to help evaluate the latest technology. *Aircraft Maintenance Technology* magazine went along.

Our interest was in discovering emerging technologies for detecting corrosion. Interestingly, most of the technology being introduced is technology that has been employed in other industries for years, but has never been adapted for use on aircraft.

The following three NDI methods are examples of emerging technology that show promise for locating corrosion and for being readily available to the aviation industry in the near future:

Compton X-ray backscatter imaging

Probably the most promising technology for the inspection of corrosion is a process called Compton backscatter imaging (CBI). CBI is a re-emerging near surface NDI measurement and imaging technique which can detect critically imbedded flaws such as cracks, corrosion, and delaminations in metal and composite aircraft structures. The technology offers exciting possibilities because it provides a two- or three-dimensional density map of the inspected area. This means that the data can actually measure the depth and extent of corrosion and/or cracks in subsurface layers.

According to a report from AANC called Emerging Nondestructive Inspection for Aging Aircraft, the information provided by this technology can be presented in forms ranging from a simple accept-repair gauge for corrosion-induced aircraft skin thinning to the more sophisticated three-dimensional tomographic-like digital flaw image displays.

Compared to conventional radiography where the whole area is flooded with X-rays, backscatter imaging equipment has a highly collimated, pencil-like X-ray beam and detector geometry and exposes only a very small volume of the inspection area. This significantly reduces the stray radiation and minimizes additional shielding requirements for maintenance personnel.

The technology is currently available from Philips Electronics Instrument Co., Industrial Automation Division, Norcross, GA. Philip's unit, called the ComScan, can determine, both in depth and size, first- and second-layer corrosion, honeycomb impact damage, density variations in carbon composite sandwiches, water entrapment in honeycomb structures, and the detection of delaminations and cracks in stabilizer stringers in all-carbon reinforced wings which have escaped detection by traditional X-ray inspection methods.

CBI is unaffected by variations in liftoff, surface roughness variations or paint, metal conductivity, delamination, and air gaps. The technology can inspect solid aluminum or composite materials to depths of 2 inches or more. And if the structure is layered with intervening air gaps, information can be obtained at greater depths.

Up until now, CBI use in the aviation industry has been virtually nonexistent because cost-effective equipment has not been developed for the industry. However, according to the report, cost-effective systems can be developed.

Airframe Technology

Magneto-optic eddy current imaging (MOI)

Nearly as promising as CBI is a process called magneto-optic eddy current imaging. MOI does real-time imaging of airframe fatigue cracks and corrosion. Although it is best for searching for fatigue cracks on aircraft skins, it also has some possibilities for corrosion detections as well.

The unit consists of a hand-held imaging scanner head, TV monitor and power unit, and 30 feet of interconnecting flexible cable. The lightweight imaging head contains the eddy-current inducing system, a magneto-optic sensor element, and a TV camera.

The MOI images result from the response of the magneto-optic sensor to weak magnetic fields that are generated when eddy currents induced by the MOI interact with defects in the inspected material.

Unlike conventional eddy current in which crack detection is based on graphs, needle movement, or other indications in which interpretational skills are required, MOI offers visual images that closely correspond to flaws.

Use of MOI by airlines and aircraft manufacturers has shown that inspection of large areas for cracks and corrosion is rapid, the inspection can be reliably performed through paint and airline decals, the inspection results can be videotaped for test documentation, the unit is portable, and it requires little training to become proficient in its use.

Airlines and aircraft manufacturers currently using this system report one-eighth to one-tenth of the time needed for conventional eddy current inspections is needed for inspections using the MOI.

Boeing Commercial Airplane Group in Wichita, for instance, has reported that the template-guided surface eddy current inspections of pressure cabin skins that took 32 hours to complete were completed in three and one half hours using the MOI.

MOI procedures have been developed by both Boeing and McDonnell Douglas for use on their aircraft. Boeing published an all model procedure in March 1992 and has included it in their aircraft NDI manuals. The final procedures from McDonnell Douglas are soon to be published. Lockheed is currently involved in writing procedures.

Advanced eddy current techniques

Eddy current equipment and techniques are becoming somewhat commonplace in aviation maintenance. But despite the advancements in equipment, conventional techniques are typically limited to inspection of only a single layer of material or skin.

Advance techniques such as dual- or multiple-frequency techniques and pulsed eddy current techniques are emerging as viable, cost-effective methods to detect cracks and quantify corrosion in single- and multilayered metal skins.

These techniques allow the detection of second-layer corrosion and cracks and liftoff variations are reduced.

Additionally, computer programs combined with automated scanners have been developed to process eddy current data. C-scan images (two-dimensional pictures) of eddy current inspection data over an area of a test piece can be made using digital encoders.

Eddy current C-scan imaging has the potential to improve inspections for many aircraft applications that require quantitative information of crack length and corrosion assessment.

Multifrequency and pulsed eddy current instruments are on the market. The development stage of the instruments is such that they can be implemented in the field immediately after applications are identified and procedures are developed to fully realize its capabilities.

According to AANC, pulsed eddy current techniques for corrosion assessment could be developed and ready for field use in two to three years.

Any further questions about any of the aforementioned technologies and/or possible applications can be directed to Patrick L. Walter, Sandia National Laboratories, NM, (505) 844-5226.

Forms of corrosion

Many forms of corrosion can be found on aircraft. The forms are as varied as the types of metals used. And not all corrosion is bad for the aircraft. Aluminum, for instance, forms a corrosive shield on its surface which prevents further corrosion. Yet, other metals are more susceptible to corrosion due to granular characteristics.

The following, from ASA's Aviation Maintenance Technician Series, by Dale Crane, is a brief overview of the most common forms of corrosion seen on today's aircraft:

Surface corrosion

Anytime an area of unprotected metal is exposed to an atmosphere that contains industrial contaminants, exhaust fumes, or battery fumes, corrosion will form on the entire surface and give it a dull appearance. Contaminants in the air react with the metal and change microscopic amounts of it into the salts of corrosion. If these deposits are not removed and the surface protected against further action, pits of corrosion will form at localized anodic area. Corrosion may continue in these pits until an appreciable percentage of the metal thickness is changed into salts, and in extreme cases, the corrosion may eat completely through the metal.

Pitting corrosion shows up as small blisters on the surface of the metal. When these blisters are picked with the sharp point of a knife, they are found to be full of a white powder.

Intergranular corrosion

Aluminum alloys are made up of extremely tiny grains of aluminum and alloying elements, and they may be hardened by heating them in an oven to the temperature at which the alloying elements go into a solid solution with the aluminum metal. When this temperature is reached, the alloy is taken from the oven and immediately quenched in cold water to lock all of these alloying elements to the tiny grains of the aluminum.

When the metal is removed from the oven and begins to cool, the grains begin to grow, If quenching is delayed, for even a few seconds, these grains will reach a size that will produce the anodic and cathodic areas needed for corrosion to form.

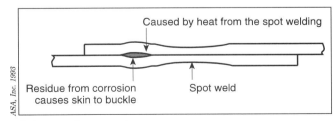

Intergranular corrosion

Spot welds and seam welds can also cause the grain structure inside an aluminum alloy to grow until they are large enough for the metal to be susceptible to intergranular corrosion.

Intergranular corrosion is difficult to detect because it is inside the metal, but it often shows up as a blister on the surface. However, the fact that there are no surface blisters does not assure you that there is no intergranular corrosion.

Ultrasonic or X-ray inspection is needed for a good inspection for intergranular corrosion. Once intergranular corrosion is found, usually the only sure fix for it is the replacement of the part.

Galvanic corrosion

Galvanic corrosion occurs any time two dissimilar metals make electrical contact in the presence of an electrolyte. The rate at which corrosion occurs depends on the difference in activities of the two metals.

Galvanic corrosion can form where dissimilar metal skins are riveted together, and where aluminum alloy inspection plates are attached to the structure with steel screws.

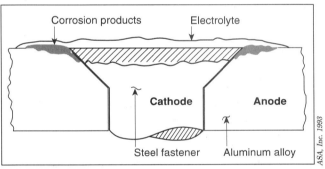

Galvanic corrosion

Fretting corrosion

Fretting corrosion forms between two surfaces which fit tightly together, but which move slightly relative to one another. These surfaces are not normally close enough together to shut out oxygen, so the protective oxide coatings can form on the surfaces. However, this coating is destroyed by the continued rubbing action.

When the movement between the two surfaces is small, the debris between them does not have an opportunity to escape, and it acts as an abrasive to further erode the surfaces. Fretting corrosion around rivets in a skin is indicated by dark deposits streaming out behind the rivet heads. These dark deposits give the appearance of the rivets smoking. By the time fretting corrosion appears on the surface, enough damage is usually done that the parts must be replaced.

Airframe Technology

Filiform corrosion

Filiform corrosion consists of threadlike filaments of corrosion that form on the surface of metal coated with organic substances such as paint films. Filiform corrosion does not require light, electrochemical differences within the metal, or bacteria, but it takes place only in relatively high humidity, between 65 and 95 percent.

The threadlike filaments are visible under clear lacquers and varnishes, but they also occur under opaque paint films such as polyurethane enamels, especially when an improperly cured wash primer has left some acid on the surface beneath the enamel.

Exfoliation corrosion

Exfoliation corrosion is an extreme case of intergranular corrosion that occurs chiefly in extruded materials such as channels or angles where the grain structure is more layerlike, or laminar, than it is in rolled sheets or in castings.

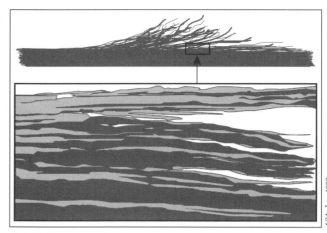

Exfoliation corrosion

Exfoliation corrosion occurs along the grain boundaries, and causes the material to separate or delaminate. By the time it shows up on the surface, the strength of the metal has been destroyed. *July/August 1995*

Window care and repair

Andrew Geist, general manager for Aeroscope, Inc. in Broomfield, CO, says that with the worldwide increase of crazing to acrylic aircraft windows, the airline industry is spending literally millions of dollars for research and development of new techniques to extend the service life of their current windows. The OEMs are also hot to find answers to the industry's problems with new formulas for their acrylics and special coatings.

Geist says that acrylic goes through a natural aging process that eventually causes crazing, even sitting on a parts shelf. It is all the other factors affecting acrylic that causes the rapid acceleration in crazing. There are numerous types and causes of crazing, which include volcanic, chemical, and age crazing. The most prevalent type seen at the corporate and airline level is caused by volcanic crazing. "Crazing," he says, "can be defined as fine micro-cracks on or beneath the surface of the acrylic. It is easily identifiable as a glaring refraction of light from the surface of the window.

"When the combination of the outside applied forces to the window surface exceeds the tensile strength of the plastic, the surface cracks and crazing begins. Simply put, acrylic stresses out. Any additional forces applied to the window just accelerate the rate at which the window crazes. These environmental forces result from pressurization, thermal variations, age, and water disorption. As the transparency is exposed to the dry environment at altitude, the transparency loses water through disorption, shrinks, dries, and cracks. It's believed by the industry that the culprit of this increased rate of crazing is caused by changes to our atmosphere, primarily the increase in sulfuric acid aerosol."

He explains, for example, the Mount Pinatubo eruption on Luzon in the Philippines in 1991 dumped millions of tons sulfuric dioxide gases into the atmosphere. Sulfuric dioxide from volcanoes combined with water vapor in the atmosphere makes sulfuric acid aerosols.

"Over the last few years," Geist continues, "the effects of gravity have brought the dioxides down to the level that most transport and corporate aircraft operate. This sulfuric acid is theorized to cause a rapid acceleration in the drying process.

"Naturally," says Geist, "there are many other contributors to this crazing problem such as ultraviolet light, improper maintenance, neglect, and chemical damage. Something as simple as a razor cut around the radius of the window can cause shelling to the panel that may eventually deem it unserviceable. Other enemies such as plastic and glass cleaners which have not been tested for use on aircraft transparencies should be avoided. Certain types of masking tapes used during painting and nonapproved window sealants can also cause crazing. The easiest way to determine if your aircraft suffers from crazing damage is to view the acrylic in direct sunlight or use a high-intensity light held at a 45-degree angle to the surface. Then move your head around to change your visual angle relative to the surface, and look for a reflection of light.

Technician removes crazing damage on Citation business jet.

"Cleaning after each flight," he says, "is extremely critical for the prevention of crazing. It should be completed as often as possible, to remove any sulfuric dust from the panel surface. The best recommendation for cleaning an acrylic window is to first remove any rings or watches so as not to scratch the windows. Then wash the window down with water using your bare hands only, so careful attention can be given and you are less likely to scratch the surface. Dry the window with a very soft cloth. Next polish the window with known products that are recommended by the manufacturer.

"Besides the obvious cosmetic reasons to have a damaged window repaired, there are other factors that could lead to more serious consequences if not addressed," says Geist. "Any damage to the window surfaces will drastically reduce the structural integrity of acrylic panel. If not addressed, a structural failure of a transparency can radically change your day while flying at FL400."

He explains a damaged aircraft window develops stress areas known as stress risers, resultant from the pressure applied to the uneven damaged surface. An average cabin window from a business jet at FL400 can experience from 1,200 to as much as 1,700 pounds of total pressure on the window surface from the combination of loads experienced in flight. Each individual stress riser will be affected by the combination of the loads.

Airframe Technology

Geist says, "All is not lost when your enclosures suffer from any type of damage which may degrade their appearance. Your options are to replace the window with a new unit, which is usually very costly, or effective restorations can be completed saving your operation a considerable amount of time and expense. Crazing and other damage can be removed and the optical clarity restored through procedures approved by the FAA and the manufacturer."

Typical crazing reflection.

Geist explains that technicians at repair facilities or on site will remove uniform amounts of acrylic through the use of various levels of sanding materials and polishes. Additional nondestructive precision equipment such as optical micrometers, ultrasonic testing equipment and prisms are used to measure window thickness to keep repairs within manufacturer's approved tolerances.

He warns, however, that improperly performed transparency refinishing can create excessive distortion and actually weaken the structural strength of the surface. "When the correct combination of techniques and materials are applied, window damage of almost any type can be repaired successfully," says Geist.

Rethinking routine care

Don Moyer of Norton Performance Plastics Corporation, Composites Operation in Ravenna, OH, says, "Making the most of the window repair option requires an industrywide rethinking of routine care, maintenance, and inspection procedures.

"The first step," he continues, "is to minimize the minor damages that make windows more vulnerable to their punishing environments. Next, catch the damage early enough to make repair possible. True, a single scratch can be enough reason for a cockpit window to be pulled. However, why then scrap and replace when it can be inexpensively repaired to like-new condition instead?

"To a certain extent, taking full advantage of cockpit window repair involves changing habits of the flight crews and maintenance people, through upper management, including purchasing agents and engineers. Think of cockpit window repair as a strategic area of cost reduction. It's especially important to educate the crews who conduct the inspections and actually have authority to pull windows. Better to pull it sooner and fix it than wait until later and pay five times as much to replace it," explains Moyer.

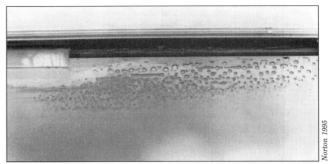

If bubbles are spotted, pull the window as soon as possible to avoid irreparable damage.

Types of damage

Moyer says, "Surprisingly, much cockpit window damage arises from the inside. A major cause, believe it or not, is banging and scraping by clipboards placed on the cockpit glare shield. Another culprit is crew members' rings hitting the windshield when they lean forward to reach for a clipboard or other item. Also, automatic seat belt retractors on the co-pilot's side cause some problems when the buckle flies up during retraction," he says.

"Of course," explains Moyer, "exterior factors also cause cockpit window damage. Hail, birds, sand kicked up from the runway during takeoff and heavy rain all contribute to chipping, pitting, scratching, and seal erosion. Under severe thermal and barometric cycling coupled with exposure to 600-mph winds, tiny cracks called crazing can impair pilot visibility.

"Another common type of damage is delamination, the separation of the acrylic or glass pane from the P-membrane. This is usually the result of age and constant thermal and barometric cycling that occurs in pressurized aircraft.

"Yellowing of the window is also a frequent problem. It also results from aging and moisture encroachment. It may not affect window functionality or truly impair visibility, but it can become unsightly and prompt maintenance personnel to question its serviceability," says Moyer.

"Milkiness may also occur. Again the cause is age and moisture encroachment into the PVB layer between the panes."

He says that in passenger widows, the most common problem is micro crazing—the appearance of hundreds of tiny surface cracks. Initially, the problem is more cosmetic than functional. However, it is often a source of passenger concern and complaint.

"These same problems also can affect the lenses protecting exterior lighting. Lenses are not normally in a crew's constant line of sight, making lens damage less noticeable. But, if lens damage goes unnoticed too long, there is the risk of burning out the electric circuitry or contaminating it with dirt or water."

Moyer offers the following description of the types of repairs that are possible at Norton's window repair facility:

Relamination—to get rid of bubbles—indicative of delamination, the window is disassembled, cleaned, and relaminated in a large autoclave to new window specifications. In Norton's case, it is to FAA-approved specifications. The relamination process is straightforward and much like the original manufacturing cycle. Each delaminated transparency is repaired and cured in a special vacuum package. Pressure and heat are applied in an autoclave. A log generated during the repair process documents the time-temperature-pressure cycle for each step.

After relamination is completed, transparencies are polished and matched to their original frames and reassembled.

Milkiness removal—This is caused by moisture at the interface between pane and membrane. The window is dried and, in most cases, relaminated as well to eliminate air spaces so no new moisture can enter.

Surface scratches—If the scratch is shallow, pane thickness permitting, it may be ground or polished out. Deeper scratches may require replacement of one of the two panes that make up the entire window.

Crazing—This problem is usually solved by grinding and polishing, assuming acrylic thickness and OEM specifications permit.

Early detection is the key to easy repairs

Early recognition and action, emphasizes Moyer, will yield major savings over the life span of the aircraft.

He offers this look at the early warning signs for each type of common problem:

Bubbles—These are signs of delamination. The pattern may be "champagne bubbles"—lots of little ones—which indicate a possible controller problem that causes overheating of the window. Large edge delamination bubbles indicate age and use. These bubbles can impair visibility, but the window can be repaired. The window should be pulled and repaired as soon as bubbles are noticed and before irreparable damage sets in. Watch for the bubbles around the edges first.

Crazing—The signs are obvious. The trick is to spot—and act on—crazing while it is still small. Crazing does not heal of its own accord and can spread in an instant. And if deep enough, the crazing can provide a pathway for less repairable damage. Remember to inspect the outside for crazing. Crazing can be repaired.

Milkiness—Again, the visible sign is obvious. The good news is milkiness can often be corrected.

Yellowing—Though the signs are obvious, the prognosis is not as good. At present, there is no known way to correct yellowing short of replacing the pane. While primarily cosmetic, it is nonreversible.

Micro crazing—This looks like a web of tiny cracks on the exterior surface. It is the most common problem in passenger windows. Spotted early, it can be corrected. Left unchecked, it can lead to early replacement of the window.

Distortion lines—These are signs of the deterioration of heating elements due to fatigue breakage. The windows are deiced by wire grids or heating films in the vinyl interlayer. Repeated fight cycles and associated deflections may cause individual wires to break and cease working. Therefore, the window is partially heated, resulting in fogged areas and localized poor visibility. Thermal stresses in the glass also can lead to more serious damage. Some wire breakage may be tolerated unless the pilot finds his vision impaired, but heating elements should be monitored closely. Distortion lines may also be caused by localized repairs, for example, where a maintenance worker sands only one scratch or pit for local repair.

Loose connections—When you inspect cockpit windows, also carefully check all electrical connections to the defoggers and window heating elements. Proper heating and defogging can go a long way toward reducing thermal cycling stresses and preventing electrical arcing and other damage.

Airframe Technology

A short list of certificated window repair companies

Aeroscope Inc.
11750 Airport Way, Hangar #B12B
Broomfield, CO 80021
(303) 465-4414

Aircraft Window Repair
2207 Border Ave.
Torrance, CA 90501
(310) 212-7173

LP Aero Plastics
Rd. #1 Box 201B
Jeannette, PA 15644
(412) 744-4448

Norton Performance Plastics
Composites Operation
335 N. Diamond St.
Ravenna, OH 44266
(216) 296-9949

Perkins
2300 W. 6th St.
Fort Worth, TX 76107
(800) 880-1966

The Glass Doctor
2390 26th Ave. N.
St. Petersburg, FL 33713
(813) 821-1761

AMT May/June 1995

Airframe Technology

Finding solutions for environmental concerns at paint shops is not easy
Some facilities face extinction in face of high cost

By John Boyce
Contributing Editor

While you might be able to make a case that environmental and safety regulations governing aircraft painting facilities are overly restrictive, you can't deny that they exist.

There are many painting facilities that are ignoring the regulations, but it is only a matter of time before they will face a regulator audit and the consequent fines for noncompliance (as high as $25,000 per day). They will also face the decision of whether or not spending the money to install systems for compliance is worth it or is indeed, feasible.

It appears it is smaller facilities that are slipping through the regulatory net, and it is the smaller facilities that face extinction because the capital outlay for systems to capture and dispose of hazardous waste is considerable. For instance, it costs Duncan Aviation in Lincoln, NE, "somewhere between $1,500 and $2,000 per paint job to take care of all the hazardous waste."

Ranger Aviation in the west Texas town of San Angelo, a self-described "small paint shop," was audited by the Texas Natural Resource Conservation Commission (TNRCC).

"We were doing some of the stuff," says Ranger owner John Fields, "but we had no idea what it was going to take to comply with the regulations." Although it crossed his mind to abandon aircraft painting, Fields decided to stick with it so that he could continue to provide full service at his FBO. It wasn't easy and it isn't cheap.

Installing systems

Fields installed a system for capturing paint stripper and the stripped paint which is then put in barrels and transported to an incineration site at a cost of between $350 and $650 a barrel. Fields also installed a water treatment system, and because of the toxicity of the main stripping agent, methylene chloride, painters wear respirators and full rain suits. Additionally, because the TNRCC said his concrete floors are porous—and thus give access to the underlying earth—and consequently pose a danger to ground water, no hazardous mat could be allowed to contact the floor. As a result, Ranger had to strip and paint on a plastic floor liner.

"Literally," says Fields, "I now spend about 50 percent of my time addressing environmental concerns. It's mind-boggling. I'm sure there are some smaller paint shops that are not going to be able to do all this. It's been difficult on us because we're a small paint shop."

Because of the hazardous nature of methylene chloride (operators suspect it will be banned eventually), Fields and other painters have investigated other forms of paint stripping, particularly with dry media. The results have been mixed although dry media manufacturers are quick to defend their products.

A variety of dry stripping media exist, including plastic "beads," wheat starch, and carbon dioxide pellets. But many operators report they don't work well on modern, multicoated paint jobs, and they are apt to damage the skins of general aviation aircraft. In addition, dry stripping methods are not approved by aircraft manufacturers.

"We have always used chemicals," says B.J. Wagner of Cypress Aviation in Lakeland, FL, "but we have tried dry stripping and haven't had good luck with it; it can warp the aircraft skin."

Jed Heastrup, sales manager for Aerolyte Systems, which makes equipment that uses lightweight media, says, "I'm not aware of anybody who is concerned about the use of plastic. You have to be trained. Even the most experienced blaster can damage an aircraft. We don't recommend blasting with any greater than 40 psi at shallow angles."

Some shops have experimented with acid-based strippers and, indeed, some such as Beechcraft San Antonio (TX) are using them. However, they have to be used with great care because the acids can eat away hangar floors and corrode fasteners on aircraft.

Concern for the air

While stripping paint seems to get the major share of attention in terms of environmental concerns, air pollution from the paint spray is covered and will get greater attention in the future.

At the moment there are limits on the amount of air pollutants that any one facility can emit into the atmosphere. But operators expect that aircraft will have to be painted in sealed booths and any hazardous air pollutants captured and remediated. In the meantime,

Airframe Technology

OEMs are continuing to develop high solid paints which use less solvent, high volume low pressure (HVLP) guns, and electrostatic applicators.

"If people don't have HVLP or electrostatic applicators," says John Stewart, owner of P&J Aircraft in Spring, TX, "they're going to have to have them real quick. HVLP is a learning process after using a syphon-type gun. It's much different. But it's to lessen the amount of overspray going out into the air."

Jobbing out and liability considerations

Full-service facilities like to offer painting because it brings in other business; if you must get a paint job, then you might as well get other modifications and repairs done during the downtime. However, the stringent regulations and their attendant costs prompted some operators to think of getting out of the paint business and just jobbing out that portion of their service. An extensive search failed to reveal a facility that had actually abandoned painting, but it is a strong testament to the regulatory climate that many operators had at least considered such a move.

"It crossed my mind," says John Fields at Ranger Aviation of giving up painting, "but I wouldn't job out paint jobs on a customer's aircraft. Painting is labor intensive and in order for the end product to be good, the preparation has to be correct. I wouldn't surrender control of that."

In another effort to avoid having to deal with regulations, there has been talk of operators in border states sending aircraft to Mexico, where the environmental regulation is thought to be less stringent. Beside the obvious fact that damaging the Mexican environment is still damaging the natural environment, there is a big question about the general quality of work south of the border.

"I'd be very careful about sending work to Mexico," says one Dallas (TX) area operator. "The quality is in question. You can get good paint jobs down there but it's not a given. If I were going to do that, I would go down and look at the facility very carefully, and I'd get a list of past customers and talk to them.

"In fact, looking over the facility and checking with customers is what I would do if I were jobbing out work anywhere, in the United States or anywhere else. It's good policy."

As with anything else in aviation, liability looms large as a consideration. Some shops will tell you the customer can be held liable for a portion of any damage to the environment arising from noncompliance at a paint facility.

"My understanding," says Jeannine Falter, manager of aircraft modification and completions at Duncan Aviation, "is that because of the cradle-to-grave concept the EPA works with, you are responsible for any hazardous waste that you generate. It (EPA) will always go for the entity that has the greatest ability to pay." Meaning that the operator and the customer could be held jointly responsible for any damage.

Because of this and because they don't want to be perceived as unconcerned about the environment, corporations are asking operators about their EPA compliance.

Regulation likely to increase

Environmental compliance is likely to become more difficult if a new EPA rule for the aerospace industry is adopted as expected some time this summer. The proposed regulation is known as National Emission Standards for Hazardous Air Pollutants for Source Categories: Aerospace Manufacturing and Rework, or NESHAP.

While the rule appears aimed at commercial and military painting operations, general aviation facilities are included. One estimate says that a facility that strips and paints 35 large business aircraft per year is likely to be covered by the regulation.

The rule calls for the virtual elimination of methylene chloride, which will force operators to use dry media or acid-based strippers. As noted earlier, there is little confidence among general aviation operators that dry media works on relatively thin-skinned GA aircraft. However, developments in stripping technology are ongoing. Stay tuned.

NESHAP is likely a harbinger of things to come—regulation is apt to get tighter and tighter in the future. Many operators don't like it; as one says, "It's getting ridiculous." But unless there is a groundswell of protest, which doesn't seem likely, the tide of regulation is rising and will continue to do so. *March/April 1995*

Pitot-static system testing

By Richard Floyd

Prior to beginning any testing on pitot-static systems, it's imperative anyone performing these tests be in compliance with FAR 91.411 and familiar with the specific aircraft system.

It's equally important the test equipment used to perform these procedures be equivalent to that specified in Advisory Circular 43-2B and Advisory Circular 43-203B, and the personnel performing the tests observe all the precautions given in the related FARs and advisory circulars.

Prior to beginning any tests, ensure the aircraft has not been flown, or the static system or the altimeter subjected to testing for at least three hours. Also verify that the temperature of the aircraft altimeter and the test equipment are within 5°F for at least 15 minutes. Lastly, make sure all hoses, adapters and test equipment are leak checked and all equipment is within its calibration interval.

Pitot system inspection

Although testing the pitot system is not required per the FARs, such tests become necessary if the aircraft airspeed indicator has an indicated range of 150 knots or less, or any static system testing will require a decrease of indicated altitude of more than 1,000 feet above ambient static pressure, or any static system testing will require a decrease of indicated altitude below ambient static pressure.

If any subsequent static system tests would cause the range limits of the airspeed indicator to be exceeded, it will be necessary to connect the pitot and static systems together. Therefore, it's important to test the pitot system for leaks prior to connecting to the static system; otherwise, leakage observed during the static system leak check could be in the pitot system.

To begin the test, connect the test equipment to the aircraft pitot system. Using the test equipment, increase pressure until airspeed reaches approximately 75 percent of full range. Then close off the pressure source and allow one minute for the system to stable. Record the airspeed reading and time for one additional minute. The reading must not change by more than 2 knots.

When testing is complete, return the system to ambient. If more than one pitot system exists, repeat the check for each.

Static system inspection

It's a good practice to first perform a visual inspection of the ports, plumbing, accessories, and instruments connected to the static system, and repair or replace

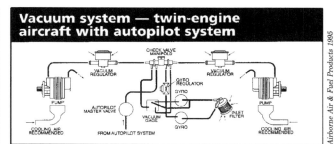

those parts which are defective. Purge the system, if necessary, to remove foreign matter which may have accumulated in the tubing, and check the static heater to assure proper operation.

When an aircraft has more than one static system, test each system separately and ensure the leak rate for each system is within tolerance.

Connect the test equipment directly to the static ports, if practical. If not, connect to a static system drain or tee connection and seal the static ports. If the test equipment is connected to the static system at any point other than the static port, make sure that it's at a point where the connection can be readily inspected after it's returned to its normal configuration. Be sure to remove all static port seals after completion of the static system test.

Static system leak testing

Unpressurized aircraft

Using the test equipment, evacuate the static system at a rate not to exceed the maximum reading of the aircraft climb indicator until the altimeter increases indication by 1,000 feet.

Then slowly begin to close off the vacuum source while watching the VSI, and ensure that a severe leak which could cause descent at a rate greater than the VSI range is not present. If a severe leak does occur, use the vacuum control to bring the system back to ambient and

Airframe Technology

correct the leakage problem before proceeding. Now close off the vacuum supply and allow one minute for the system to stable. Record the altimeter reading and time for one additional minute. The reading must not change by more than 100 feet.

When testing is complete, return the system to ambient at a safe rate. If more than one static system exists, repeat the check for each system.

Pressurized aircraft

For pressurized aircraft the test is similar, but you must first determine the *maximum cabin differential* pressure for which the aircraft is type certified.

Having preset the aircraft altimeter barometer to 29.92, evacuate the combined pitot-static system at a rate safe for the aircraft VSI until the predetermined maximum cabin differential pressure has been achieved (as indicated by the Cabin Differential Pressure indicator or an accurate vacuum gauge connected to the static system).

Then slowly begin to close off the vacuum source, while watching the VSI, to ensure that a severe leak (which could cause descent at a rate greater than the VSI range) is not present. If a severe leak does occur, use the vacuum control to bring the system back to ambient and correct the leakage problem before proceeding.

Now close off the vacuum supply and allow one minute for the system to stabilize. Record the altimeter reading and time for one additional minute. The leakage rate shall not exceed 2 percent of the indicated altitude, or 100 feet, whichever is greater.

When testing is complete, return the system to ambient at a safe rate. Again, if more than one static system exists, repeat the check for each system.

Altimeter tests

It's wise to begin with a visual inspection to check the general appearance of the altimeter for obvious defects such as a bent or broken setting knob, cracked or loose glass, cracked or broken case, bad port threads and/or fittings, peeling paint on dial or pointers and evidence of corrosion.

Next check that the knob, pointers, and baro scale turn smoothly without binding. Also, check that it has sufficient drag so the setting will not shift with normal vibration and that pulling or pushing on the knob while turning will not cause disengagement of either the pointers or barometric scale.

If the altimeter is being bench tested, a case leak should be performed next. To do this, connect the altimeter to a controllable source of vacuum. With the baro set for 29.92, reduce the pressure at a rate not to exceed 20,000 fpm until the altimeter reaches 18,000 feet; then close the vacuum supply and allow the pressure to stabilize. The leakage rate at 18,000 feet shall not exceed 100 fpm. Lastly return the pressure to ambient at a rate not to exceed 20,000 fpm. Any excessive jumping or sticking during ascent or descent will be cause for rejection.

Next, perform the *Barometric Scale Error Test*. With the altimeter or system vented, no test equipment connected, set baro to 29.92. With normal vibration, record the indicated altitude. Set the baro scale to each of the values specified in Table IV of Appendix E, Part 43. Again with normal vibration record the indicated altitude for each test point. The altitude difference from that recorded for the original 29.92 setting shall agree at each point ± 25 feet.

The next test is the *Barometric Correlation* (altimeter setting). This requires connecting the test equipment to the altimeter for bench testing or to both the pitot and static systems on the aircraft.

With both the master and test altimeter baro set for 29.92, adjust the vacuum control of the test set to produce zero feet altitude, as indicated by the master altimeter. Allow at least one minute for the pressure to stabilize; then carefully set the barometric correlation to produce a zero feet indication with a baro set to 29.92.

It should be mentioned that the FAA does not condone resetting of the altimeter barometric scale unless the person conducting the test is qualified to make the necessary barometric correction.

We are now ready to perform the *Scale Error and Friction Tests*. If testing on aircraft, remember that although rate of ascent is not critical, it is specified as 5,000 to 20,000 fpm during descent for the hysteresis tests which follows this test. If this exceeds the aircraft VSI, then it will be necessary to isolate the VSI. Also some provision is required to isolate the altimeter from the static system to perform the case leak test.

To begin the *Scale Error Test* ensure that both the master and test altimeter are set to 29.92. With the system vented to ambient, record the test altimeter reading and the current barometric pressure (altimeter setting). These values will be used later for the *After Effect Test*.

With the tester properly connected to the altimeter, or static system, operate the tester valves or controls to establish the first test point (-1,000 ft). Now, apply the correction factor from the calibration card to the master altimeter. Maintain the test pressure for not less than one, nor more than 10 minutes.

Apply light vibration to master and test altimeter. Check that test altimeter is within allowable ±20-foot tolerance at -1,000 feet; then increase altitude to zero feet, per corrected master, and allow minimum one, but less than 10 minutes for pressure to stabilize. Next, apply light vibration to master and test unit. Check that test altimeter is within allowable tolerance.

The friction test may be combined with Scale Error Test by increasing altitude at 750 fpm while vibrating the master altimeter only. Then when the master reaches 1,000 feet (first friction test point), maintain the pressure and note the test altimeter reading.

Airframe Technology

Apply vibration to the test altimeter. The test altimeter after vibration shall not exceed the ±70-foot tolerance at 1,000 feet. Continue the combined scale error and friction tests in this manner for each specified test point until reaching 18,000 feet.

After completing the scale error test at 18,000 feet, close the vacuum supply to the test altimeter. Isolate the altimeter from the static system if the test is being conducted while installed in the aircraft. After allowing the pressure to stabilize, observe the test altimeter for one-minute. Leakage shall not exceed the 100 feet in one-minute tolerance. Return the system to normal operation and continue the combined scale error and friction test until reaching the desired maximum test altitude.

The *Hysteresis Test* must be performed within 15 minutes after reaching the desired maximum test altitude of the scale error test.

To accomplish this, begin increasing pressure (descending altitude) at a rate not less than 5,000 fpm nor more than 20,000 fpm (isolate aircraft VSI if required). Take care not to overshoot the test point. When reaching a point 3,000 feet above the first hysteresis test point, reduce the descent rate to approximately 3,000 fpm.

Observe the master altimeter while applying moderate vibration, and stop the descent when the master reaches the approximate hysteresis test point with proper correction card.

After not less than five minutes, nor more than 15 minutes, while applying moderate vibration to both master and test altimeters, the test altimeter reading shall not differ from the reading observed by more than the allowable 75-foot tolerance.

If any hysteresis check point is not within 500 feet of one of the scale error points, it will require a notation of the test altimeter reading. Since the master altimeter may not have a correction value for this point, the master should be stopped at the appropriate check point without correction and the test altimeter reading noted.

Continue descent to second hysteresis check point in the same manner. With the master altimeter stable, at the second check point with appropriate hysteresis correction, after not less than one minute nor more than 10 minutes, with moderate vibration, the test altimeter shall not differ from the reading observed during the scale error test by more than the allowable 75-foot tolerance.

To accomplish the *After Effect Test* continue descent in the same manner as described in the hysteresis test until the system has reached and is vented to ambient pressure. After reaching ambient pressure, within five minutes of completion of hysteresis test, determine the current barometric pressure (altimeter setting) and compare current barometric setting with that recorded prior to start of scale error test. If barometric pressure has changed since beginning of test; determine amount and direction of change and change setting of barometric scale of the test altimeter from 29.92 by the same amount and direction.

Then with vibration applied to the test altimeter, observe the reading. The difference between this reading and that recorded prior to the start of the scale error test, shall not exceed 30 feet.

Completion of testing

Upon completion of satisfactory testing disconnect the test equipment from the altimeter or aircraft If testing on aircraft, remove all static port seals and restore systems to their original status. Comply with all maintenance record entry requirements.

Failure of any of the tests will necessitate correction and retest. *January/February 1995*

Richard Floyd is a field engineer for Barfield in Atlanta, GA.

Airframe Technology

Engine-driven fuel pumps
Reliable, but still in need of TLC

In the complex world of fuel injection systems, the fuel pump is the device most often overlooked because of its simplicity. "Generally, because of their simple design, they are very reliable," says Charles Chapman, shop manager for Southeast Fuel Systems in Rockledge, FL, "but fuel pumps are susceptible to contamination problems and can fail or cause serious problems at a moment's notice."

The best way to care for them is to prevent contamination of the systems and to know how they operate so that you can quickly and effectively troubleshoot them.

Chapman explains that "the most common problems we see at our repair station are the result of foreign objects in the pump, particularly in the relief valve area which causes the pump pressure to drop into the lower rpm ranges.

"The introduction of foreign objects to the pump relief valve can be eliminated, however. Careful installation, fuel line hose inspection, and proper application of Teflon tapes (anti-seize) to threaded fittings will go a long way toward this, and checking your filters and strainers regularly will also help reduce pump failures."

Chapman explains that they always recommend regular inspection and cleaning of fuel filters at the tank, and at the fuel injector.

Following are some specific items that should be looked at along with brief descriptions of the critical operating features of different fuel injection pumps.

Continental fuel pumps

Scott Rivenbark, shop supervisor for Southeast Fuel Systems, says, "Anytime you have a problem with a Continental fuel injection system, go to the pump to inspect it—regardless of what fuel component is failing. The pump is almost always involved in some way with the problem."

Rivenbark explains that the Continental fuel injection system on a normally aspirated engine meters its fuel as a function of the engine rpm, and doesn't use airflow as a metering force.

The engine-driven pump is the heart of the system and provides the fuel metering pressure. That's why you should always look at the pump first in the event that problems develop.

This pump is a vane-type, constant displacement with special features that allow it to produce an output pressure that varies with engine speed.

This system works well for flows in the cruise or high power range, but when the flow is low, as in idling, there's not enough restriction, he explains, to maintain a constant output pressure. Therefore, an adjustable pressure relief valve is installed in the line.

During idle, the output pressure is determined by the setting of the relief valve, and the orifice has no effect. While at high speed operation, the relief valve is off its seat and the pressure is determined by the orifice. As in any fuel injection system you must have vapor-free fuel in the metering section. A special function of the pump is to remove all vapor from the fuel and return it to the tank.

An additional feature of this pump is a bypass check valve around the pump so fuel from the boost pump may flow to the fuel control upon starting. When the engine pump pressure becomes higher than that of the boost pump, the valve closes and the engine pump takes over.

Turbocharged engines

Rivenbark continues that turbocharged engines, have a unique problem during acceleration. If the fuel flow increases before the turbocharger has time to build up to speed and increase the airflow proportionately, the engine may falter from an overly rich mixture.

Airframe Technology

There is a special aneroid valve for these pumps which will vary the output fuel pressure proportional to the inlet air pressure. In layman's terms, when the throttle is opened and the engine speed increases, rather than immediately supplying an increased fuel pressure to the control, the aneroid holds the orifice open until the turbocharger speed builds up and increases the air pressure into the engine.

Lear-Siegler pumps

Bill Rivenbark, president of Southeast, explains that Lear-Siegler pumps are positive displacement, vane-type units, designed mainly for use in reciprocating aircraft engines fuel injection systems.

He says that "although similar in construction, each pump is tailored to meet the specific requirements of the fuel system in which it has application. Engine and fuel system features such as direction of rotation, mounting considerations, interface with engine accessory drive, and fuel system performance demands, affect relief valve components and adjustment, drive shaft and spline configuration, and may require casting modifications to accept special fittings."

He continues, "Lear-Siegler uses a Romec developed, vane-type structure that's enclosed in an aluminum housing and driven by an accessory drive gear. Protection of the fuel system is provided by a built-in pressure relief and pump bypass valves.

"While the pump is operating, its discharge pressure maintains the bypass valve in a closed position. Should the pump become inoperative, fuel pressure from auxiliary pumps overcomes the valve's light spring load, and fuel is allowed to flow through the disabled pump."

Scott Rivenbark explains that a line mechanic is really only authorized to perform a few adjustments. He says that "with most pumps that are constant pressure, only the relief valve can be adjusted. Frequent adjustment after prolonged use can be a sign of trouble on the way, and possibly a relief valve assembly is hardening or the valve seat is showing signs of wear.

"Other types of variable pressure pumps will have a tapered pin and variable orifice or aneroid adjustment. But as before, only a minimum amount of adjustment should be necessary on installation," he says.

Typical fuel pump troubleshooting chart

Trouble	Probable Cause	Remedy
Air leakage	Cracked castings, faulty gaskets, ruptured diaphragms	Dissamble and inspect parts in proximity of leakage area. Replace faulty parts as necessary.
	Loose mounting screws, scratches, grooves or other surface irregularities to mounting or mating surfaces that afford a path for leakage	Tighten screws to prescribed torque value. Disassemble and inspect parts in proximity of leakage area. Dress out or replace parts as necessary.
Low or pulsating flow	Air leak in test system	Trace out plumbing circuit. Make sure all plumbing connections are clean and tight. Bleed air from system.
	Worn pumping element components	Disassemble and inspect pumping element components. Take necessary action to correct condition.
	Faulty instruments	Make sure only calibrated instruments are used in tests. Instruments must be accurate within ±2 percent of reading.
Unable to obtain pressure setting	Relief valve hang-up	Disassemble and inspect condition of relief valve poppet and seat. Inspect valve spring and associated parts for improper assembly or damage. Repair or replace parts as necessary.
	Faulty instruments	Make sure only calibrated instruments are used in tests. Instruments must be accurate within ±2 percent of reading.
Noisy operation; unit runs hot	Foreign matter in pumping mechanism; improper assembly	Completely overhaul pump.

Airframe Technology

"The line mechanic should also be alert for dye stains and for the presence of fuel leaking from the overboard drain. Any leaking in a fuel pump is unacceptable.

"Also, prior to installation of any pump, check all fuel lines to the pump for deterioration, and use torque values as specified by the pump manufacturer. If fittings are to be installed on your pump, be careful not to contaminate the pump with tape or a thread compound.

"And finally," he says. "if your aircraft is going to be down for any length of time, be sure to drain your pump of fuel. Also, avoid the use of compressed air because it can introduce moisture in your pump and cause corrosion. Cap all open fittings with recommended caps."

Contamination can lead to a damaged valve seat and cause fluctuating pressure or the inability to attain proper operating pressures.

More fuel pump tips

Welton Maynard of Consolidated Fuel Systems in Montgomery, AL, says they see contamination of fuel pumps as being the biggest problem. He recommends that during the installation of pumps, all lines to the pumps should be blown through with clean shop air or flushed to assure they are clean.

He warns that during the installation of the fittings into the pump, the metal is frequently pulled from the threads of the pump housing which then causes damage to the valve and seat area of the pump. "Because of this, pumps overhauled by Consolidated Fuel Systems are sold with the fittings installed. This assures that any metal pulled from the thread is cleaned out before the assembly of the pump is completed," he explains.

He also suggests the following:

- Take care not to depress the drive shaft on installation or in handling even though it will spring back into place. A drive seal leak may result.
- Proper torque on all fittings and the use of Teflon paste will help prevent cracked bosses.
- When removing and installing fittings, if you must clamp the pump in a vise, do it by the mounting flange—not the body.
- At periodic inspections, you should pressurize the pump with the boost pump and the mixture in idle cutoff. This way you can check for leaks on the pump body and drive seal.

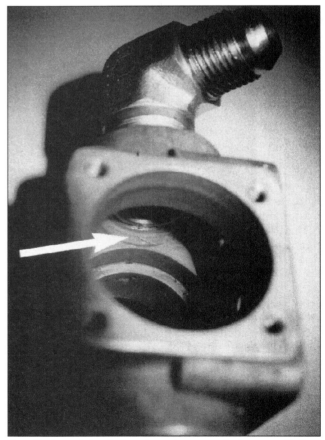

Installation of the fittings into the pump body can pull the threads and result in metal contamination. This must be cleaned prior to running the pump.

Airframe Technology

What every technician should know about turbine fuel contamination

Fuel contamination in jet fuel can lead to many problems. Most of the time, contamination is easy to control by changing filters and adhering to maintenance procedures which stress cleanliness.

Unfortunately, most maintenance personnel aren't aware of the fact that fuel systems can be contaminated by microorganisms, and this contamination is difficult to control. It cannot be seen by the naked eye, and once it contaminates the aircraft's fuel system it will continue to exist unless it's controlled.

The following Q & A, offered by Hammonds Fuel Additives Inc., describes the effects of microorganism fuel system contamination in jet fuel and discusses methods for detecting and controlling it:

Q. What are microorganisms or bugs?
A. The scientific names for the types of organisms that live in petroleum products are Cladosporium resinae and Pseudomonas aeruginosa. The organisms are either air- or waterborne and contaminate fuel systems by entering through vents, standing water in sump bottoms, dissolved "free water," or trash incurred during transport or delivery. They grow at incredible rates with some varieties having the ability to double in size every 20 minutes.

Q. What do these microorganisms do?
A. These slimy bugs live and multiply in the fuel/water interface. They actually exist in the water and feed off the hydrocarbons in the fuel. They are referred to as "hydrocarbon utilizing microorganisms." As they grow, they form mats that are dark in color and appear gel-like. Their waste produces water, sludge, acids, and other harmful byproducts. Microorganisms will also consume rubber gaskets, O-rings, hoses, or tank linings and coatings.

Q. How do I know I have microorganism contamination?
A. The signs of microbial growth can vary. Some of the obvious signs are clogged filters, loss of engine power due to fuel starvation, plugged lines, contamination on tank bottoms, fuel with a sulphur smell and tank access lids with green or brown slimy formations. A more definite appraisal may be made with the use of a microbe detector kit.

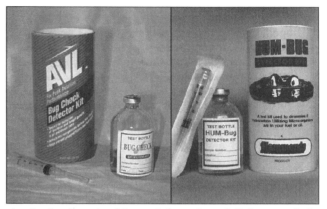

A microbe detection kit can help make a definite appraisal of whether or not fuel is contaminated. Above are two detection kits currently available.

Q. Why do I need to control microbes in fuel systems?
A. A sterile fuel system is one with lower maintenance costs. That translates to reduced equipment disruptions or downtime due to fuel system complications. The elimination of bugs in both storage and equipment tanks will also reduce the chances of fuel tank damage due to corrosion from the microorganisms' acid waste products.

Q. Isn't good fuel management and "proper housekeeping" enough to avoid problems?
A. The highest standards of housekeeping will not ensure the absence of microbial infestations and associated problems. Microbial spores can exist in a dormant stage for extended periods of time waiting for trace amounts of water or improved growing conditions.

Q. What should I do if I find I have a problem?
A. An additive is recommended (such as Hammonds Biobor® JF) to exterminate the infestation. The remains might appear as coffee grounds in the bottom of the fuel tank.

For more information on microorganism detection, contact the following:

Hammonds Fuel Additives Inc.
P.O. Box 38114-407
Houston, TX 77238

Aviation Laboratories
5401 Mitcheldale #B6
Houston, TX 77092

November/December 1994

Airframe Technology

The importance of cleanliness in hydraulic maintenance
A little attention to a clean work environment can mean the difference between failure and success

By Greg Napert

Hydraulic systems require relatively little maintenance on today's aircraft. For the most part, hydraulic fluids are capable of withstanding a wide range of operating temperatures, and the loads placed upon the systems are relatively light.

So it's not in operation that these systems usually break down; instead, it's more often than not that foreign substances are unwittingly introduced during scheduled maintenance. And it's the introduction of these contaminants that work on the system and degrade it.

"Ninety-nine percent of the key to hydraulic maintenance is cleanliness," says Paul Finefrock, president of Thunderbird Accessories Inc., in Bethany, OK. "When you work on a hydraulic component, line, or whatever, you've got to treat the area like a surgical operating room.

"Cap all ends immediately, clean all your tools, and the surrounding area, and do your work in a clean environment.

"And don't leave hydraulic fluid uncovered in an open container. Every bit of dust in the air will stick to the surface of the fluid, and the surface will begin to oxidize."

Joe Lundquist of Pall Corp., a manufacturer of hydraulic filtration systems, agrees. "We've conducted tests and have found that with hyperfine filtration (below 1 micron) bearings just don't wear out. This ties in the need for better filtration. And it's more important especially with engines that they are trying to get more and more horsepower out of. We also find that better filtration can extend the life of the oil itself, the change intervals. We have been doing some work in this area on helicopters and have found our programs to be very successful.

"If we can improve the cleanliness of the hydraulic system, you're looking at such things as improved component life and a savings on hydraulic fluid where instead of having to change it occasionally, you can get indefinite life out of the fluid.

"We're finding also that if you properly maintain a fluid by removing water (with vacuum distillation purification units), filtering particulate contamination,

Cleanliness is absolutely essential in a hydraulic pump due to very tight clearances of the pump mechanism.

and you don't stress it by shearing it or subjecting it to high temperatures, the fluid will not oxidize and will remain stabile."

Lundquist says Pall likes to recommend that maintenance facilities beginning a program like this use a company that does spectrometric oil analysis. He suggests monitoring the oil closely until they are comfortable with how long they should use hydraulic fluid and how often they should filter it.

Finefrock says that hydraulic components have become outrageously expensive; you're talking between $14,000 and $15,000 for a new Skydrol® type pump.

The criticality of proper cleanliness and proper clearances in a pump are exemplified by the fact that if you install gears in a pump that produce a clearance of over one and one-half thousandths, the pump won't pump fluid.

Airframe Technology

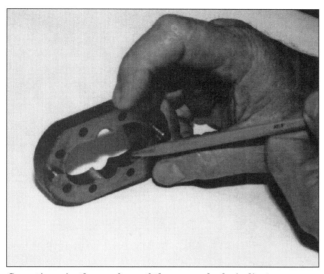

Serrations in the surface of the pump body indicate excessive contamination of the hydraulic fluid in this pump.

Source of contamination

Finefrock says that you need to be aware of contamination from sources like rubber gaskets and seals. A piece of rubber, although the reservoir strainer will take it out most of the time, can take out the hydraulic pump very quickly. Because there's only between .001 to .002 clearance in the pump, a piece of rubber can jam between the gears and seize the pump, which will shear the shaft. We've also seen where rubber works its way between the flat face of the gear and the housing and builds until it pushes the opposite side against the housing and causes metal to metal contact. And this results in the gear seizing.

Most contamination is primarily metal as a result of normal wear and tear of components in the hydraulic system, however, and this slowly wears away at the gears and bearings in the hydraulic pump.

Hydraulic fluid will oxidize and turn sticky where it's exposed to the air. And if you have a suction leak and cause foaming in the reservoir, it will cause the hydraulic fluid to break down.

You've got to remember also that when you attach the hydraulic mule to the aircraft to cycle the gear or other hydraulic systems for maintenance purposes, you are changing out the fluid with the fluid in the mule at that time. You need to be aware of this and make sure that your fluid in the mule is fresh, clean, and properly filtered preferably with a 1-micron or better filter.

Some maintenance shops will go through many aircraft maintenance checks with their mule and not even change the fluid in the mule once. Also, you've got to be aware that if you use the mule to purge a system and filter contaminated hydraulic fluid, you should clean the mule and change the fluid on the mule before using it on another aircraft.

In fact, the mule is a good way to filter the hydraulic system occasionally if it has a good filter on it.

We've also seen some of the rubber hoses in hydraulic systems break down after a period of time and begin contaminating the system. Solid lines aren't a problem. But rubber/synthetic hoses become brittle after a while and small chunks of the liner break off and contaminate the system. That's why it's important to change the hoses out at the slightest sign of them becoming brittle or damaged. There are too many aircraft flying around with 20-year-old rubber hoses that have never been changed.

Inadequate filtration systems

Finefrock says that, "one of the major problems with aircraft hydraulic systems is that there is no real good filter or filtration system on them. Basically, one reservoir services the entire aircraft. And the only filter on the system is a strainer that is on the inlet side of the pump This is a fairly coarse mesh strainer, and I swear to God that coffee grounds will go through. All it will take out is some large particle, like two microns.

"A really good idea for protecting your hydraulic system is to install a suction side filter on the aircraft that can filter particles down to 1 micron. As a matter of fact, I've installed one on mine. All you need to do is install a filter housing that's equipped with a bypass, and a flow gauge to monitor the condition of the filter (you'll need this so that you don't cause cavitation of the pump), and a filter, and you'll protect the pump from ever receiving any kind of contamination. We highly recommend putting this on the suction side of the pump between the reservoir and the pump. You could put it on the return side before the fluid returns to the tank, but then you wouldn't prevent contamination that's introduced into the reservoir from getting to the pump. Basically, you're installing the filter to protect the pump, which is the most critical part of the system and most susceptible to damage."

Finefrock continues, "Kits are available for this purpose from various automotive or commercial sources for a relatively low cost, and a field approval should be relatively easy to get.

"Some of our customers install a temporary filter to run on the system after finding contamination or after a component failure. They'll then run the system for six to eight hours with the filter on to clean out the hydraulic oil and then remove it for flight.

"Another option, of course, is to use a mule with a good filter on it and to purge the system with a mule. But you need to do it right. You need to put it up on jacks and run the gear up and down and cycle all of the hydraulics to assure moving all of the oil through the system.

"In any case, you really need a filter that is capable of removing all contamination down to a 1 micron size. With this level of filtration, the system will operate properly, you won't have any problems, you won't break pumps, and you won't be sending them back to us for warranty," explains Finefrock.

Some product improvements

Finefrock says that one of the changes they have made to accommodate the inevitable introduction of contaminates to the hydraulic system is a to replace the bronze bearings on certain hydraulic pumps with needle-type bearings. Contamination in the hydraulic fluid causes conventional bronze bearings to elongate. The result is too much play and eventually the gears begin binding on the gear faces until it shears the drive shaft.

The modification involves boring out the existing brass bushing, replacing it with an oversized aluminum bushing, and then pressing in a needle bearing.

"The cost of performing this modification on an 105HBG pump, for example, is approximately $1,100, where buying a new one is $5,600," he says.

"Additionally, we've found the need to increase the diameter of the shear section of the drive for this same pump. We feel that it was originally designed to fail at too low a pressure.

"A discovery that we made, quite by accident, is that shafts are often shearing due to malfunctioning accumulators. What happens is as a result of the check valve in the accumulator sticking, the accumulator bangs in rapid succession, and this hammers at the drive shaft and eventually shears it. This happened to us once on our test bench.

"It really doesn't take much hammering either. It only has to happen one time and it cracks the shaft, or it can also shear the key that engages the driven gear of the pump.

"When we receive a pump in for overhaul that has a sheared shaft or key, we now call the customer immediately and tell them that they may have a problem with the accumulator," says Finefrock.

He explains that if the accumulator isn't working, it can also blow the seal out of the intake side of the pump. "It happens because you shut the engine down and the back pressure forces its way through the gears and puts reverse pressure on the seal," he says.

"Many times technicians assume that it's a problem with the pump, but if you blow out a seal, it's usually a problem with the accumulator.

"If the seal is blown, you have to take the pump apart to replace the seal, which means you have to put the pump back on a test bench to make sure it works.

"When we see this, we immediately alert the maintenance facility that they need to repair the accumulator."

Finefrock says that another way to damage a pump is to fail to purge the air out of the hydraulic systems after performing maintenance on them. If you don't purge the system and get the air out of it, the pump cavitates and then surges, and this frequently shears the drive, he explains.

Modified pump for the purpose of making it more resistant to wear due to contamination involves replacing the original brass bearing with a needle bearing. First, the brass bearing is machined out (left), an aluminum bushing is installed (middle), and the needle bearings (right) are installed.

"What we do is use a cap that is adapted with an air regulator and gauge on it. We attach it to the reservoir and put 2 1/2 pounds of pressure on it. We then take the outlet line from the installed pump and pull the prop through by hand or motor it to turn the pump until we have clear fluid coming out of the discharge line, no bubbles. Sometimes it takes a gallon or 2, but it's worth it to save the pump. You'll never have a problem if you do that. Directions for this are in most aircraft maintenance manuals," he says.

"We had one company put three pumps on down in Georgia that had air in the system, and he called accusing us of problems with the pumps. We told him if he read the directions, that he would purge the system and that was the problem."

More frequent maintenance please

Finefrock says he's observed that too many people are trying to make their accessories last to TBO before overhauling them. "We never use to do that. We use to overhaul all of our accessories at 50 percent of TBO, and we never had any problems. And I guarantee that it costs very little to have a unit that's in good condition overhauled, compared to replacing something that's broken, especially where TBOs on engines are being extended longer and longer. People are expecting the accessories to last as long as these engines, and the accessories rarely went to TBO when they were much lower.

"Not to mention that if you wait till something in the hydraulic system fails, there's a good chance that you'll get contamination from the parts that have failed, such as metallic chips, gaskets, shavings, etc., and end up damaging other components in the system. At the very minimum, you'll have to flush the system thoroughly."

September/October 1994

Auto pilot INOP: Oh no!
Key to troubleshooting autopilots is in asking the right questions

By Jim Sparks

Automatic Flight Control Systems (AFCS) are categorized by how many axes of flight they can control: The single axis autopilot, found in many small aircraft, is a device for holding the wings level. Dual axis autopilots can sometimes coordinate aileron and rudder, or, in other systems dual axis will work with an on-board "yaw damper" to provide pitch and roll control. Many systems today are three-axis, and integrate pitch, roll, and yaw.

Isolating malfunctions

When malfunctions in the AFCS exist, it becomes necessary to associate the fault with a specific axis of flight:

The pitch axis is the vertical mode and control information supplied from air data systems, radio navigation, vertical navigation computers, or flight management systems. Autopilot response to commands from the vertical mode can be made by elevator deflection, changing the angle of attack of the horizontal stabilizer or even by moving trim tabs. If the flight control cannot respond to a specific condition, the autopilot can interpret this as a failure.

A common malfunction in the vertical mode of auto flight is "porpoising." This is defined as "a low frequency oscillation in the pitch axis." The usual image of a porpoise is jumping in and out of water and not maintaining a level path. There are various situations that can result in porpoising. For instance, if cables are used to connect the autopilot servo motor to the elevator system and cable tensions are not correct, overcontrol is probable and the aircraft will porpoise. Also, several types of aircraft make pitch changes by changing the horizontal stabilizer angle of attack. Most use an electric motor for this. If the motor operates too slowly, the autopilot will always be lagging and this will cause porpoising. In an attempt to prevent this, manufacturers sometimes incorporate "brakes" to ensure an immediate stop and prevent overtravel.

However, the brake is often manufactured from carbon. With wear the carbon dust accumulates between the friction disc and rotor. This dust works as a lubricant and significantly reduces the stopping power of the brake. The result is stabilizer overrun and again, porpoising.

Not all of these low frequency oscillations can be blamed on mechanical components, though. Faulty information supplied to the auto flight system may result in the autopilot commanding the porpoise.

The information driving the autopilot is available from one of two sources: the pilot using a manual autopilot input, or more often, a "flight guidance" or "flight director system."

Flight directors will compile all flight data available and summarize it using priorities issued by the pilot. The result is a pictorial display using "V" bars or "crosshair pointers," typically on the attitude indicator. The pilot can manually fly the aircraft using this display or can allow the flight director to communicate directly with the autopilot.

By interrupting this communication or by switching the autopilot to a second flight director, the input data problems can be isolated. If the porpoising stops with an alternate flight guidance system supplying the information, a high likelihood is that the problem is with the primary flight director or one of its information providers.

Operation of the autopilot in "basic mode" is usually a good place to start fault isolation. "Basic mode" is the condition in which the autopilot will function without commands from the flight guidance system. Pitch hold and vertical speed hold are the most common "basic vertical modes." If the autopilot is engaged without selecting a flight director mode, the autopilot will most likely hold the aircraft just as it was when activated.

When a porpoising problem is encountered, you should determine if the porpoising is also occurring with the autopilot in its basic mode. If it is, then the problem may be associated with the basic mode sensors, air data, attitude reference, or with the autopilot to airframe interface. If the porpoising does not occur in the basic mode but only with ALTitude Hold engaged, then the sensors that supply the aircraft altitude reference and the altitude selection may be a good beginning in the fault isolation process.

Find out how the system performs with other vertical modes, which are: air speed hold, mach hold, vertical speed hold, and approach. If the porpoising disappears during approach with the automatic flight system receiving its information from a navigation receiver vs. an air data computer, common sense dictates that you check the air data systems.

Other common vertical mode discrepancies include "pumping," a low frequency control wheel movement back and forth, "stick bump," where controls give a quick

Airframe Technology

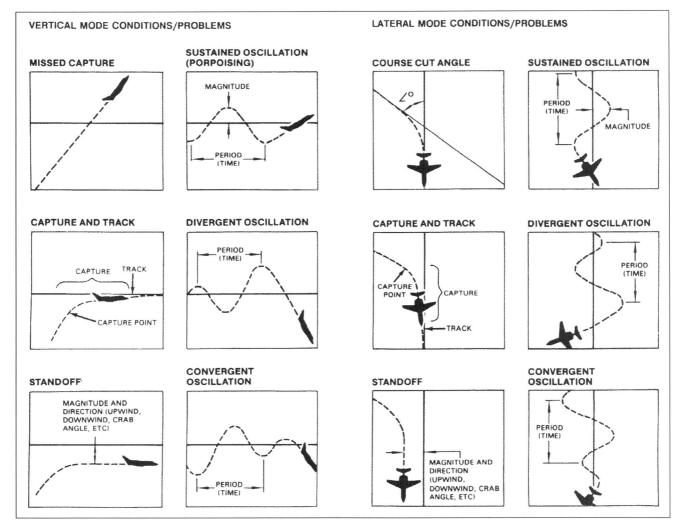

moderate movement with little aircraft reaction, and "stick buzz," a high frequency small movement of the control wheels with no aircraft response.

Lateral mode auto flight discrepancies can also prove quite challenging. The basic lateral mode is "bank hold," or in some cases "heading hold." These modes are if the aircraft is presently making a turn and the auto pilot is engaged the aircraft will continue to turn until the autoflight system receives a command that has priority over the basic mode. Other common lateral modes include HeaDinG, NAVigation, and APProach.

The HeaDinG mode will use information provided from the Horizontal Situation Indicator. The actual aircraft heading is compared to a heading reference selector. Anytime the HDG mode is activated and actual aircraft heading does not agree with selected heading, the AFCS will command a bank to the selected heading. The bank angle is preprogrammed and can be automatically altered, depending on aircraft speed or altitude. A failure of the directional sensing system will usually prevent the autopilot from being engaged. The directional system also can provide rate of turn information.

For the NAVigation mode to function a navigation receiver has to be operational. When maneuvering with VOR (very high frequency Omni directional Receiver) information, not only does the NAV receiver have to be appropriately tuned, but a course also has to be selected for the specific station. The VOR station transmits omni directional signals in a full 330-degree arc. Each degree of the arc is referred to as a VOR Radial.

The Approach function operates in a similar manner as NAV. The gain or system responsiveness might be modified with the aircraft in a landing configuration since airspeed is usually much lower than during cruise. This means the autoflight system will have to use greater flight control deflections.

This sensitivity adjustment is made using airspeed information or radio altitude. Localizer gain programming can also be activated by tuning a NAV receiver to an Instrument Landing System frequency. Malfunctions in the gain programming can cause problems such as oscillations. The autoflight system continually overcorrects and the aircraft cannot properly track the NAV signal.

The yaw axis is usually controlled by a rudder autopilot servo or yaw damper. This function is used to ensure turn coordination and to prevent slip. During high speed flight, if airflow over an outboard wing is disrupted, a wing dip could result. If uncorrected, this asymmetric airflow could result in a dutch roll. A yaw damper can provide immediate rudder response to prevent the dutch roll tendency.

The primary source of information for rudder control is a yaw sensing gyro. Some systems will use the normal inertial reference system for this information. Other manufacturers will use an independent yaw gyro. Yaw damper authority, or the amount of rudder deflection, can be regulated by airspeed. In many cases discrepancies with the yaw damper can be related to airframe or engine problems. If a flap system is misrigged and one flap is extended slightly, this will cause an asymmetry in wing airflow and the yaw damper will have to compensate. In some systems a continuous yaw damp command signal will result in a yaw damp disconnect. When the yaw damper can be engaged and consistently disconnects, a good place to start troubleshooting is with checking flight control rigging, engine alignment or asymmetric thrust, and even possible fuel imbalance.

In most AFCS the yaw damper can be engaged without the autopilot, but the autopilot will not engage without the yaw damper.

A good guideline to follow when faced with an autoflight discrepancy is based on two rules:

1. Before taking any action, determine conditions when the problem exists.
2. Get the flight crew to give a complete description of the problem.

Asking questions

There are many questions to be asked for accurate diagnosis. Here are a few to keep in mind:

- Are any fault annunciators illuminated?
- Which flight axis is experiencing the problem?
- In what modes or conditions does the problem occur?
- Is the problem related to flaps, landing gear, or airbrakes?
- Does it occur only at certain engine power setting?
- Is airspeed a factor?
- Is altitude a factor?
- Does the problem occur in more than one flight guidance mode?
- Could the problem occur with a specific sequence of mode selection?
- Do only certain radio frequency selections give the problem?
- Does the problem only occur when using a radio transmitter?
- Does the weather radar have an effect?

Remember, even the configuration of electrical power sources (alternators, generators, inverters) may introduce malfunctions.

Built in test equipment

In AFCS using digital technology it is common to find Built In Test Equipment BITE. This is used to continually monitor the AFCS and record any faults. This memory can then be accessed and the displayed fault code translated to locate the failure. These "flight fault" memories, in many cases, can be erased by removing power from the system. It is essential to inform the flight crew to leave the system powered so any fault codes can be recovered. Prior to troubleshooting any AFCS discrepancy and after a thorough crew debrief, it's beneficial to obtain an operating manual for the specific installation and read about how it's really supposed to work. The airframe manufacturer's maintenance manuals, *Airline Transport Association*—Chapter 22 "Auto Flight" and Chapter 34 "Navigation"—can add depth to the text of the operating description. This sometimes is sufficient to make a knowledgeable judgment as to faulty components. If the problem can't be diagnosed at this level, it may be necessary to get avionic manufacturers' manuals and installation diagrams. *July/August 1994*

Jim Sparks is an instructor for FlightSafety International in Little Rock, AR. He has over 13 years of maintenance instruction and holds an A&P and FCC certificate.

Airframe Technology

Large aircraft air conditioning
An overview

By Jim Benson

To transport passengers and crews from point A to point B comfortably, an aircraft's environment must be designed to satisfy any demands that will be encountered during a flight. To do this air inside the cabin is modified with a process called air conditioning.

Air conditioning is the process of treating air so as to control simultaneously its temperature, humidity, cleanliness, and distribution.

The Garrett Corporation, a division of Allied Signal of Phoenix, AZ, furnishes about 90 percent of environmental systems for the aircraft produced in the free world, and today, virtually all modern commercial aircraft use an air cycle system called an air-conditioning pack (short for package). The basic supply of air for an air-conditioning pack while in flight comes from the aircraft's engines. When parked, it comes from a built-in auxiliary power unit called an APU.

The air is commonly referred to as pneumatic air and is bled from the compressor sections of the engines or APU. Newer air-conditioning systems have been designed to run so efficiently they use less of this pneumatic air source, thereby, allowing more air to be used by the engines or by smaller more efficient APUs.

Improvements in the design and operation of air-conditioning systems have led to less maintenance, higher efficiency, and better troubleshooting. Computers not only monitor and control the operation but store fault information and data for current and future interrogation to help keep the system running at peak performance. This translates directly to lower maintenance and fuel expenses.

The air-conditioning pack

A basic air-conditioning pack is made up of a pack flow-control valve, two heat exchangers, a bypass valve, anti-ice valve, air cycle machine, and a water separator. The air is typically bled from the engine at about 45 pounds of pressure and 400°F. The bleed air enters the pack through the flow-control shut-off valve. This valve can be a combination shut-off and flow-control valve, or separate shut-off and flow-control valve. Its purpose is to start or stop the pack and control the volume of air based on the area to be heated or cooled.

The flow-control valve is basically a venturi with a valve. When closed, it can act as a shut-off valve and when open it is modulated to control the flow of air. If we measure the air pressure at either end of the venturi and compare it to the air pressure at the throat, we find that the air pressure at the throat of the venturi is lower than air pressure at either end.

This difference in pressure is affected by the amount of air flowing through it, so the greater the flow of air, the greater the difference in pressure. This difference in pressure is used to control the movement of the valve, thus controlling the amount of air flowing to the air-conditioning pack.

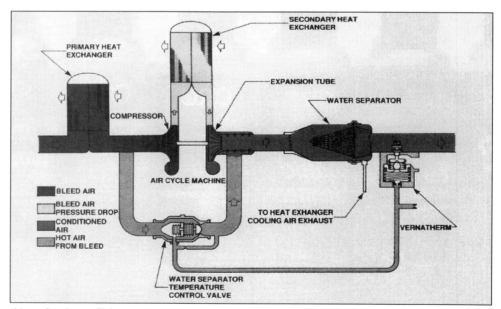

Air-cycle air conditioner

After the air passes through the flow-control shut-off valve, it's routed through a heat exchanger. The heat exchangers are a radiator-type device that uses ram air or a cooling fan (or both) as a heat sink to cool down the air before it enters the air cycle machine.

The typical air cycle machine (ACM) consists of a centrifugal-type compressor and turbine. When the cooled air enters the ACM's compressor, the air is compressed, thereby, increasing its velocity and temperature. This increase in temperature brings the compressor output temperature back to what it was when it entered the heat exchanger, but the air velocity is now increased. The compressor outlet temperature increase is closely monitored and if it is too high, the pack will be signaled to shut down. This feature protects the subsequent heat exchanger in the pack. If the air is too hot, it will warp the cooling fins and destroy the exchanger, plus it will be too hot for a sufficient pack output temperature.

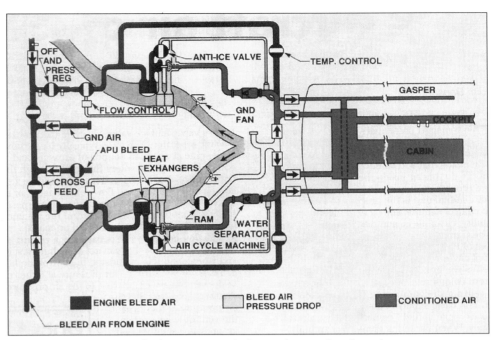

Air-conditioning system of a jet transport airplane using an aircycle system.

The next path for the air depends on the type of pack installed in the aircraft. In older traditional pack systems, the air is routed to the turbine section of the air cycle machine where the air is expanded. This results in lower velocity air with a cooler output. The moisture content of this expanded air must now be removed with a water separator.

This brings us to one of the weaknesses related to this type of system. With this style of air-conditioning pack, if the temperature drops below 35°F, the water will freeze as it passes through the water separator and block the pack output. To prevent this, an anti-ice valve is installed, which bypasses some of the hot air upstream of the air-cycle machine to prevent the output from going below 35°F. This type of air-conditioning system, therefore, is limited to a low pack temperature output of 35°F.

New generation packs

In the newer advanced air-conditioning pack before the air is routed to the turbine it will pass through two additional heat exchangers. They will drop the temperature below the dew point thus forcing the water into suspension. The moist air goes through a water separator which will remove the water before entering the turbine of the ACM. Now the pack output can go below 35°F without freezing anything because there is not enough moisture left in the air. The two additional heat exchangers are sealed units with two passageways. As the air passes through, the first heat exchanger will remove some heat and the second will pick up the heat that was extracted on the first pass.

The first exchanger is called a reheater and the second one is called a condenser. The condenser is cooled by the pack output and thus has the most effect in cooling the airflow below the dew point. The second pass of the air through the reheater absorbs the heat removed on the first pass to raise the air above the dew point before entering the turbine. The second pass through the condenser absorbs the heat removed on the first pass which will raise the pack output temperature. But because of the low output temperature of the ACM turbine will produce a colder output than in a traditional system with the 35°F limited output.

This colder pack output means that less pack air is needed to satisfy the temperature demands of the cockpit and cabin. This results in fuel savings because less air is bled from the engines to run the pack.

The newer systems and some of the older systems have another improvement which involves the air cycle machine itself. Older ACMs have oil bearings which need servicing to make sure there is enough oil to prevent bearing failures and the destruction of the ACM itself. Newer units have air bearings which nearly eliminate servicing requirements.

A common procedure with old style ACMs with bearings is to bypass the ACM in flight to slow it down. To do this the ram air doors are opened to allow more outside air to flow across the heat exchangers and make up for the idling ACM. This results in more drag on the airframe, however, and a higher fuel burn. With the newer ACMs and a resulting cooler pack output, the ram air doors can be kept closed for better overall efficiency.

Controlling pack output

There are various methods of controlling the pack output temperature. The first method is by allowing more air to flow across the heat exchangers to give a colder pack output. Another method involves controlling the speed of the ACM. A faster turning ACM results in a colder pack output, so it follows that by controlling the ACM speed, you're also controlling the pack temperature.

This is accomplished by opening the bypass valve which will bypass warmer air to the pack output to raise the pack output temperature. Because of the parallel path that the air takes, it will slow down the ACM, which will further raise the temperature. This gives the operator a wide range of control over the air-conditioning pack output.

Because there might be some conditions where not all of the moisture is removed, another device called an anti-ice or a low limit valve is installed to prevent the condenser from freezing up. This unit is normally a pneumatically controlled valve that monitors the pressure drop across the condenser. If the pressure drop becomes greater because of the freezing of air passages, the valve is commanded to open and add warm air to melt the ice.

The final part of the air-conditioning system is a mixing chamber. This is where the pack output air is mixed with warmer air for temperature control. This is typically used on some older systems like DC-9 and 727 type aircraft. This is not a desirable type of an air-conditioning system because there are two separate packs that receive temperature information and demands from two different sources. One source is the passenger cabin and the other is the cockpit. The result is two separate packs feeding a common mixing chamber with two different outputs, which is like having two packs fighting each other.

Newer aircraft, 747, 757, DC-10, A320s etc., employ a mixing chamber with pack inputs, and recirculate cabin and cockpit air via recirculation fans. This cuts back on the amount of pack air required to accomplish the same job performed with the older systems. Normally, these new systems result in a 50/50 mixture of fresh air and recirculated air. Again, the result is more fuel savings.

Think of the controls of aircraft air conditioning like the thermostat in your house. You input a demand by turning a dial to a desired temperature. This signal is sent to the furnace to tell either heat or the air conditioning to turn on. When the house temperature is met, the thermostat signals the heating or cooling to stop.

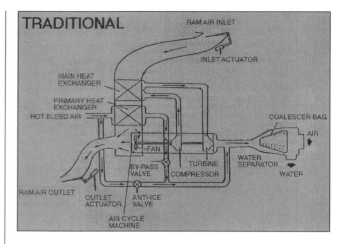

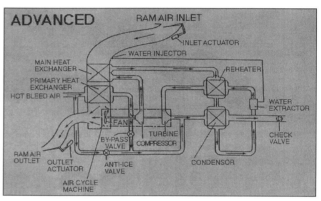

It will then cycle on and off to keep the house at the desired setting. Essentially the same events occur on an aircraft.

On newer aircraft, the air-conditioning packs are commanded to feed their outputs to the mixing chamber along with the recirculated air (if so equipped) to satisfy the demands of the lowest selected zone. The warmer demands of the remaining zones are satisfied by adding some hot air called trim air. This is more efficient than older aircraft where the packs try to satisfy individual zones. Unlike the example of the house, the packs keep running to keep the temperature constant. This is necessary because the air is being dumped overboard through an outflow valve to satisfy the aircraft's pressurization system.

Sensors are installed in the systems that feed information to computers to monitor and control the packs. Overheat sensors will monitor such things as bleed air, which will shut off the bleed air supply if it exceeds the design limits. Sensors will also signal a pack to shut down if the compressor outlet temperature or the temperature at the inlet to the turbine gets too high. Also if the pack output temperature gets too high, the pack will be commanded to a colder setting, and if that doesn't help, the pack is shut down.

There are also sensors in the ducts going up to the cabin and cockpit so that if too much trim heat is added, the pack is commanded to a colder output. These sensors also turn on fault lights and warning lights in the cockpit and can send messages to CRTs in the cockpit which can aid in troubleshooting or recommend alternate means of controlling a faulty system.

Other sensors help the computer to control the valves and ram air doors. They also control the pack outputs and monitor the mixing chamber and ducts for proper operation of the trim air and hot air valves. Finally there are temp sensors and temp bulbs in the cabin and cockpit zones. These sensors feed information to the computer to signal when hot or cold air is needed.

Although some of the newer systems are provided with built-in troubleshooting, it's still as important as ever to understand the basics. Information obtained from the maintenance and troubleshooting manuals is a great aid in the repair of these systems.

May/June 1994

Jim Benson is a training instructor for Northwest Airlines in Detroit.

Airframe Technology

Keeping corrosion at bay
The key is in knowledge and prevention

With the recent focus by the general aviation industry on corrosion control, there is a definite need in the industry for more education on corrosion.

A thorough knowledge of corrosion: why it occurs, how to limit its effects, how to identity it and how to correct it can significantly reduce the costs and downtime associated with its repair.

Whether or not they have formal corrosion inspection programs, airplane manufacturers would have to agree, corrosion is a threat to any aircraft, regardless of make or when it was built.

The basics

As materials corrode, they attempt to return to their original elements. This is a natural phenomenon that occurs to all material, but it can be particularly destructive to metal.

In the 1950s and early 1960s, the two primary considerations in selecting materials for aircraft were ultimate tensile strength and light weight. 7000 Series aluminum alloy treated to the T6 temper became widely used by jet air-frame manufacturers.

The industry employed these alloys for many years. However, as the first jet aircraft came of age in the 1960s, manufacturers and owners discovered that the tempered alloys showed an increased susceptibility towards stress corrosion cracking and in heavy section structures to general corrosion. From that time to the present, manufacturers have been faced with the challenge of finding extremely strong, lightweight materials which will resist corrosion.

By the early 1960s, metals experts had developed a new heat-treat process called "overaging," which included the T7 and T8 tempers for 7000 Series alloy. Overaging is a process which yielded alloys that were more resistant to stress corrosion cracking and, in heavy section structures, to general corrosion. Unfortunately, this process also lowered the tensile strength of the alloys.

Because many aircraft had been designed to employ T6 alloys, adoption of the weaker "overaged" alloys would have meant redesigning entire aircraft.

So manufacturers waited until the advent of its next aircraft model or next major redesign before using the overaged materials. So it was really the mid 1960s before the new alloys were adopted.

Over time, however, the industry has responded by developing new alloys that offer even better corrosion resistance and strength.

With each new aircraft designed, aircraft makers tried to employ the latest and best that technology had to offer.

Corrosion of metal is a primary concern as it eats away at the structural integrity of the aircraft. Chemical or electrochemical corrosion over time will change metal into metallic compounds, such as rust.

Corrosion to metal accelerates in the presence of an electrolyte. Electrolytes can be substances like water or other contaminating materials that allow electrical current to flow from a positively charged area to a negatively charged area.

This flow of electrons results in metallic deterioration and compounds like oxide that are formed from the metal. The longer the metal is in contact with an electrolyte, the more likely it is that corrosion will occur.

Corrosion takes many different forms. Pitting, for example, is a localized form of corrosion confined to small areas of metallic surfaces. Pitting usually occurs after the paint surface or other protective film has been removed or penetrated. Exfoliation is a severely destructive form of corrosion where sections of metal actually leaf out from the rest of the metal. By the time exfoliation is detected, the strength of the part is impaired.

Attempts at prevention

It's no secret that many alloys, including aluminum alloys, can corrode within a few years unless coated with an effective finish system.

During the 1950s and early 1960s, and today, manufacturers typically applied a three-part system consisting of a chemical conversion coating, primer, and, in the case of exterior components, a topcoat. Each component of the system plays a different role. The chemical conversion coating provides both a surface to which the primer can adhere and some corrosion protection. The primer gives additional corrosion protection, and is a substrate for the topcoat. And the topcoat gives color to the aircraft, while protecting the primer from environmental and mechanical damage.

Early finish systems, which utilized anodizing or alodining, a chromated alkyd resin primer and a nitrocellulose topcoat, provided moderate topcoat protection, but required frequent stripping and repainting. In the mid 1960s, the airframe industry adopted a number of finish improvements, including the use of Epoxy polyamide primer and polyurethane as a topcoat material.

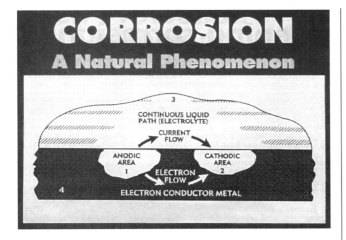

Epoxy polyamide proved to be tough, durable, and virtually impervious to water and aircraft fluids. Epoxy-based primers were also available with corrosion inhibitors such as zinc chromate mixed into the primer. Although not as durable as pure epoxy, chromated epoxies offered the added ability to mend minor breaches in the finished surfaces.

Today, manufacturers can choose from among numerous chromated primers, some employing zinc, barium, calcium or strontium chromate or combinations thereof. They may also choose non-chromated polyurethane finishing systems. Interestingly, due to today's environmental regulations, the industry stands at a crossroads in developing new water-based nonchromated aircraft coatings.

Many of these various finishes, both old and new, provide an effective barrier to the elements. However, none last forever and all of them must be rigorously preserved through regular maintenance.

Many factors affect the level and rate of corrosion which occurs to a specific aircraft or part. In addition, geographic location and environment are significant factors. Aircraft located in areas that frequently experience adverse conditions, such as saltwater areas, heavy industrial areas, soils and dust in the air and tropical climates, can encourage corrosive attack.

How an aircraft is used and how often can also influence the rate of corrosion. The FAA in one of its advisory circulars (43-4) says, "To postpone inspections or corrective action because something new in corrosion control is 'just around the corner' is ultimately an expensive proposition."

Some answers

Preventive maintenance of the aircraft on a regular schedule is the only sound practice. It minimizes the cost of labor and productive time loss. It also puts both of these costs on a predictable basis, and removes uncertainty and guesswork on the actual equipment.

Removal of electrolyte (cleaning the aircraft) and contaminating materials like moisture, salt, dirt, or grease is the most important preventive procedure that maintenance personnel can perform on the aircraft.

As the FAA reported as far back as 1973, the primary approach to corrosion detection is visual. There are areas, however, that are obstructed from view by structural members or equipment installation. Ingenuity should be encouraged as long as the improvised inspection methods are thorough.

If corrosion is detected, it should be treated in accordance with accepted repair practices. If questions arise during the course of inspection or repair or part replacement, it's prudent to contact the manufacture, and get involved in the repair scheme.

In an attempt to provide detailed information about preventive maintenance procedures, Sabreliner Corporation introduced its Corrosion Control Inspection Guide in 1989. This guide identifies additional areas on the aircraft that should receive more than a cursory inspection. With input from the Sabreliner operator's advisory panel, the guide was expanded into a complete corrosion inspection program.

Sabreliner consolidated the inspections in the guide and published them in 1990 as part of the required recurrent maintenance inspection in Chapter 5 of the Sabreliner maintenance manual.

Compliance with the inspections became mandatory for all Sabreliners being operated under Chapter 5 within 12 months of publication of the manual. During this initial 12-month compliance period, Sabreliner was able to acquire a significant amount of data relating to corrosion as it inspected nearly 100 Sabreliner aircraft, representing all models.

Examples of some findings include:

Camp Card 53.022 calls for a visual inspection of the left- and right-hand splice between the center skin and fuselage stations 99 - 156 to detect any distinctive bubbling or pillowing of the skin or flaws in the paint or sealer. Bubbling or pillowing would indicate that corrosion is present between the two materials. In more severe cases, rivet heads begin to cup, as the hidden structure begins to exfoliate. In the most severe cases, rivet heads actually pop loose.

Airframe Technology

The skin was removed to further investigate the area.

Cracked or deteriorated sealer is also visual evidence that there may be corrosion. Anytime that you identity such a condition performing any of the inspections, the sealer must be scraped away to take a closer look. Assuming that nothing further is needed after identifying cracked or deteriorated sealer is always improper. This is particularly true in window areas. There is case after case where corrosion is identified on the window frame after the sealer is removed.

There are many cases where minor corrosion can be removed, treated and primed and the parts replaced without replacement parts.

To generalize, when major structural part replacement is necessary, the use of specialized tooling is required to assure proper alignment of the fuselage.

When corrosion is discovered, it's critical that corrective action be taken to repair or replace the affected area in accordance with appropriate practices. Postponing corrective action is not only an expensive proposition in terms of dollars but, more importantly, it compromises the integrity of the airframe.

Corrosion is a continuing process. The safety of any aircraft ultimately depends on a thorough ongoing corrosion control and inspection program.
March/April 1994

The preceding information is from Saberliner Corporation's Corrosion Education Control and Prevention videotape.

Airframe Technology

Going beyond the requirements in a helicopter airframe inspection

By John Boyce
Contributing Editor

SAN ANTONIO, TX—Kerby Neff is like many aircraft technicians; he doesn't get a lot of calls to explain for publication how and why he does or does not do things.

Neff, the director of maintenance at Helicopter Specialists, Inc. just outside this south central Texas city, has been working on helicopters for over 13 years. During that time he has learned that with some items on an inspection checklist, you don't go by the book, you exceed the book. It's just that he's so used to doing things rather than explaining them that the things that experience has taught him don't necessarily jump to mind.

"I guess if I thought for a while I could come up with some things that go beyond what the manual tells you," Neff says. "But the 100-hour inspection on this ship is pretty simple. The checklist covers just about everything. It didn't used to—but now I think they've just about got everything covered."

This ship is a Schweizer 269C helicopter owned by the San Antonio Police Department. It's at Helicopter Specialists for a 100-hour inspection. The 100-hour inspection checklist on this aircraft runs to 40 items, each of which is clearly distinguished from the others with a carefully lined box and a place for the technician's initials after each item. The boxing was Neff's idea.

"I've done it with all the inspection checklists," Neff says. "It makes it easier to read and there's no confusion over which item you're reading. The feds like it because they can quickly read it and see that each item is inspected."

Tail rotor cables

In inspecting the Schweizer 269C Neff takes extra care with the tail rotor cables. The checklist does dictate inspection of the cables for fraying, chaffing, broken strands and corrosion. However, Neff and Helicopter Specialists' general manager Harry Geyer have found that simply checking the cables as dictated by the checklist is not enough.

"The last major problem we had was a tail rotor cable break on a customer's machine," Neff says. "We inspected the tail rotor cable system as dictated by the 100-hour checklist. That basically tells you to inspect it with reference to the HMI (Handbook of Maintenance Instruction). The HMI tells you to visually inspect everything; however, it doesn't tell you to remove the cables and look at them much like you would on some fixed wing setups; draw a rag over it and see if you can catch a loose strand without doing it with your fingers.

"The tail rotor cable setup on this ship is pretty simple; it only runs from the bell crank in the front of the floor to the aft section on the center frame and its push-pull tube and bell crank in the back.

"There are actually only two sections of cables because in the middle they have a switch-ball; then it goes over the bell cranks on the front and the back. They hook together with two turnbuckles.

"The cable makes a hard 90-degree turn right under the co-pilot's seat and goes over two pulleys. It's in a box section underneath the floor and it's very hard to see. If you can get a rag in there, you can really only get the outer section (of the cable) that goes in the groove. You can't get the part that runs in the groove of the pulley. That's where we had the break."

Consequently, Neff now physically pulls those pulleys to get enough slack in the cable for close inspection and cleaning.

"The way that engine generates airflow, being close to the ground and landing in dusty environments, those cables take a beating," Neff says. "The HMI doesn't say to dismantle that assembly but I always do now after that one snapped. That cable probably had about 2,400 to 2,600 hours on it."

"When we started inspecting this way," says Geyer, "we found another three or four of them that were fixing to do the same thing. Some had a lot more time on them, some had less so you can't go by the time on it; you have to look at it closely."

Tail rotor control and drive systems

The entire tail rotor control and the tail rotor drive systems in all Helicopter Specialists' inspections come under close scrutiny because they are two of the most critical systems in the helicopter. "On any of the helicopters I've ever worked on," Neff says, "the tail rotor drive system and the tail rotor control system are the parts you really have to worry about. They are probably the weakest links in the machines. If you lose them they can get you into the most trouble. I have always had a tendency to keep a close eye on those two systems—above and beyond what the book says to do.

The Best of Aircraft Maintenance Technology Magazine

Airframe Technology

"You don't know if somebody has serviced some of the old KP bearings or not. They don't tend to get a lot of attention. Nobody ever takes them out and greases them, and they need it. Eventually, you can tell there is something wrong if you get a lot of play in the control system, but you really can't tell the state of the grease in each one of those bearings without removing each and everyone of them. Most of the bearings down there are like that and those down at the tail rotor pedals.

"Those bearings are down under the ship and they pick up a lot of dirt. On one that we inspected, the tail rotor pedals were stiff and the customer had gotten used to it and never mentioned it. Then they got 'notchy' and all it turned out to be was the cross-shafts and the bushings just for the mechanisms that make the pedals work had gotten full of grit and dirt. So those items we check real close."

Landing skids

Helicopter Specialists has a lot of experience with helicopters that are used in ranching applications. This means, many times, the aircraft are transported by trailer, which is particularly hard on landing gear skids. As a result, Neff has developed the habit of carefully checking the bearings in the landing skid of any helicopter he works on, particularly ones that have the older friction dampers rather than the newer elastomeric dampers.

Neff says he physically unloads those bearings to check them by jacking up the aircraft to do a shake test.

"The trained eye can actually look and tell if you've got wear, but somebody that doesn't know where those parts should sit in relation with the other assemblies, the next higher assembly, would not be able to tell if there was wear because you can grab hold and shake it while it's sitting on the ground and there won't be any physical play. We usually jack them up and do a shake test on it. Then you can really tell what kind of looseness and play you've got in there."

Looking at everything

Neff says that the clutch actuary cable or lower coupling drive shaft should receive normal service at every inspection, but he has been removing the lower pulley for years because that is one sure way to determine if there has been an engine overspeed.

"I've had people overspeed the engine and not even tell us about it," Neff says. "It's the quickest place you can get to on normal maintenance where you can tell if the engine has been oversped."

Although it isn't mandated, Neff removes the console between the two seats of the Schweizer 269C because it gives him an "excuse to go in there and look at all of those controls that go through that area. It doesn't tell you to physically remove that cover, but I want to see everything under there. There's a throttle control and lower longitudinal rod and those control rods go up to the main rotors. It's just a good idea to take a look at those.

"I've found problems just from one 100-hour inspection to the next in some of the bell cranks. They have a ball-type bearing and those things will seize up from lack of grease and actually cause the outer race to ruin the casting of the bell crank that it's pressed into. Those little bell cranks are quite expensive."

January/February 1994

Getting a grasp on hydraulic systems

By B.J. High

It was a dark and stormy night. I was wishing I was on Snoopy's dog house, but, no such luck. Fact was, I was easing out of the wheel well of a modern corporate jet aircraft, soaked in hydraulic fluid, and trying to remember where I had gone wrong. Yeah, I hadn't been on this type of aircraft for all that long and I had already put in my eight hours.

Then with the flash of a nearby lightning strike, an apparition appeared and I heard a voice from the darkness.

"What?" I said. And the voice came back, "I make laws, send you to school, buy you books, and even send an old hand around from time to time to give you a clue and look at you. You're a mess, and so is the airplane."

"What do you mean make laws," I said. "Who are you, anyway?"

"Pascal's the name. I'm the guy who put all the effort into figuring out how fluid works under pressure and put it into words so most can understand. I did this so you wouldn't have to get wet when you break a connection on a hydraulic system without bleeding down the accumulator."

"Oh," I said, realizing that I had not bled the accumulator off before I broke the B-nut on that line. "Oh well, it's bled now," I said.

That was a long time ago and I wasn't the first nor the last maintenance technician to forget his place in the physical world. But the laws of hydraulics still apply, and they apply not only to aircraft, but looking around our world we see heavy equipment, manufacturing equipment, farm equipment, to name a few, heavily involved in the application of Pascal's law.

The law

Pascal's law says that if pressure's applied to any part of a confined liquid, the pressure's transmitted with undiminished intensity to every other part of that liquid.

To an engineer, this means lots of mathematics and formulas in order to design a system that performs the work the way we want it to. But to us technicians, it means that we can identify, troubleshoot and repair hydraulic systems without getting wet and with only having to do it once. So if you know the normal operation and the basic design of a hydraulic system, it will make the job much easier.

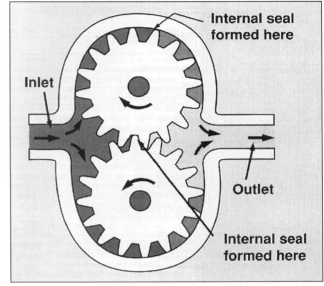

Typical gear-type pump.

Basic systems

Systems range from simple to complex. Simple being a hydraulic jack, and a complex system being the hydraulic system on a jumbo jet aircraft. And yet these have a common bond. They apply pressure to a fluid, and cause work to be done.

In order to stay out of the engineer's domain, we will keep it simple but with the idea that if we understand the way it works, we won't get wet, won't have to work so late and will fix it right the first time.

An aircraft hydraulic system (like many systems used in aircraft) consists of pumps, valves, pressure regulation devices, controls and indicating systems. Because these components are often arranged within the aircraft in a manner so that the technician can't get a finger on them, much less a wrench, a review of components is in order.

Reservoirs

I love it when we speak French. But it's only a tank. Sometimes it's a tank which stores and supplies fluid for the system and is sitting alone, high in the aircraft, someplace where any air in the system will work its way up and be bled off through venting or a manual bleed valve. And some reservoirs are pressurized to prevent foaming at altitude, ensuring the pump gets fluid and not air.

Airframe Technology

Lines		Pumps		Miscellaneous Units (cont.)		Valves	
LINE, WORKING (MAIN)	——	HYDRAULIC PUMP FIXED DISPLACEMENT		FILTER, STRAINER		PILOT PRESSURE REMOTE SUPPLY	
LINE, PILOT (FOR CONTROL)	- - -	VARIABLE DISPLACEMENT		PRESSURE SWITCH		INTERNAL SUPPLY	
LINE, LIQUID DRAIN	------	**Motors and Cylinders**		PRESSURE INDICATOR		**Valves**	
FLOW, DIRECTION OF HYDRAULIC / PNEUMATIC		HYDRAULIC MOTOR FIXED DISPLACEMENT		TEMPERATURE INDICATOR		CHECK	
LINES CROSSING		VARIABLE DISPLACEMENT		COMPONENT ENCLOSURE		ON-OFF (MANUAL SHUT-OFF)	
LINES JOINING		CYLINDER, SINGLE ACTING		DIRECTION OF SHAFT ROTATION (ASSUME ARROW ON NEAR SIDE OF SHAFT)		PRESSURE RELIEF	
LINE WITH FIXED RESTRICTION		CYLINDER, DOUBLE ACTING SINGLE END ROD		**Methods of Operation**		PRESSURE REDUCING	
LINE, FLEXIBLE		DOUBLE END ROD		SPRING		FLOW CONTROL, ADJUSTABLE— NONCOMPENSATED	
STATION, TESTING, MEASUREMENT OR POWER TAKE-OFF		ADJUSTABLE CUSHION ADVANCE ONLY		MANUAL		FLOW CONTROL, ADJUSTABLE (TEMPERATURE AND PRESSURE COMPENSATED)	
VARIABLE COMPONENT (RUN ARROW THROUGH SYMBOL AT 45°)		DIFFERENTIAL PISTON		PUSH BUTTON		TWO POSITION TWO WAY	
PRESSURE COMPENSATED UNITS (ARROW PARALLEL TO SHORT SIDE OF SYMBOL)		**Miscellaneous Units**		PUSH-PULL LEVER		TWO POSITION THREE WAY	
		ELECTRIC MOTOR		PEDAL OR TREADLE		TWO POSITION FOUR WAY	
TEMPERATURE CAUSE OR EFFECT		ACCUMULATOR, SPRING LOADED		MECHANICAL		THREE POSITION FOUR WAY	
RESERVOIR VENTED / PRESSURIZED		ACCUMULATOR, GAS CHARGED		DETENT		TWO POSITION IN TRANSITION	
LINE, TO RESERVOIR ABOVE FLUID LEVEL		HEATER		PRESSURE COMPENSATED			
BELOW FLUID LEVEL		COOLER		SOLENOID, SINGLE WINDING		VALVES CAPABLE OF INFINITE POSITIONING (HORIZONTAL BARS INDICATE INFINITE POSITIONING ABILITY)	
VENTED MANIFOLD		TEMPERATURE CONTROLLER		SERVO MOTOR			

Typical symbols used in fluid power diagrams.

Some reservoirs use a standpipe inside to allow a remaining amount of fluid in the pipe to supply an emergency system in case the main volume becomes depleted due to a malfunction downstream.

An emergency hand pump might get its fluid directly from the standpipe while the main pumping station (like an engine-driven pump) may get its supply from the main volume of the reservoir.

Pumps

The pump delivers fluid at a rate needed to do the work. If the load offers resistance to the fluid flow, pressure will build up in all of the places where the fluid flows. This can be felt in a hand pump as your arm gets tired. Pascal's law branches out here and says, "The greater the resistance, the quicker your arm gives out," or something to that effect.

There are complete technical training courses developed for pumps alone. There are so many different types. But we'll limit this discussion to a prominent few.

The easiest to understand is the hand pump. It has a piston, two ports and a check valve. When the lever on the piston moves the piston in one direction, it draws fluid into its little cylinder from the inlet port. If your arm has not become too weak from that, you can move (push) the lever the other way and a check valve (ball and spring) prevents the fluid from going back out the inlet, and makes it go out the outlet.

If you push hard enough, and there's resistance of a load present, a pressure will result within the entire flow stream. If the piston on our pump has an area of 1 square inch, and it takes 5 pounds of force to move the load, you could cause a pressure gauge (if you had one included in the circuit) to indicate 5 pounds per square inch, or 5 psi, the pressure applied throughout the pressure system.

Engine-driven pumps are used in most aircraft as the main pressure source and they too can be of varying types. There are gear-, vane- and piston-type pumps which also vary in their delivery style.

Variable delivery pumps use an internal compensating unit to control the flow and pressure demands placed on the pump. These pumps typically idle in free-flow when no systems are requiring work to be done.

Constant delivery pumps move the same amount of fluid all the time, regardless of the demand. If several systems need pressure at the same time, somebody has to wait. (Ever see a DC-3 gear go up?)

Piston pumps are intricate in their design, but, like many things in life, they're more design than substance. They are typically expensive, hard to fix and don't work as well as gear-type pumps.

Airframe Technology

Accumulators

This is the dude that will get you. Just when you think you have no pressure on the system (no pumps on, no electrical power on, what could go wrong?). Well, this is the unit that can store fluid under pressure. By doing that, it's storing energy. Energy that can do work if you direct it to the right place. But if you break a line that's in the pressure path, you're going to have Pascal's law all over you, the hangar floor and the airplane.

The accumulator has two chambers in it. One for hydraulic fluid, and the other has a pro charge of nitrogen of a pressure ranging in excess of 1,500 to 2,000 psi, depending on the designed operating pressure. Some units separate the two chambers by a diaphragm, while some are a cylindrical design with a floating piston separating the chambers. To allow maintenance downstream of the fluid side of an accumulator, the remaining trapped system pressure must be reduced to zero by relieving the pressure back to return until the piston (or diaphragm) has bottomed out inside the accumulator housing, leaving only the precharge of nitrogen remaining. Most systems using accumulators provide a manual valve for this purpose.

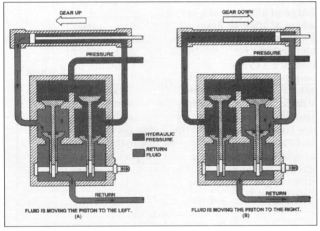

Closed center, four-part, poppet-type selector valve.

Control valves

Control valves direct the pressure to where you want it to go. That is, you can use the same pressure source (pump) to operate many different actuators, doing many different jobs about the aircraft. Someone must tell the valves when they are to open or close because without us technicians, they are stupid. Some valves are hand-operated (like with a selector lever), some are electrically controlled (solenoid or electric motor driven), and some are operated by the hydraulic fluid under pressure (shuttle or sequence valves). Obviously, some valves are smarter than others. Many selector valves provide a port to direct the flow of pressure by use of a ported cylinder or piston (spool valve), and the ported selector will usually provide a second ported path to allow for return flow from the actuated unit.

Actuators

Actuating cylinders convert energy from the hydraulic fluid into motion. These are the guys that actually displace the load when the pressure gets to its piston. Pascal says, "Because of the pressure on it, it'll move." For, you see, the pressure is undiminished at all points. So, with a push rod connected to the piston, they will move together against a load.

Single-action actuators apply force in only one direction, while double-action actuators will apply force on both the extension stroke or retracting stroke. When troubleshooting hydraulic systems, we must remember that when the piston inside the actuator is moving in one direction (such as the extending stroke) something must replace the area it's moving away from. It will not move very far if it puts a giant suck on the part of the cylinder it's vacating. On a single-action actuator, it may be vented to atmosphere. On a double-action unit, hydraulic fluid will flow into the evacuated space and prevent liquid lock.

Condiments

You need salt, pepper and lots of catsup to allow the real food to form a great meal. So, you need several little items in the hydraulic system to make it work right. Check valves, thermal relief valves, pressure-regulating valves, system relief, all make the real stuff in the system perform smoothly.

Check valves have a neat little ball and spring that will allow flow of fluid in only one direction. Some also have a drilled port called an orifice, and will allow just a tiny bit of back flow. This might be used to prevent trapped pressure you don't want.

Thermal relief valves are much like a check valve except they act as a vent for any over pressurization of a section of line, or a component due to landing late on a cool night, and tomorrow's sun causing the fluid to expand so as to possibly rupture a line, or something. (More rag wrenches, please.)

Pressure-regulating valves do just that. Some do it by being placed where any fluid pumped must pass through it and get regulated to a constant designed pressure value. Others do it by controlling the angle of the rotating plate in a piston-type pump and thereby control the pressure directly at the source.

System relief valves are the pop-off valves of the system. Any excessive buildup of pressure will be relieved as in the thermal relief valve, but will allow for much more flow.

Airframe Technology

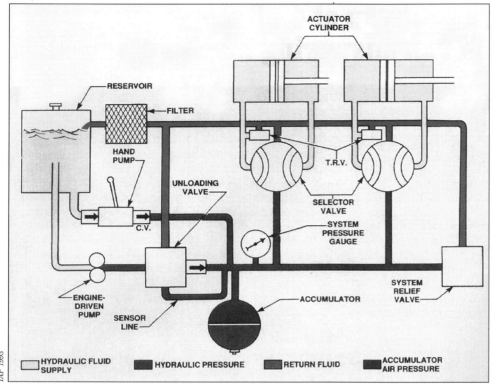

Complete basic aircraft hydraulic system using both an engine-driven pump and a hand pump.

Hydraulic fluid

None of the above mentioned components can do much without a transmitting medium. That medium is a fluid that can meet the requirements of our aircraft system. It must be fire safe (have a high fire point and flash point). It must be chemically stable in order to withstand the extreme conditions of altitude and temperature variations, and the viscosity must be compatible with the condiments and pumps and other components in the system.

A major concern is compatibility of fluid and seals. If you have the wrong seals with a harsh fluid, guess what? The seals turn to mush. So, what are the types of fluid commonly used these days?

There are three types commonly used:

Vegetable oil base fluid (MIL-H-7644) is not used much any more but some of the older aircraft systems may be using it. Being a mix of castor oil and alcohol, you will recognize it from the odor. Also, it's dyed blue so you can tell it from the other alcohol products you may have stashed. Any fool knows not to eat or drink anything blue. Natural rubber seals are found in a system with this fluid.

Mineral base fluid (MIL-O-5606), is a red-dyed petroleum based fluid of which most of us are familiar. Synthetic rubber seals are found in these systems. Another mineral base fluid gaining wider use is BRACO (MIL-H-83282). While it's considered to be compatible with 5606, you should mix only when needed to service up the system on a temporary basis.

And then there's Pascal's favorite, phosphate ester (she may have been Pascal's daughter). Known to most of us as Skydrol or Hyjet, this fluid (MIL-H-8446) is potent stuff and the maintenance manuals of all the aircraft that use it have a listing of safety precautions about it. Be sure you follow them. Skydrol 500 is blue, 500A/B are purple, and 7000 is light green. These systems use butyl rubber seals.

So, here we are at the hangar, 10 o'clock at night, I am missing the continuing saga of Maxwell Smart reruns on Nick at Nite, and guess what? An early launch and a hydraulic selector valve to change. But now I have arisen in status to director of maintenance, corporate aviation department, which means now it's me and one new guy that's working on his A&P ticket.

"I'll go to the supply room and get the parts while you start removing the valve," I said.

About the time I got to the supply drawer, I heard that dreaded sound. The sound of an accumulator piston bottoming out at warp speed, followed by that "waterfall sound, and then followed by a bolt of lightning with a voice coming out of the darkness muttering about schools, laws, books and that which I had heard years before.

"It wasn't me this time," I said, pointing my finger out toward the hangar floor. "He did it."

And the voice of Pascal came forth again and said, "I'll take care of him, I want to talk to you about management and responsibility."

Well, he did that (which is another story). I thanked him, asked him how his daughter, Ester, was doing, and with a flash, he disappeared.

November/December 1993

B.J. High is an aircraft technical instructor for Citations at SimuFlite Training International in Dallas, TX.

Airframe Technology

Taking command of composites
Knowing the basics is a good start

By Cindy Foreman

Inadequate or incorrect repairs can result in further damage to a composite component causing it to exceed repair limits. Thus, a very expensive composite component can be ruined if the proper techniques are not applied.

It's not difficult to complete an airworthy repair to a composite structure. However, the techniques, materials and tools used are different than those which are used on conventional repairs.

In an actual repair situation, the manufacturer's structural repair manual must be consulted regarding such information as operating environment, damage size limits, repair proximity limits and other information pertaining to the specific reps.

General repair procedures for advanced composites

The first thing that needs to be accomplished when confronted with a composite repair is to evaluate the extent of the damage. Check the damaged area for entry of water and/or foreign matter and for delamination around the damaged area. Always consult with the manufacturer's structural repair manual to determine whether the damage is within repairable damage limits.

Remove surface contaminants and paint around the area to be repaired. Paint strippers should not be used on any composite structure. Strippers can remove surface layers of resin and expose fabric. Paint should be removed by mechanical means. Typically this is done by sanding. Care should be taken not to sand into the structural fiber layers.

If damage has occurred to the core material of a sandwich structure, the damaged core material must be removed prior to step cutting the laminate. This procedure is performed by using a router to remove the damaged core. Caution should be taken not to damage the core materials that remain in the structure.

To accomplish the proper step cuts in the laminate, each successive layer of fiber and matrix must be removed without damaging the underlying layer. Great care must be exercised during this portion of the repair procedure to avoid damaging the fibers surrounding the area being removed. Sanding is the method that's usually used to remove the plies with the most control.

Sand each layer down about 1/2 inch all the way around the damaged area. The idea is to sand 1/2-inch-wide concentric circles (assuming a round repair) which

The damaged core is routed out, and each layer is sanded down in concentric circles around the repair area.

tapers down to the core material. Start sanding at the outermost mark and work down toward the center, removing one layer at a time.

All repairs must be cleaned after sanding. Removal of sanding dust and any oil on the surface is essential for proper structural adhesion of the repair plies. The strength of a bond is directly related to the quality of the surface preparation.

Next, gather together the materials needed for the repair. The structural repair manual will list the materials you need. Cross reference the area on the aircraft where the damage occurred with the manufacturer's description of the composite which was used to fabricate the part. The materials used should be identified by:

- Material type, class and style
- Number of plies, orientation of warp and fill, and stacking sequence
- Adhesive and matrix system
- Type of core, ribbon direction, core splicing adhesive and potting compound

Be sure that all resins, adhesives and prepregs are within their usable shelf life.

Identify the manufacturer's recommended cure system and ensure that the proper tools are available (e.g., hot patch bonding machine, heat blankets of the proper size, vacuum bagging equipment and materials, etc.).

The warp compass is used to align the fiber orientation of the patches for the repair.

The patches for the repair must carry the same stress loads that were manufactured into the part. The ability to endure these stress loads is dependent on the way in which the fibers are oriented into the repair area.

Airframe Technology

There should be one fabric bonding patch ply of the same thickness and ply orientation for each damaged ply removed.

Some manufacturers call for the use of prepreg materials for use in certain repairs; however, other repairs can be accomplished by allowing the technician to fabricate and impregnate resin to raw fabric. Since prepregs may not be available for a repair, it's often necessary to impregnate the fabric at the time of the repair. Be sure to use the proper resins and weigh and mix it properly. The fabric is impregnated by using a squeegee to work the matrix into the fibers. A plastic backing is used to keep the materials clean prior to being installed into the repaired area.

The core and all patches are ready for installation into the sanded part.

Once the repair patches have been impregnated with resin and cut to shape, the repair plies of fabric can then be laid into the step cuts.

Lay the patches over the sanded area and remove all plastic backing. Be sure to place them with the correct ply orientation and in the right sequence—following your structural repair manual.

Once the repair is made and the patches are in place, cover the area with a parting film or a parting fabric (peel ply). This allows the excess matrix to flow through to the upper surface and into the bleeder material. Parting film is easily removed after the cure is completed, and it prevents other materials, such as bleeder material, from sticking to the repair. The parting film or parting fabric also feathers in a seam, or overlap, of fabric to produce a smooth surface.

Some release fabrics can be used instead of a parting film to provide a final rough surface (slightly etched) suitable for painting.

A bleeder material is an absorbent material that's either placed around the edges or on top of the repair to absorb the excess matrix. A breather material is placed to one side of the repair to allow air to flow through it and up through the vacuum valve. Bleeders and breathers can be made of the same material and can be used interchangeably in many cases.

Next, attach sealant tape around the circumference of the repair area. Sealant tape, in conjunction with the vacuum bagging film, is designed to produce an airtight seal which can be removed from the surface after the repair is made, without taking the paint off.

If a heat blanket is used to cure the repair, place a layer of parting film over the repair with a thermocouple to control the heat output of the blanket. Then, place the heat blanket over the thermocouple and cover the entire area with vacuum bagging film. Next, work the vacuum bagging film into the sealant tape to produce an airtight seal.

The repair area is vacuum bagged and air is evacuated to produce pressure on the part during the cure.

Attach a vacuum hose to the valve and evacuate the air. Air leaks in the vacuum bag will produce a hissing sound. Leaks often occur where the sealant tape overlaps, or where wires pass through the tape, and can be eliminated by pressing the bagging film down into the tape until the hissing sound stops.

Vacuum bagging is probably the most widely used and recognized method of applying pressure for use on advanced composite repairs.

Once the repair is vacuum bagged and sealed, it's ready for a controlled heat cure.

The most widely accepted method of curing structural composites employs the use of resins which cure at higher temperatures. These adhesives and resins require elevated temperatures during their cure in order to develop full strength and reduce the brittleness of the cured resin.

When a part is to be cured with heat, it's not enough to simply apply heat at the final cure temperature. It's also important that the resins be allowed enough time to flow properly before they go through their curing process. If this isn't allowed, a resin rich area may result. "Ramping" (raising in increments) the temperature up to the final cure temperature slowly over a period of time allows the slow heating process which is critical in the curing of the composite.

The temperature is held constant at the final cure temperature for a determined amount of time. It's also important to allow a repair to cool at the proper rate. Composites gain much of their cure strength in the cooling down process. The heating and cooling process is known as a "ramp and soak" profile.

After completing a repair, the part should be painted. For most aircraft, the same type of paint that is used for the metal portions of the aircraft is suitable for use on the composites. *September/October 1993*

Cindy Foreman is the chief executive officer of Composite Educational Services and CES Composites. She is also the author of Advanced Composites, *published by International Aviation Publishers.*

Pneumatic system maintenance
And the role of the dry air pump

Like any other mechanical system, the aircraft pneumatic system deteriorates with age. Loose hose connections, dry-rotted hose, oil contamination, dirty filters, malfunctioning pneumatic components and inadequate maintenance procedures will invariably take their toll over the years.

In many cases, this is the reason why the first air pump installed in aircraft at the factory seems to operate forever, but each succeeding pump installed lasts for fewer and fewer hours.

However, a knowledgeable maintenance staff performing scheduled pneumatic checks can result in every air pump operating hours well in excess of the warranty.

Back to basics

All pneumatic system problems can be solved by getting back to the basics. Let's look at a few points and see if we can generate more in interest in checking aircraft pneumatic systems more closely.

The aircraft gyro vacuum gauge (as shown in Figure 1) is connected across one of the pneumatic flight instruments. As air is drawn through the instrument, a pressure differential is created, causing the gauge to indicate in a suggested range (usually 4.7 to 5.2 inches hg). This value only indicates that there is sufficient airflow through the gyro flight instruments to assure proper operation. *The gauge does not tell the pilot or technician the operating pressures (or vacuum) of the dry air pump or pneumatic system.*

In a properly operating single-engine pneumatic system, if you were to measure air pump inlet vacuum, and the gyro vacuum gauge in the instrument panel was indicating 5.0 inches hg, the highest reading you should obtain at the air pump is 6.5 inches hg. This 1.5-inch hg increase at the air pump is the additional vacuum required to make up for what we generally call "system line loss." Line loss for a single-engine aircraft should not exceed 1.5 inches hg—no exceptions.

The pneumatic system on a twin-engine, vacuum instrument aircraft is covered by basically the same rule. However, due to the longer distance between air pump and gyro panel, a maximum of 2 inches hg line loss is allowed. For example, if the twin-engine aircraft gyro vacuum gauge is indicating 5.0 inches hg, the desired maximum vacuum developed at the inlet to the air pump should be 7 inches hg.

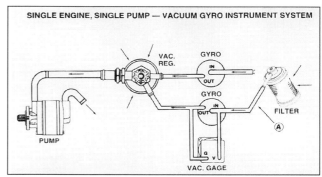
Figure 1.

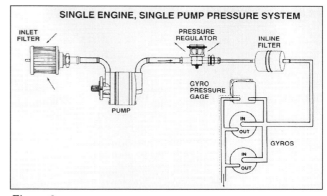
Figure 2.

Pilot/owner error

Let's look at a typical scenario:

The new aircraft leaves the factory with a completely leak-free pneumatic system.

Over the years, hoses deteriorate, connections loosen and clamps aren't replaced following that new panel improvement. One day, the pilot/owner happens to look at his gyro vacuum gauge to discover it indicating 3 or 4 inches hg. With his or her knowledge of pneumatic systems, the problem can be corrected by adjusting the aircraft vacuum regulator.

The adjustment is made, the gyro instrument vacuum gauge increases to the "green arc" and the pilot/owner pats himself/herself on the back for correcting the problem without the help of a technician—saving several dollars in the process.

Because the aircraft vacuum regulator has been adjusted toward the closed position, less air enters the regulator, and more air has to be drawn through the inlet filter and gyro instruments. This loads down the air pump.

Airframe Technology

With more load on the air pump, the vacuum at the air pump inlet will increase causing increased pump operating pressures and shorter air pump life—so much for saving dollars.

Testing

Every aircraft, engine and/or pneumatic system combination subjects the air pump to a different environment, so testing of the pneumatic system and its components must be accomplished using an approved pneumatic system test kit.

The system cannot be tested with the aircraft's air pump. You've got to use a source of pressure that is provided from an external source.

For example, Parker Hannifin, Airborne offers a 343 pneumatic test kit, a kit that's called out in the majority of aircraft maintenance manuals. If you have one of these kits, make sure you have the latest laminated instruction manual. Also, it's good to know that if you don't have a test kit, Airborne's customer support department makes these available on a "loan" basis.

Where has all the carbon gone?

There's a misconception about what happens to the rotor and vane carbon particles that are generated when an air pump fails during operation. To understand what will happen in this instance, you've got to understand the conditions of the system during normal operation.

During operation of a typical pneumatic system (refer to Figure 1), air enters the inlet filter, flows through the gyro instruments and moves on toward the pneumatic regulator and air pump. The following conditions should exist under normal conditions:

- If the inlet filter is clean, there will be no vacuum measured at point A.
- With the gyro producing a vacuum of 5.0 inches hg, vacuum should be measured at all points between the gyro "out" connection and the inlet to the dry air pump.
- Pressure at the outlet port of the air pump will be a positive pressure, normally engine compartment pressure.

Under these conditions, any carbon contamination caused by a sudden stoppage will be drawn (by vacuum) "upstream" toward the vacuum regulator.

So if you change an air pump after a sudden stoppage, but don't check the inlet lines and vacuum regulator for foreign material, the contamination may cause instant pump failure. Or if the foreign material happens to be wedged in the vacuum regulator, it might be a long time before the material is dislodged and drawn into the air pump.

This same situation can occur in a pressure system (refer to Figure 2). In a pressure system, line pressure of 5 inches hg (2.5 psi) is present all the way back to the "inlet" port of the gyro instruments. Pressure at the inlet filter is basically outside ambient pressure, or in some cases engine nacelle pressure. With an air pump failure, contamination may flow toward the inlet filter.

Because of this possibility, Airborne is installing a screen in the air pump inlet port area of new pumps. This design provides extra protection against premature pump failure caused by foreign object ingestion. This same screen also helps to prevent the larger carbon particles from leaving the air pump and entering the pneumatic inlet lines during failure.

Despite the new design, however, its impossible to retain all contamination within the pump, so always clean the inlet lines before installing the next pump.

This vice ain't bad

Most aircraft technicians are not aware of one of the most helpful indicators of air pump failure.

That little red and white anti-vise sticker provided on every Airborne dry air pump should be examined not only after a pump failure but whenever the cowling is off and the pump is visible.

With 200 Series air pumps, if the pump creates higher pressures than normal due to dirty filters, kinked or blocked hoses and misadjusted regulators, pump body (housing) temperatures will increase. This increase in body temperature added to the already hot environmental operating temperatures will cause the anti-vise label to darken in color.

A pump label darkening in color is an indication that there's something drastically wrong with the system. Ignore the label warning or the aircraft system condition, and the label will get darker. Eventually the increasing pressure and temperature will result in premature failure of the pump.

With 240 and 400 Series pumps (cooling finned pumps), as pressure within the air pump increases due to faulty deice valves or blocked filters, internal temperatures will rise. Eventually pump housing temperatures will reach a point that can't be dissipated by the cooling fins.

Overheated cooling fins will start to burn dark lines through the back side of the anti-vise label.

Lines through the anti-vise label are an indication of pneumatic system problems, that if not corrected, will cause premature pump failure.

An ounce of prevention

Twin-engine aircraft and the winter months bring on a whole host of problems with deice systems.

Although each aircraft model and pneumatic system may be different, there are several common problems that continue to receive attention.

To prevent problems with deice systems during the winter months, maintain the systems during the summer months.

The cause of most inoperative systems is usually traced to a sticking deice control valve (engine compartment), or a deflated valve made inoperable by freezing water (holes in the boots) or plain old rust.

During normal flight, vacuum is applied to the boots for holddown via the system deflate valve. In most twin-engine aircraft this deice deflate valve is located under the cabin floor, at a lower level than the wing deice boots.

Deice boots with holes or failing boot patches provide a sure means for water to enter the system and be drawn into the deflate valve. Here, the moisture causes corrosion that results in problems during flight at altitude. These problems often disappear when the technician tries to troubleshoot the system in a warm hangar.

Holes in deice boots also cause pressure problems.

Several aircraft deice systems cycle to the off position only after maximum boot pressure is reached. With boots that have multiple holes, activation pressure is never reached, and the air pumps struggle to satisfy the requirements.

A word to the wise

Failure of the air pump or any other component of the pneumatic system during IFR flight in Instrument Meteorological Conditions (IMC) can lead to spatial disorientation of the pilot and subsequent loss of aircraft control.

Filter Change Schedule

Air filters must be replaced at each air pump replacement and at the intervals specified below.

Frequency	Vacuum System	Pressure System
100 hrs./ annually	Vac. Rg. Garter Filter (P/N B3-5-1)	Pump Inlet Filter (P/N B3-5-1) (P/N D9-14-5) (P/N D9-18-1)
500 hrs./ annually	Central Gyro Air Filter (P/N D9-14-5) (P/N D9-18-1) (P/N 1J10-1)	Inline Gyro Filter (P/N 1J4-4) (P/N 1J4-6) (P/N 1J4-7)

Use of single-engine aircraft in IMC is increasing. Many single-engine aircraft do not have a backup pneumatic power source or backup electric attitude gyro instruments. In aircraft without such backup devices, the pilot, due to added workload, may not be able to fly the aircraft with only "partial panel" instruments (that is, turn and slip indicator, altimeter and air speed indicator) in the event of primary air pump or pneumatic system failure.

Air pump or pneumatic system failures can and do occur without warning. This can be a result of various factors, including but limited to normal wear of components, improper installation or maintenance, premature failure or use of substandard overhauled components.

Therefore, a backup pneumatic power source for the air-driven gyros or a backup electric attitude gyro instrument, must be installed in all aircraft which fly IFR.

Airframe Technology

Air pump pointers

Keep the following points in mind when troubleshooting and maintaining air pumps and pneumatic systems:

- If you are consistently experiencing short pump life (or decreasing pump life) from an air pump, this is an indication that there's something seriously wrong with your aircraft pneumatic system.
- Twin-engine aircraft equipped with deice boots have additional pneumatic system components which require periodic maintenance. Operators consistently experiencing short pump life, (or decreasing pump life), on Airborne 400 Series dry air pumps should perform a thorough check on the aircraft pneumatic system and its components.
- Functional checks of the aircraft pneumatic system with the engines operating will not identify a serious system problem.
- The gyro vacuum (or pressure) gauge located in the aircraft instrument panel does not indicate pump vacuum (or pressure).
- The need for adjustment of the pneumatic system regulator normally indicates a system problem.
- Pump failure during engine operation will probably cause carbon contamination of the inlet hose.
- Examination of the air pump "anti-vise" label may provide important clues to the cause of premature pump failure.
- The aircraft deice boot system should be activated at random intervals during the summer months.

July/August 1993

Airframe Technology

This is not your father's fabric
State-of-the-art coverings for aircraft

By Greg Napert

ELK GROVE VILLAGE, IL—Although fabric covering on aircraft isn't typically considered state-of-the-art, its use in general aviation continues to be widespread. In fact, fabric is used on a regular basis to restore and maintain older aircraft and is quite popular in the kit-building industry.

Dip Davis, technical adviser for Superflite, a supplier of fabric, dopes and finishes for use in aviation, says that the bulk of the fabric used on aircraft today is synthetic.

Synthetic materials are stronger, lighter, longer lasting, and offer advantages that are just plain superior to cotton and linen. For instance, Davis says that synthetics can have a service life of 15 to 20 years or more, with virtually no care at all, whereas, you're lucky to get eight to 10 years of use out of cotton if you keep it covered and indoors.

If there's one advantage to cotton or linen, it's that many aircraft that were produced before synthetics were introduced had cotton or linen listed in the aircraft's type certificate. So using these original materials requires no STC.

According to Davis, though, that really isn't a problem because there are plenty of STCs available to cover most installations of synthetic fabrics. Superflite, he says, has over 115 STCs for various applications. "Virtually every type of aircraft is covered under these. And if you have an airplane for which there isn't an STC, we can probably have an STC for you before you're ready to cover the aircraft."

Davis says that synthetic Dacron® or Du Pont polyester fabrics were initially introduced in the '50s, and that products such as Ceconite® and Eonex® were the first on the market. These products were virtually indestructible. "In fact, we're finding people around with aircraft that are 30 to 35 years old or more, and the fabric is still punch-testing good. But the earlier products were very heavy-bodied materials that were stiff and difficult to work with. Modern synthetics, are lighter weight (2.7 pounds/square yard, as opposed to 3.8 pounds/square yard for Ceconite and 4 pounds/square yard for Grade-A cotton) and similar in texture to cotton."

Davis says that dope finishes were originally developed for cotton fabric which had a nap or "fuzz" to which the finish could adhere. Synthetic fabrics, how-

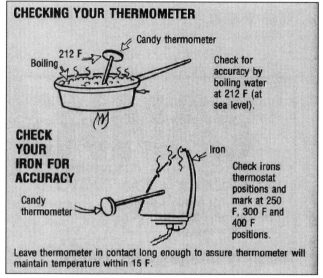

Tautening fabric requires accurate temperature settings with the iron.

ever, have a hard, silk-like texture, and if special primers aren't used, the dope will not properly adhere. The special primers penetrate the synthetic fabric and provide a surface to which nonshrinking dope can be applied.

Another alternative is the use of polyurethane finishes. Superflite, for example, has developed a finishing system which uses polyurethane from start to finish.

Davis says the polyurethane finishing system is superior to dope because of the flexibility and durability of the coating. Also, says Davis, as little as three coats of polyurethane can be used to finish the aircraft, as opposed to the many coats required to finish an aircraft with dope.

Working with synthetic materials

"Unlike cotton and linen, synthetic fabric is very easy to work with," says Davis. "With cotton, you've got to shrink it and then when you put dope on it, it loosens. With linen, you've got to worry about the fabric shrinking as the finish dries.

"When you heat-shrink synthetic coverings into place, they remain taut even as you apply the finishes.

*The Best of Aircraft **Maintenance Technology** Magazine*

Airframe Technology

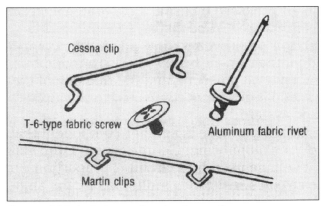

Types of fabric fasteners for aluminum ribs.

"To shrink the fabric, use a standard household iron to mark the thermostat positions to indicate 250°F, 300°F and 400°F. Begin at the 230°F setting and iron the entire surface keeping the iron moving much in the same manner as ironing a shirt. Increase the heat setting, and go over the entire surface again using the same technique. One more pass at 400°F should produce the desired results."

Davis says to keep an eye on the seams as the fabric tightens so that the seams align with the trailing and leading edges, and to apply heat more liberally as needed to center the seams. It's also important, he says, to make sure that the heat doesn't exceed 400°F or you can melt the material.

Davis doesn't recommend using air heat such as a blow dryer or heat gun because the heat cannot be controlled as well. He says that it's common to burn through the fabric with a heat gun, and also the heat is not as evenly applied. It's important to apply the heat over the entire surface without missing any spots, he emphasizes,
because an area that's not heated will sag as the outside temperatures change.

Davis says that virtually all of the fabric that's sold today is sold in presewn envelopes that slide over specific parts of the aircraft. Sewing has been virtually eliminated and replaced with overlapped and glued joints, and mechanical fasteners are often used in place of rib stitching.

There are envelopes for virtually all standard category airplanes.

When you place the envelope on the aircraft, you put it on like you're slipping on a sock. "There will be a large variation in the way that it fits," he says. "Remember that synthetic fabric should be installed rather loosely. Superflite fabric has a heat shrinkage potential of approximately 10 percent, or 5 inches in a typical chord of 50 inches, and it's desirable to use the majority of this potential.

"After you've slid it on, all that's left is to cement the edges down, cement it shut and iron it," he says.

"If you don't wish to use envelopes, sheets of material are available. Orientation of the weave of the fabric may be parallel in either direction," says Davis.

He explains that pinked edges aren't necessary using synthetics. "We do provide pinked edge tape for people who want it to look authentic. On synthetics, you don't have to worry about the material unraveling. As soon as you glue the edges, the adhesive prevents the material from unraveling. Also, tapes are cut with a hot knife which seals the edges and prevents them from unraveling.

"Surface tape should be applied as recommended over ribs, seams and trailing edges even if no seams exist at that point," says Davis, and is applied using adhesive or dope as specified with the system that's being used.

Tape can also be applied over leading edges to provide additional surface protection. On curved surfaces, it may be difficult to get the tape to lie flat against the surface. In this case, a small amount of heat may be applied locally with the tip of an iron to get the tape to lie flat.

Finishing

Synthetics can be finished with a number of finishing systems that are available on the market. The closest finishing system to conventional dopes is the use of a primer (typically a nitrate dope that's thinned to penetrate the weave of the material) and a nonshrinking butyrate dope for successive coats.

Other systems include resins and polyurethanes that offer many advantages over conventional dope finishes. Polyurethane finishes, for example, are not temperature sensitive or moisture sensitive. The temperature does need to be about 65°F to cure, however.

The polyurethane comes as a two-part finish, the color and a catalyst. You mix them together to apply them. The number of coats required to finish the aircraft will vary, but generally, all that's needed are three base coats of primer and two topcoats of the color you choose.

The primer is typically applied in three successive coats with a few minutes of drying time allowed between each coat. When the final primer coat is applied, it's allowed to sit overnight to dry. The next day, joint areas can then be blended by sanding with a medium grit wet or dry paper. If further coats are needed to "bury the tapes," more coats of primer are added and sanded after they are dried.

The finish coats are then applied in the desired colors by spraying a light "tack coat," which is allowed to set for a few minutes, and then applying a full, wet coat.

Davis says that the proficiency of the person applying the finishing system will determine how many coats are used. Those who are really proficient can produce a desirable finish by using only two coats of primer/filler and one cross coat of color.

Davis cautions that there are many fabrics around, and not all of them are approved for use on aircraft. There are lighter fabrics around that are used on ultralights that can be easily confused with the fabrics certificated for use in general aviation.

Make sure that you're using fabric that's marked with FAA-PMA markings or other FAA certification markings. And if you're not using the original materials as specified in the aircraft's type certification, be sure you comply with the STC or approval for the system that you're using. *May/June 1993*

Airframe Technology

Aircraft fluorescent lighting systems

By Gerald R. Stoehr

The interior cabin lighting is a system on the aircraft that presents an interesting problem to the aircraft technician. Although it is not as important to the actual airworthiness of the aircraft as the avionics, flight control or power systems, it's one of the most visible systems to the actual users of the aircraft, the passengers. The technician has to make sure this system is working at its best to assure the complete satisfaction of those passengers.

Fluorescent lighting systems and benefits

Incandescent lighting was once the choice for aircraft interior lighting, but fluorescent lighting has become the preferred alternative. A fluorescent lamp provides the equivalent amount of light as a 100-watt incandescent lamp while using just 20 watts of electricity. Fluorescent lighting systems also provide significant benefits in reduced weight, heat and power consumption while maximizing safety and design and installation flexibility.

Within the category of fluorescent lighting there are choices. The three types of fluorescent lighting that are used in aircraft are cold cathode, hot cathode and low-voltage instant start. Cold cathode systems use a sustained high voltage to start and operate the lamps. These systems require minimum wiring but they need special wire, connectors and tools to control the high voltages. Hot cathode systems operate at a lower voltage but, do have some design and installation constraints. Low voltage instant start systems offer low operating voltage and easy installation and maintenance.

All of these systems are made up of an inverter which changes the 28v DC to AC, and a ballast or power unit that generates the voltage level required to drive the lamps. When these elements are contained in the same component it's referred to as an inverter/ballast or power supply. All of these devices are collectively called primary components. The other element of the system is the lamp.

Fluorescent lamps come in a variety of sizes and shapes. They are usually either cool white or warm white in color temperature. High color rendering index (CRI) lamps are also available for use where maximum color saturation, brightness and clarity are needed. An important benefit of fluorescent lamps is that they have a long life.

Two of the most important lamp life factors are the voltage applied to the filament and the purity of the lamp's internal environment. If either of these factors is not optimized, the lamp life will decrease. If both are not optimized, lamp life will be very short. As lamps reach

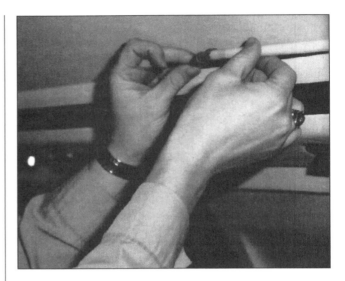

the end of their life, they require a higher operating voltage and begin to decrease in light output. This is a normal function of fluorescent lighting in general, as is a gradual darkening at the lamp ends near the filaments.

Defining and achieving optimum installation

Gaining the optimum performance from any fluorescent system is a result of proper installation as much as it is of an effective program of preventive maintenance. Quite often the lighting system manufacturer's installation instructions are not followed closely by the completion center, making it even more difficult for the technician to obtain the optimum results from the system.

When the aircraft is newly completed or refurbished, it's a good idea to check the lighting system for some simple criteria to assure that the system has been designed and installed correctly. Of course, the specific manufacturer's instructions should be followed, but there are some general guidelines that should be addressed.

The installation can be easily checked to ensure that these general guidelines have been followed. Components of the system that perform the inversion of the 20v DC to AC should be installed on metal surfaces to allow heat dissipation. These units typically produce heat as part of the inversion process.

Lamp holders should be placed so that they support the entire length of the lamp, not just the ends. The lamp holder should be constructed of a material that will allow easy installation and removal of the lamps while maintaining a tight grip.

If the lamp is less than 3 feet long, the lamp holders should be placed one-sixth of the lamp length from the ends of the lamp. If the lamp is longer than 3 feet in length, the lamp holders should be placed one-quarter of the lamp length from the ends of the lamp. In all cases, the lamp holders should never be closer than 2 inches from the end of the lamp. This will keep them away from the lamp filaments and will protect the lamp holders from premature aging.

System wiring should meet the lighting systems manufacturer's specification and should be in line with the guidelines contained in FAA advisory circulars for wire gauges, voltage drops and wiring run lengths. Documentation that shows which components are responsible for lighting each section of the system and where they are located on the aircraft is very important for subsequent troubleshooting and maintenance. System controls should be conveniently placed and should be easy to understand and operate.

Troubleshooting and maintenance

In the late '80s, the FAA instituted a series of Airworthiness Directives (ADs) for aircraft interior fluorescent lighting systems. The directives came in response to a prevailing concern over the high voltages associated with fluorescent lighting and the possibility of arcing and its resulting problems. The ADs required inspection and maintenance of the systems and/or disabling and placarding the lighting system.

Since those ADs were initiated, there have been significant strides in fluorescent lighting system design, both from the standpoint of safety in the form of engineered solutions that deactivates the system under certain high voltage conditions, as well as in overall system performance.

As with any system, certain areas are more prone to problems than others. By taking basic precautions, potential faults can be minimized. For instance, making sure that lamp connections are secure and that the lamps are connected according to the manufacturer's instructions will eliminate a major source of problems. Also, check to ensure that the system has sufficient input voltage.

Troubleshooting can best be accomplished by starting at one end of the system and methodically working toward the other. If a single lamp is out, you would start at that lamp and work toward the power units or ballasts that are driving that lamp. If all of the lamps in a section are out, you would start at the lighting controls and system input power and work toward the individual power units or ballasts.

The input voltage to a 28v DC system can be easily measured, and should always be checked at the input connector. But, the voltage within the system is usually a high-frequency AC potential and requires a special meter capable of measuring high-frequency AC voltages. Usually the best way to work is to replace a suspected

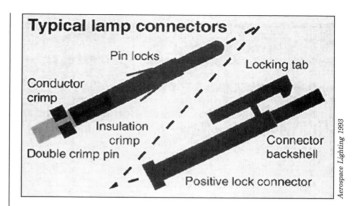

component or lamp with a spare to determine if it's the problem. You can use the following troubleshooting sequence to find a fault in most of the lighting systems in use today.

If a lamp does not light, the odds are pretty good that it has reached the end of its life. Changing the lamp should solve the problem. It's a standard rule of thumb for fluorescent lamps that are connected in series with a common power unit or ballast that they should all be changed at the same time.

It has been proven that the performance of a new lamp will be adversely affected if it's paired with an old lamp. Lamps should be replaced according to the lamp manufacturer's specification, and they should be replaced before they stop working. When replacing lamps, it's always important to wear protective eyewear in case the lamp fractures during the removal or installation process. It's also a good idea to inspect the lamp holders and connectors during lamp replacement.

The lamp connectors should be well-maintained and well-mated. There should be no indication of breakage or chipping of the connector bodies, and there should be no broken wire strands where the conductor is crimped at the connector. Connectors that provide a double crimp, on the conductor as well as the insulation, will require less maintenance and will be more reliable over time. If several lamps are out, but not the entire section of light, you could be facing one of several problems. If the lamps are being driven by a common power source such as a power unit, ballast or inverter ballast, the problem may be localized in the output circuit of that component.

In the case of a protected system, the problem could be that one of the lamps is faulty, or there could be a problem that has caused an excessive voltage and triggered the protection circuit. In this case, the best method is to replace the lamps on that circuit, and to check the wiring and connections visually to determine if there are any wiring problems.

When an entire section does not light, the problem is usually In the inverter, inverter/ballast or power supply that drives that section of light. You should check to see if an input voltage (usually 28v DC) is being provided to the component. If the correct voltage is present, then the

Airframe Technology

component is probably at fault. If there is no voltage or a very low voltage present, you should check the lighting system controls and power circuit for problems.

Primary components such as inverters and power units are designed to work together; therefore, it's not a good idea to mix and match different manufacturers' parts. In addition, many of the components produced by different manufacturers look similar. During maintenance, extra care should be taken to make sure that you have the right part. The manufacturer should be able to answer any questions you may have. Usually inverters, inverter/ballasts and power supplies are repairable. Lamps, ballasts and power units are typically not repairable. If a component is repairable, the repairs should only be accomplished by an FAA-authorized repair center that can perform the appropriate testing and issue a return-to-serviceability tag for the component.

Keeping the interior lighting system operating at its highest efficiency requires no more effort than any of the other systems on the aircraft. A system that has been correctly installed and maintained will give you long-term, trouble-free performance. If a system is not installed or properly maintained, it will create a multitude of problems for the aircraft technician. It's important to keep the lighting system manufacturer in mind when working on the system, as it is often the best place to turn for information and help solving lighting system problems. *March/April 1993*

Gerard R. Stoehr is director of marketing and sales for Aerospace Lighting Corporation. Prior to joining ALC, Stoehr spent 16 years in the electronics industry.

Airframe Technology

Lead-acid battery servicing tips

by Greg Napert

There are two types of lead-acid batteries used in general aviation today: vented cell (also called dry-charged) and sealed (also referred to as maintenance-free or recombinant gas batteries).

The options available today combined with the technological improvements in lead-acid batteries in the last few years, are resulting in batteries that are more powerful and require less maintenance. In fact, many lead-acid batteries are now being sold to replace nickel-cadmium battery installations in some business jets. The new batteries offer decreased maintenance in some cases, along with reduced cost.

Although vented cell and sealed lead-acid batteries are similar in appearance, they're quite different in terms of storage, maintenance requirements and charging.

Additionally, sealed lead-acid batteries (currently manufactured by Hawker Energy in the United Kingdom, and Concorde Battery Corporation in West Covina, CA) have only been available in general aviation for about the last five years. Although there are sealed batteries approved for a wide variety of aircraft—to include singles as well as business jets—the manufacturers are still developing the list of approvals for installation in various aircraft.

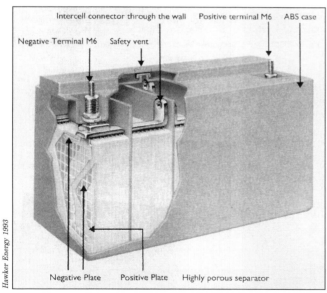

Cutaway of Hawker SBS 30 showing sealed construction.

Aircraft batteries differ from automobile batteries in that they are lighter in construction and typically the active material is higher density which requires a stronger electrolyte. Aircraft batteries use a 1.285 SG (specific gravity) electrolyte, whereas, automotive uses a 1.265 SG electrolyte. This is an important point to consider, because it means that handling and servicing are much more critical.

Aircraft batteries should be kept in a full state of charge at all times for two reasons: First, the batteries need to be airworthy, and are only so when they can deliver at least 80 percent of their capacity for emergency or essential operation. And second, during a normal discharge period, the batteries build up sulfate on the plates. Unless removed through regular charging cycles, the sulfate forms on the plates, becomes hard and crusty, and ultimately reduces the efficiency of the battery.

Vented cell

Vented cell batteries are dry-charged at the factory. According to Dan Rankin, sales manager for Teledyne Battery Products, the term dry-charged means that the plates are formed, or charged, at the factory, dried out in vacuum driers, and then assembled into cells that are placed into a battery case. Storage seals are placed into each cell to prevent moisture or condensation from getting into the cell that would react with the active material and degrade the life and performance of the battery.

According to Teledyne, the batteries can be stored indefinitely if kept in a clean dry place at normal ambient temperatures. However, the company also says that the length of time required for charging after activating with electrolyte will vary with the length of storage time and temperature. Longer storage periods and higher ambient temperature during storage will cause greater loss of the dry charge.

Probably one of the most critical things that can affect the life and performance of a battery is how the battery is initially placed into service.

Airframe Technology

"The vent caps should be firmly in place while charging," he says "because one, it keeps the electrolyte in the cell where it belongs and two, keeps electrolyte off the top of the battery which could provide a path for electrons to flow between terminals and cause self-discharge. If self-discharge is suspected, take the negative probe of a voltmeter and place it on the negative terminal of the battery. Take the positive probe and slide it around the top of the battery between the cells. If voltage is present, the battery is self-discharging."

Heavy-duty, hot-shot or fast chargers shouldn't be used on aircraft batteries, he explains, because aircraft batteries are built with thin plates to obtain high cranking power. These chargers will destroy the plates and actually drive the active materials from them. Dark or brown electrolyte is a sign that this is occurring. If you notice the electrolyte turning dark, it's a sign that you should reduce the ampere rate and voltage used to charge the battery.

"The recommended charger to use is a constant current type," says Rankin. This type of charger is equipped with ampere rate adjusting capability, has a timer and is capable of charging in constant current, or constant potential modes.

The best way to determine the state of charge in a vented lead-acid battery is to measure the specific gravity of the electrolyte. Automotive or industrial hydrometers that contain color balls or floats can't be used, as they are weighted for 1.265 electrolyte. A hydrometer specifically designed for aircraft batteries is recommended.

Any time the specific gravity of the battery falls below 1.260 SG it should be recharged. The battery is fully charged when the specific gravity stabilizes at or above 1.285 SG after three consecutive hourly readings. "Remember," says Rankin, "it'll take time to reach this point. Never rush a battery into service."

Sealed lead-acid

Skip Koss, vice president marketing for Concorde Battery Corporation, says that sealed or recombinant gas (RG) batteries are quickly gaining popularity because of their low cost and reduced maintenance.

"Conventional vented lead-acid band nickel-cadmium batteries generate hydrogen from the negative plates and oxygen from the positive plates at the end of the charge cycle. Problems with these batteries such as corrosive fumes and electrolyte leakage have been resolved with the RG battery," he says.

"The percentage of hydrogen gas generated on overcharge with the RG design is less than is required for flammability in air. The design features of the battery minimize hydrogen generation by maximizing the plate surface contact, which allows oxygen generated at the positive plates to diffuse through the glass mat separator and to recombine efficiently and safely with hydrogen generated on the negative plates," he explains.

"Also, the RG batteries don't have any free electrolyte; the electrolyte is absorbed and immobilized within the micro-fiberous glass mat separators. This makes the batteries fully aerobatic," he says.

Although the term "maintenance-free" is used in connection with sealed lead-acid, the batteries do require some maintenance, primarily when the batteries are in storage and when they're low on charge.

In storage, it's recommended that the sealed battery be boost charged every 90 days.

RG batteries should be charged when their open circuit voltage is below 2.08 volts per cell (12.5 for 12-volt batteries, or 25.0 for 24-volt batteries). Koss says that the batteries should be charged with a constant potential or constant voltage charger regulated at 2.35 volts per cell (14.1 volts for 12-volt batteries and 28.2 volts for 24-volt batteries). The battery is charged when the charge current diminishes to approximately 0.5 amperes for one hour, he explains.

These batteries also need to be checked periodically for reserve or emergency capacity if they are kept in service for more than one year or 600 hours of operation. The first check should be done after 12 months or 600 hours of operation and every three months or 100 hours of operation after that.

"To test the battery for capacity," says Koss, "you must first make sure the battery is fully charged. Then, with the battery temperature above 59°F, discharge the battery for one hour according to manufacturer's recommendations for the model in question.

"The minimum end point after one hour of discharge," says Koss, "must be 9 volts for 12-volt batteries, and 18 volts for 24-volt batteries. If the end point voltage is below minimum, emergency or reserve capacity, the RG batteries are to be "conditioned" with a constant current charger at the C/10 rate (C is the one-hour capacity as determined by the manufacturer) for 14 to 18 hours. The battery should sit for one hour and then be retested. If the battery passes the second capacity test, it should be recharged with a CP charger that has at least a C-rate for three hours.

Technicians can't jump

Never jump-start an aircraft from another power source for the following two reasons:

First, the battery that's being jumped is not an airworthy battery. It takes approximately three hours to recharge a fully discharged battery using the aircraft's generating system. In order for the battery to be airworthy, it must have the capacity to operate the aircraft electrical system, avionics, and be able to crank the engine in the event of an emergency during flight.

Second, active material on the positive plate will expand and the fast recharge from the charging source can damage the adhesion of the active material. This will result in premature battery failure.

And finally, aircraft batteries utilize relatively thin plates to create more surface area for higher performance. But these thin plates are susceptible to damage from excessive charging current applied during a typical jump-start.

For more information on servicing batteries, contact the following lead-acid battery manufacturers.

Concorde Battery Corp.
2009 San Bernardino Rd.
West Covina, CA 91790

Hawker Energy
Stephenson St.
Newport, Gwent, NP90XJ, UK

Teledyne Battery Products
210 Interstate N. Pkwy.
7th Floor
Atlanta, GA 30339

January/February 1993

Airframe Technology

Helpful tips for handling flexible hose

By Greg Napert

Hoses do more than route fluid from one location to another; they allow hydraulic fluid, oil and fuel to flow while allowing movement and vibration of various components. And when used in high-pressure applications, they absorb shock.

It's only when you consider what they're capable of doing, that you realize hoses need as much attention as the propeller that drives the aircraft, or the wheels you rely on for landing. But unfortunately, hoses are an often neglected item of an aircraft. They tend to be overlooked in favor of items that are seemingly more important.

Don Meadows, hose shop supervisor for Superior's Aeroquip hose shop in Dallas, TX, confirms this common hose neglect when he explains how his facility receives many hoses quite regularly that have been in service longer than they should have; hoses that have been exposed to extreme temperatures, kinked, cracked, corroded and chaffed.

"We also see some hoses in here that were built using industrial hoses," says Meadows. "Often they'll work just fine, but the fact is that these hoses are non-FAA approved."

General information

Most of today's hose assemblies are constructed internally of either rubber or Teflon™ and are categorized in basically three pressure ranges: low pressure (up to 200 psi), medium pressure (200 to 1,500 psi), and high pressure (1,500 to 3,00 psi).

Hoses are constructed in a variety of ways depending on the pressure application or operating environment. All hoses, though, have an inner tube, reinforcement layers and an outer cover to provide protection from fluids, heat and chaffing.

Identification markings typically are printed directly on the outer cover or may be printed on an identification tag if the cover happens to be stainless. The identification marks typically include: the Mil-Spec number, manufacturer's name, part number, size, date of manufacture and hose manufacturer's code. If a hose is not clearly identified, it shouldn't be used.

Hose fittings can be the reusable type or the compression crimp type. The compression crimp fittings require special equipment to crimp the fittings into place, which then become a permanent part of the assembly. The fact that compression crimp fittings are not reusable reduces the overall service life, which typically reduces their value. Reusable fittings can be used over and over, making it necessary to only replace a hose

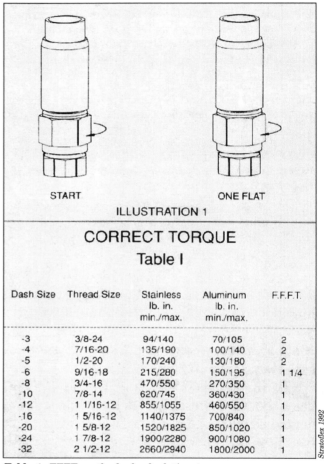

ILLUSTRATION 1

Table 1. FFFT method of calculating torque.

CORRECT TORQUE
Table I

Dash Size	Thread Size	Stainless lb. in. min./max.	Aluminum lb. in. min./max.	F.F.F.T.
-3	3/8-24	94/140	70/105	2
-4	7/16-20	135/190	100/140	2
-5	1/2-20	170/240	130/180	2
-6	9/16-18	215/280	150/195	1 1/4
-8	3/4-16	470/550	270/350	1
-10	7/8-14	620/745	360/430	1
-12	1 1/16-12	855/1055	460/550	1
-16	1 5/16-12	1140/1375	700/840	1
-20	1 5/8-12	1520/1825	850/1020	1
-24	1 7/8-12	1900/2280	900/1080	1
-32	2 1/2-12	2660/2940	1800/2000	1

instead of the entire assembly. But don't always assume that reusable fittings are serviceable. A damaged sealing surface means that they must be removed from service.

Installation of hose

Improper installation of hose can be a major contributing factor in how long a hose will be in service.

According to Aeroquip, a flexible hose manufacturer, "it's advisable to restrain, protect or guide the hose to protect it from damage by unnecessary flexing, pressure surges and contact with other components or structures."

The first item of concern regarding installation is the bend radii of the hose. Generally speaking, bends should be made as gradual as the installation will allow.

Manufacturers recommend that bends do not begin too near a coupling or fitting. The recommendation, generally, is that the hose should extend at least one and one-half times the hose diameter before starting a bend.

Other areas of concern:

- Make sure there's a small amount of slack in a line to allow for growth and contraction in the line due to temperature and pressure variations.

- Be sure that hose clamps fit snugly around a hose but not so tight as to constrict fluid flow. The hose clamps shouldn't restrict travel or cause the hose line to be subjected to tension, torsion, compression or sheer stress as it flexes.

- Use abrasion sleeves, support coils and abrasion pads as needed to eliminate chaffing and rubbing.

- Group and clamp hoses together to improve appearance, aid in routing and prevent chaffing, but try to avoid clamping high- and low-pressure hoses together, as they react differently in operation.

- Be sure that the hose isn't twisted or kinked during installation. Most hoses have a "lay line" to aid in the installation.

Torquing

Although many technicians get by with using torque wrenches and consulting recommended torquing charts to tighten fittings, Stratoflex, another hose manufacturer, suggests that there may be a more accurate and repeatable way to tighten fittings.

The method was developed, says the company, because variables such as surface finishes, material type, surface treatment, lubricants and others do not remain constant; therefore, measuring input force is not always accurate.

A method called Flats From Finger Tight (FFFT) was developed by the company to assure repeatable "throughput" force, or the force that's transmitted to the metal-to-metal sealing surface.

The FFFT method works as follows:

1. Finger tighten the nut (approximately 30 inch-pounds).
2. With a marker, make a longitudinal mark on the body of the hex and onto the body of the mating part.
3. Tighten the joint by the number of flats in Table 1.

This method, says the company, measures the "throughput" of the torque. The recommended FFFT values listed in Table 1 reflect the force necessary for solid, leak-free connections, says the company.

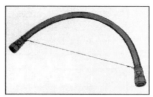

Use a tie wire to secure hoses in position before removing. Don't try to straighten hoses that have taken a set.

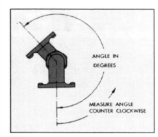

When ordering hoses that have elbows at each end, figure the relative angles by measuring clockwise from the base of one of the fittings as shown above.

Length of Service

Although most hoses are considered to be "on condition," meaning that they shouldn't be replaced unless the hose shows obvious signs of defects, manufacturers suggest shelf lives and operating lives for certain types of hoses.

For example, Stratoflex says that it doesn't allow its rubber hose to be sold as new (based on MIL-STD-1523 and SAE AS 1933) if the hose is more than eight years past the cure date. Teflon hose, and its silicone-based hoses, however, are considered to be "non-aging."

The company also recommends a maximum shelf life of five years (with extended shelf life based on inspection/test or service), and a maximum of seven years in service. (FAA AC200-7N recommends five years for general aviation in-engine-compartment service.)

If you add these years up, a hose can be up to 20 years old by the time that it has spent seven years on an aircraft! Regardless, Stratoflex says that if hose continues to pass testing and inspection guidelines per ARP1568, it can continue to be used.

The company also warns that all hose takes a "set," eventually conforming to the installed configuration. This is important to know, because by the simple act of removing a hose for any reason, the hose may become damaged, terminating the life of a hose that could have otherwise continued operating.

Handling Teflon

Although Teflon (a Du Pont trademark for Tetrafluoroethylene resin) hose offers the advantage of unlimited service life, ability to withstand virtually all types of fluids and ability to withstand wide temperature ranges, the material is somewhat delicate to work with and not quite as forgiving as rubber-type hoses.

Teflon hoses should always be handled with extra care to prevent excessive bending, twisting and kinking. Kinking of Teflon hose occurs more easily in larger sizes and in very short assemblies. It also tends to preform to the installed position of hot fluid lines. Never step, stand or set anything on top of Teflon hose.

Airframe Technology

You should also always permit the assembly to take its own "lay" when being removed or stored. When shorter bend radii are required, the assembly may need to be preformed at the factory. These factory preforms are secured into shape prior to shipping.

It's recommended that any lines that are either preformed or have taken a set be handled by using a tie wire to secure the shape of the hose.

Building hoses

Depending on the type of hose, fittings and equipment, there are various procedures for cutting, preparing and assembling hose assemblies.

Three procedures that are common, regardless of the type of hose or fittings that you're working with, are proper cleaning, inspection and proof testing.

Be careful to assure that all residue, shavings, etc., are cleaned thoroughly from the inside of the hose assembly after performing any cutting of hose or installing fittings. It's recommended that each assembly be cleaned with clean, dry compressed air prior to further inspection or testing.

The assemblies must then be inspected thoroughly to assure that there are no internal cuts or bulges in the inner tubes, and that there are proper gaps between the nuts and sockets, or hex and sockets. The nuts should be free to swivel.

Finally, the hose assembly must be tested at a ratio of 4-to-1. This means that the operating pressure to which a hose assembly is subjected should not exceed 25 percent of the minimum burst pressure. This is a general recommendation, however, and may vary depending on the type of hose, or the aircraft manufacturer's recommendations.

Ordering hose assemblies

For those who choose to order hose assemblies for a specific application instead of building the hose in-house, there are a number of hose shops throughout the country that provide the service of building and testing hose assemblies.

Meadows says that due to the great number of applications, changes that are made in the field and changes that are made during manufacture, the best way to accurately order a replacement hose is to send in the one which you wish to replace. This way, he says, the hose shop can duplicate your hose exactly, without question.

If you're ordering a replacement hose and can't send in the original for some reason, there are a few tips that Meadows offers to help you order the correct part.

When you measure the hose, remember to calculate the length of the hose to include the portion needed to attach to the fittings.

Kinked Hose
Usually is a result of Teflon* hose being improperly handled. Fluid flow is reduced. A break is in the making.

*Teflon is a Du Pont trademark.

Twisted Hose
Incorrectly installed, easy to spot. Flow is cut down. If the hose is permanently deformed, replace it at once.

Scuffed Hose
Results from abrasion against a frame, an engine component, another hose or from incorrect clamping. Hard to detect on unsleeved wire-braided hose. Sleeved hoses usually exhibit evidence of wear if this problem is present.

Brittle Hose
Rubber hose that's hardened, no longer flexible. Feel for stiffness. Wiggle and listen for cracking. Temperature and time produce this effect. Be sure you have the correct type of hose for the application.

Seeping
First appears as a slow leak. Look for wetness on the hose or dripping at the socket. Instead of disappearing, leakage will get worse. Inspect hose routing to ensure that the hose has some straight length where it exits the fitting.

Rusty Hose
Indicated when outer rubber cover is split, abraded or cracked. This means wire braid is corroding inside and cannot be seen. Check carefully.

Broken Braids
Danger signals which can be costly if they're missed. Easy to see if you take the time to check carefully. Wear a leather glove and run hand lightly along the hose. Broken braids can lead to multiple problems.

Seven ways to spot hose line problems before they cause big trouble.

Aeroquip 1992

Airframe Technology

If the nose has an elbow at each end, the angle of the fittings in relation to each other must be calculated correctly. This is done by placing one of the angles flat on a bench and measuring the fitting at the other end counterclockwise from the surface of the bench. This angle may already be marked on the identification tag on the hose, or may be an extension of the part number. The angle is critical, as many types of hose can't be flexed enough to make up for errors in the calculation.

Finally, if you order the hose by part number, don't expect the hose to be exactly like the one you removed. Many times, the manufacturer of the aircraft will have changed the routing of the hose during production of the aircraft and the length of the hose will have changes. Refer to the maintenance manuals for proper hose routing if you find that there's a problem related to hose length. *November/December 1992*

Note: Much of this information has been taken from publications and bulletins provided by Aeroquip Corporation and Stratoflex Aerospace/Military Connectors Division.

Airframe Technology

King Air five-year landing gear inspection tips

By Greg Napert

Once every five years, King Air owners roll their aircraft into the hangar for an examination of the aircraft's landing gear.

The inspection, which is applicable to most all King Air Series, is due every five years, or within a specified number of cycles (whichever comes first). It involves complete disassembly of the main and nose gear shock absorber strut assemblies, main and nose gear drag brace assemblies, axle assemblies and torque knees. Additionally, where applicable, inspection of the retract gearbox and clutch, and gear actuators is also required.

Dave Fisher, technician for Elliott Beechcraft in Eden Prairie, MN, says that they approach the inspection by assigning one technician to each individual gear. That way, he explains, each gear is removed, disassembled, assembled and installed by one individual. This speeds up the process and adds to the quality of the inspection.

After the aircraft is placed up on jacks, the next step is to perform an operational check. "We perform a retraction test by operating the gear to make sure that there aren't any abnormal noises. If we hear any unusual noises, we can then investigate them right away or note the problem so that it can be checked during disassembly. Typically, he says, if there are any problems, they are found in the gearbox, with the actuator assembly, or with the nose gear retract chain assembly.

Pulling each main and nose gear off of the aircraft can be a bit tricky, he explains. The first thing that needs to be done is to retract the gear a bit so that it's not in the down-and-locked position. "There are two things that you need to keep in mind at this point," he says.

"First, don't disconnect the torque links before releasing the pressure in the cylinder or the pressure will blow the lower shock absorber assembly off of the aircraft. These struts don't have a stop mechanism to prevent the lower and upper shock absorbers from coming apart once they're disconnected. The link is the only thing that's holding the upper and lower cylinders together," says Fisher. He explains that it's best to remove both the upper and lower shock assemblies (essentially the entire shock strut) from the aircraft as an assembly.

Todd Olson, another technician at the Beech facility, says that disassembly of the gear, once it's removed, is a fairly simple process that requires minimum tooling. "One thing that can make disassembly easier is to remove the support assembly (metering tube) before you slide the lower shock absorber assembly out. This is

Carefully inspect the nose gear drag brace-to-airframe "intercoastal area" for loose rivets or damage.

because the lower assembly helps guide the support assembly out of the upper shock without cocking and making it hard to remove."

Fisher says that after the gear's disassembled, they dimensionally check the areas called out in the component maintenance manual, and inspect for wear, corrosion and cracks. "The biggest areas to inspect for corrosion are on the top brace assemblies of the main gear. The inner bore and the bolt holes of this brace are particularly susceptible to corrosion. This is because the brace is made of magnesium and is in contact with dissimilar metal.

"A certain amount of corrosion (in the form of pitting) is allowed, as long as it can be cleaned up and remains within limits dimensionally. It's kind of a gray area, however, and if you find one with corrosion, it would be wise to call Beech engineering and give them more detailed information about the extent of corrosion. They may be able to help you make the call on whether or not the part is serviceable."

If you've determined that the corrosion is within limits, you've got to remove the corrosion as best you can, and treat it with an approved corrosion preventive compound.

"What we do," says Fisher, "is clean it up with Scotchbrite® until all signs of corrosion are removed. We then treat it (the top brace and upper shock absorber assembly) with Dow 19, a corrosion preventive that works well on magnesium." All the other areas of the top brace are cleaned up as well, treated and then painted as appropriate.

Areas of the gear that are coated with epoxy aren't stripped because the coating is too tough. Cracks that develop under the epoxy will also crack the brittle but tough epoxy and be visible at the surface so it's not necessary to strip most of the epoxy-coated components. Also, you don't want to be putting strippers on some of these parts because the strippers are corrosive.

One area that does need to be stripped of paint, however, is the weld bead on the nose gear fork as specified in AD 87-22-01, R1. According to the AD, the gear must be inspected around the weld area for cracking by removing the paint and then fluorescent penetrant testing. A small amount of cracking is allowed (with recurrent inspections) according to the AD, but switching to the new style nose gear fork that's been redesigned to eliminate the weld bead precludes the inspection requirement.

Incidentally, this particular 90-Series King Air did have cracks that extended into the surrounding material. The part was rejected and a new fork installed.

An area that's particularly susceptible to corrosion, says Jay Mitchell, lead technician at Elliott Beechcraft, is under the felt pad that lays between the upper and lower shock absorber assemblies on both main and nose gear. This felt pad must be adequately oiled prior to assembling the strut. If it isn't, moisture will enter the cylinder and be absorbed into the felt. This eventually causes corrosion to form where the felt touches the inner wall of the upper shock absorber.

Mitchell says that newer designs have eliminated the felt pad and improved the O-ring seal. This, he says, has cut down significantly on corrosion.

Look for wear to take place in an egg-shaped manner, says Mitchell, due to the way that force is applied to the front of the gear upon landing.

The most common areas for wear are the bushings and pins that support the landing gear structure. Particularly, he points to the bushings in the center of the torque link.

"The torque knee pins are sometimes corroded and difficult to remove. We've found that in some cases, we've got to use a rivet gun to exert enough force to remove stubborn pins," he says.

The bottom torque knee and the top torque knee on the King Air are made of different materials. The bottom is made of steel and the top is made of aluminum. Look for dissimilar metal corrosion where the top and bottom torque knees come into contact.

"The bolt holes that secure that drag brace to the cylinders are also problem areas because of dissimilar metal corrosion. We usually inspect these areas carefully and lubricate them well," he says.

John D'Amato, another technician at Elliott, says that an important item to pay attention to during assembly of the nose gear is the installation of the nose gear straightener assembly (which guides the gear into the wheel well). D'Amato warns that it's possible to install the straightener backward.

He explains that if it's installed backward, the gear won't properly align as it's retracted into the wheel well, which can result in significant damage. The confusion usually occurs because the inner shock absorber can be rotated 180 degrees from its correct orientation and installed that way. One hint that this has occurred is that the aircraft will be hard to steer or turn.

The aircraft may operate for a while before anyone actually notices that there's anything wrong. If the gear is perfectly centered, he explains, the roller will go into the guide without causing any problems. But as soon as the nose gear is retracted with the gear rotated slightly, it'll tear the guide assembly apart and possibly jam in the wheel well.

D'Amato says that one area to inspect carefully is the steering stops on the nose gear. It's common for these to be damaged as a result of turning too sharp while towing the aircraft. Look for cracks at the base of the stops, he says. If the stops are damaged, the upper shock absorber will have to be replaced.

On the lower shock absorber assembly, the chromed areas must be dimensionally checked and inspected for damage. Grooves in the chrome, or other types of damage, that allow fluid to leak out of the cylinder, are cause for rejection.

Inspect for cracks around the weld area of the old-style nose fork per AD 87-22-01, R1.

"We sometimes try to blend out small scratches if they're questionable, but maximum allowable wear is one-thousandth of an inch, which doesn't give you much room for repair," he says.

"When we disassemble the strut," says D'Amato, "there are certain parts that automatically get replaced, such as the packings, wiper rings in the cylinder, felt and valve cores."

D'Amato says that there's a lot of potential damage that can be done during assembly of the strut that you won't know has been done until you finish the inspection and put the weight of the aircraft back on the gear.

For example, he says, the ring at the bottom of the support tube can be broken when the lower shock absorber is inserted into the upper shock absorber assembly. The lower shock assembly must be carefully centered in the ring during assembly.

Airframe Technology

Care must also be taken to liberally lubricate the packing. If installed with inadequate lubrication, there's a chance that it'll roll. The packing will then be twisted and won't seal properly.

Other pointers

- D'Amato says that you should keep a close eye on anything up in the wheel well that may be exposed to dirt and moisture.
- Inspect the gear actuating mechanisms for proper operation because they are also exposed. On the sprocket shaft, which is actuated by the chain on the nose gear mechanism, there are sprockets that have needle bearings which require lubrication—this is an ideal time to service them.
- Check the roller bearings on the shaft to the actuator for freedom of movement.
- Carefully inspect the "intercoastal area" (where the nose gear mounts to the airframe). D'Amato says that the area has been beefed up on the new King Airs, but should be inspected closely for loose rivets, regardless.
- This inspection provides a good opportunity to check the shimmy dampener for proper operation and for proper fluid level.

Finally, after servicing the struts to manufacturer's recommendations, perform a retraction check and listen carefully for sounds that indicate improper operation. If all is well, the aircraft can be removed from the jacks.

September/October 1992

Airframe Technology

Touch-up painting pointers

By Greg Napert

Although a common sense can prevent most hangar rash, it's not all that uncommon to end up with a few scratches and nicks on painted surfaces during the course of maintenance. Returning an aircraft to an owner that's free of nicks and scratches can add a touch of class to any job.

Most of this damage can be eliminated by simple touch-up techniques. Covering up damage is not only important for aesthetic reasons, but failure to tend to damaged paint surfaces can promote corrosion of the exposed surfaces, making touch-up especially important when base metal is exposed.

Types of paints

One of the most important aspects of touch-up painting is making sure that the paint you're using for touch-up is compatible with the paint that's already on the aircraft. Certain types of paints cause others to soften or lift and act like strippers which can quickly turn the need for minor touch-up into the need for a new paint job.

Randall Effinger, owner of Central Aviation Inc., a paint facility in Watertown, WI, says that lacquer, for example, will soften and lift existing coats of enamel or polyurethane.

The simplest way to determine the type of paint on the aircraft, he says, is to look in the logbook. Many times, though, nothing's mentioned about the paint that was used.

If this is the case, you'll have to test the paint on the aircraft. This can be done by a process of elimination. You should start, he says, by applying a little drop of lacquer in an inconspicuous area. If the paint softens or lifts, you know that it's not a lacquer.

If you suspect that the paint may be a lacquer but aren't sure, you can apply a small coating of engine oil. If it's lacquer, the surface of the paint will soften in a few minutes.

If you've determined that the paint isn't lacquer, it's probably either an epoxy, acrylic or polyurethane finish. To determine if it's an epoxy or acrylic, wipe a small amount of MEK (methyl ethyl ketone) on the surface of the paint. MEK will pick up the pigment from an acrylic finish, but will have no effect on an epoxy coating.

Finishes on fabric aircraft, he says, can be deceiving also. Some fabric aircraft are doped; then a finish coat of polyurethane or enamel is applied over the dope. If you use dope to touch up the polyurethane or enamel, it will cause the surrounding finish to peel or lift—as lacquer does with enamels and polyurethanes.

Preparing surfaces

Effinger says that surface preparation for minor scratches should be minimal. You need to remove any loose paint first, by scraping with a small knife, razor blade or similar item.

If you're just touching up very small scratches or nicks with a brush, there's typically no need to prepare the surface in any way except to make sure it's clean, he says. Sanding or any further preparation will only make the damage larger and more noticeable. Alodine or chromating the surface will usually stain the surface of the surrounding paint, he says, and won't allow you to match the color of the existing paint.

If you're dealing with a fairly large area, however, you should sand the surface down to bare metal first and feather the edges of the damaged paint. You then need to clean it thoroughly to remove any leftover stripper residue or oils that may be on the surface. The metal must then be etched, and properly treated before applying finish coats of paint.

The touch up

Effinger says that most scratches can be handled quite well using a fine No. 0 or No. 1 artist brush with soft bristles.

When you're applying small portions of polyurethane, you need to mix the paint with the appropriate activator to get proper curing. The paint will dry without the activator but it'll take a long time and won't be as durable.

"We typically use something like a bottle cap to measure the correct proportions of paint to activator in small quantities," Effinger says.

On areas that are subject to oil and grease, like under the cowl or on landing gear, you need to clean the area thoroughly before you do anything.

Airframe Technology

The advantages of lacquers become apparent when it comes time to blend in repairs with existing finishes. With lacquers, it's very easy to blend it into the surrounding finish by buffing. In fact when you paint an entire aircraft with lacquer, you typically have to buff it to produce a smooth finish.

Polyurethanes and enamels are difficult to buff out and blend in. Typically, the only way to make a touch-up job unnoticeable is to repaint an entire panel.

For making repairs on fabric covering, says Effinger, scratches down to the fabric need a couple of coats of silver coat to protect the fabric; then, the color dope can be applied. Dope buffs out very easily and can be blended out very nicely.

You've got to be careful here also, he says. Some fabric aircraft are doped and a finish coat of polyurethane or enamel is applied. You can't apply dope to these finishes; if you do, the surrounding finish will peel or lift, as lacquer does with enamels and polyurethanes.

Paint compatibility rules

The following rules, taken from the Airframe Handbook AC65-15A is a valuable reference in determining the compatibility of paints.

- Old type zinc chromate primer may be used directly for touch-up of bare metal surfaces and for use on interior finishes. It may be overcoated with wash primers if it is in good condition. Acrylic lacquer finishes will not adhere to this material.

- Modified zinc chromate primer will not adhere satisfactorily to bare metal. It must never be used over a dried film of acrylic nitrocellulose lacquer.

- Nitrocellulose coatings will adhere to acrylic finishes, but the reverse is not true. Acrylic nitrocellulose lacquers may not be used over nitrocellulose finishes.

- Acrylic nitrocellulose lacquers will adhere poorly to both nitrocellulose and epoxy finishes and to bare metal generally. For best results, the lacquers must be applied over fresh, successive coatings of wash primer and modified zinc chromate. They will also adhere to freshly applied epoxy coatings (dried less than six hours).

- Epoxy topcoats will adhere to all the paint systems that are in good condition and may be used for general touch-up, including touch-up of defects in baked enamel coatings.

- Old wash primer coats may be overcoated directly with epoxy finishes. A new second coat of wash primer must be applied if an acrylic finish is to be applied.

- Old acrylic finishes may be refinished with new acrylic if the old coating is thoroughly softened using acrylic nitrocellulose thinner before paint touch-up.

- Damage to epoxy finishes can best be repaired by using more epoxy, since neither of the lacquer finishes will stick to the epoxy surface. In some instances, air drying enamels may be used for touch-up of epoxy coatings if edges of damaged areas are first roughened with abrasive paper.

July/August 1992

Airframe Technology

Fire detection/extinguishing systems
No respect

By Nick Levy

How many times have you looked over the aircraft only to find an accumulator or tire needing a little servicing? Probably once a week in some flight departments, once a month in others. But how many times have you looked at the fire extinguishing system?

Fire detection/extinguishing systems are a little like Rodney. We don't give them the respect, or in this case the "inspect" they deserve considering the job these components perform. Not only do they protect the lives of the passengers and crew, they save the flight department thousands of dollars in repair costs every time they effectively put out a fire.

Basics

All corporate aircraft are manufactured with some sort of fire detection system. But it's pretty safe to say that most of these aircraft are equipped with either a "continuous loop" or "pressure-sensitive loop" detection system. While providing the most effective fire detection coverage for the dollar, these systems provide two totally different methods of detection.

Continuous loop

While the "continuous loop" system is the easiest to maintain, it sometimes becomes one of the hardest to replace. The loop is essentially a thin wall hollow tube that is connected to ground on both ends. As part of the manufacturing process, the hollow tube is filled with small beads which support a fine wire running through its center. The beads are made of a salt solution and insulate the inner wire from the case. It's these small beads that determine the operating temperature of the detection loop.

In the event a fire occurs near the loop, the salt beads "melt" and allow the voltage in the center wire to connect to ground. The ground triggers the warning circuit, and the crew is alerted with a light and aural warning in most aircraft.

Exceptional care must be taken during the installation of the loop to assure proper operation. Most important is the bend radius of the curves as the loop is installed in the compartment which is being monitored. Since the beads are the insulation for the center wire, a sharp bend will "break" a bead and result in false indications to the crew.

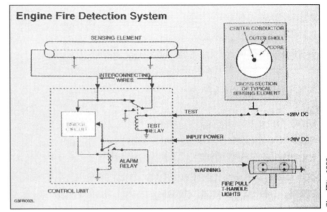

Continuous loop detection system.

Primary concern to the technician is the integrity of the system. This is ensured by attaching both ends of the center wire to form a "loop." Thus, if the loop breaks or becomes disconnected within the compartment, either end still remains connected to the detection circuit. This assures complete coverage of the area even though a fault has occurred. In order to provide a valid test, one end of the loop is electrically disconnected when the "test" button is depressed in the cockpit. This effectively tests the total length of the element by providing a continuity check through the loop.

Pressure-sensitive loop

The second type of detection element used in the corporate aviation industry is known as the "pressure-sensitive loop." This system provides an equal amount of coverage in the desired area, but uses a pressure-filled tube to detect the fire. The tube is filled with an inert gas (usually helium) and sealed on one end. The other end is connected to an electrical connector which is referred to as a responder.

The responder is merely a pressure diaphragm and a microswitch. As the temperature around the detection loop increases and the loop warms, the helium gas starts to expand. At the prescribed temperature, pressure in the loop closes the switch and the alarm is initiated in the cockpit.

To provide "spot" detection as well as area detection, the loop is made with a small core of discrete metal (usually titanium). This small core will be heated during

Airframe Technology

a fire and will release a second gas to activate the pressure switch. Thus, the loop provides "overheat" protection as well as "fire" detection.

While operation of the loop depends on a closed tube, the circuit provides adequate protection in case of failure. The "test" switch usually checks the pressure within the tube by using a second switch in the responder. If the helium gas escapes due to a crack, broken loop or bad diaphragm the test will show a fault in the system. However, the loop will still remain functional. Since the tube contains the core metal, it will release sufficient gas pressure in the tube to activate the warning system should a fire occur. In this case, the detection of a fire is ensured, while an overheat situation would go undetected.

Installation of the "pressure loop" is just as critical as the "continuous loop." Careful attention must be paid to the bend radius of the detection element and proper support within the compartment to prevent damage to the element. Inspection criteria are outlined in the applicable maintenance manual and must be followed to the letter. Most of the items are common sense such as clamps, chafing relief, insulators, connections and visual inspection. But these are the places where we tend to overlook the obvious damage.

Particular attention must be paid to the distance between the clamps. A long unsupported loop will tend to vibrate and eventually workharden which will soon result in a failure. Each clamp is also installed with a plastic or neoprene sleeve which prevents chafing with the clamp. All connections must be secured with safety wire, and any visual damage to the exterior of the loop will usually result in replacement. Be sure to consult the aircraft manual or the system manual to determine allowable limits.

Extinguishing systems

All jet aircraft are required to be equipped with an engine fire extinguisher system. While the number of engines, size of bottles, type of agent and delivery methods vary from aircraft to aircraft, all provide a greatly improved fire extinguishing system. Usually the extinguisher agent is stored in a round or accumulator-shaped bottle in the aft compartment or nacelle area. Most bottles are equipped with a pressure gauge which is used to determine proper servicing according to a temperature chart. Located on the bottom of bottle are one or two delivery ports which connect by aluminum lines to the protected compartment. (Engine extinguishing bottles usually have two ports which allow delivery to either engine.) Within the compartment are "fan-tail" nozzles made of stainless steel which assure proper delivery of the extinguishing agent within the area.

Since rapid delivery is mandatory, the bottle is delivered by the activation of an electrical cartridge. This explosive "squib" will break a ribbed disc which will allow the bottle to empty in one second or less.

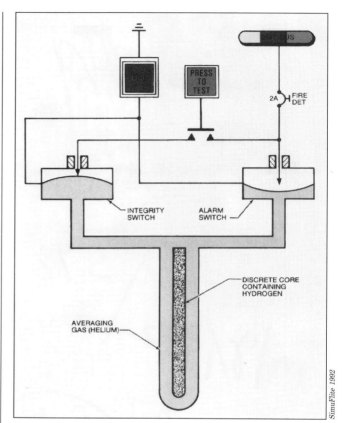

Pressure-sensitive loop detection system.

The most common type of agent used in corporate aviation is called "halon." This refers to halogenated hydrocarbons, which are made by combining carbon, fluorine, chlorine and bromine atoms. Currently Halon 1211 and Halon 1301 are the most frequently used fire extinguishing agents in corporate aviation. These agents provide excellent fire protection at a reasonable cost.

Most bottles are in the capacity range of 2.5 to 5 pounds and require annual weight checks to verify the proper quantity of halon. These bottles also require pressure testing to ensure the integrity of the container. This "hydrostatic" inspection is performed at a predetermined interval based upon the inspection cycle of the aircraft.

Since pressure is also of concern, the bottles are equipped with an overpressure relief valve or disc. These valves or discs usually relieve the pressure within the bottle if it increases to a dangerous level. At a predetermined pressure (or temperature) the bottle will be vented automatically and will be indicated to the ground crew. An obvious indication is the loss of pressure on the bottle pressure gauge. But, this may not be visible without opening compartments or panels on the aircraft. Thus the aircraft manufacturers have developed an easier method.

The most common method is to cover the end of the vent line with a colored, plastic disc. The disc will be visible when the aircraft is preflighted and serves as a visual indication to the crew. If the bottle has vented its pressure due to excess heat or pressure, the disc will be blown away and the inside of the vent line is painted a high contrasting color to make it easily identifiable.

Remember, the presence of the disc does not ensure the bottle is serviced correctly. This can only be determined by the removal and weighing of the suspect bottle.

While much has been written about the advantages and disadvantages of fire extinguishing systems, one thing remains true. If a fire occurs while an aircraft is in-flight, only the little 2.5-pound bottle of halon provides the protection necessary to save the passengers, crew and aircraft. *May/June 1992*

Nick Levy is an aircraft technical instructor for SimuFlite Training International. He currently teaches Falcon 10, 20 and 50 and Professional Troubleshooting Skills courses. Levy has over 25 years of professional aviation maintenance and technical teaching experience.

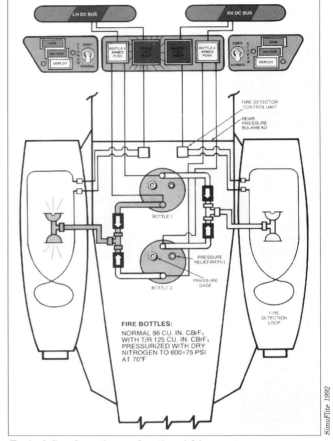

Typical fire detection and extinguishing system.

Airframe Technology

Magnetic particle inspection
Doing it right

By Douglas Latia

Magnetic particle inspection, a non-destructive testing method for locating surface and near-surface defects in ferromagnetic materials, is no stranger to the aviation industry. The process was originally used to locate cracks in the steel engine parts of reciprocating engines and is still in use today for the same purpose, along with a host of other applications.

The fundamentals of magnetic particle inspection are quite basic. Magnetic lines, referred to as "lines of flux," travel easily through ferromagnetic materials such as steel and iron. Flux lines tend to establish themselves in closed loops which never cross one another. There's a certain flux density associated with the material's permeability (the ability to accept magnetism) and with the amount of magnetizing force applied to the part.

The magnetic lines of flux will route themselves around defects, frequently referred to as discontinuities, looking for new paths. Normally the path of least resistance for flux lines is through the ferromagnetic material.

However, if the flux density is so great that there is no room for additional lines in the material, the flux lines will leave the material at the point of the discontinuity, creating a flux leakage. It's this external magnetic field formed by the flux leakage that will hold magnetic particles, making the discontinuity visible. Particles treated with a fluorescent pigment will make discontinuity indications stand out brilliantly when viewed under a black light.

Direction of magnetization is important since discontinuities running parallel to the flux lines will cause little or no disruption of the paths. Passing current through the part will generate circular magnetism, and placing the part in a coil will establish a longitudinal magnetic field.

Successful magnetic particle inspection depends upon attention to many details. As aircraft technicians, we know that the particulars of any maintenance operation can be found in the manufacturers' maintenance manuals. This is also true for non-destructive tests.

Lycoming, for example, publishes the specifications for magnetic particle testing of its ferromagnetic engine parts in a service instruction.

The direction of magnetization is given along with the amount and type of amperage. The direction of magnetization is important to provide the greatest sensitivity to cracks of certain orientation. The optimal amount of amperage will ensure that there is adequate magnetism

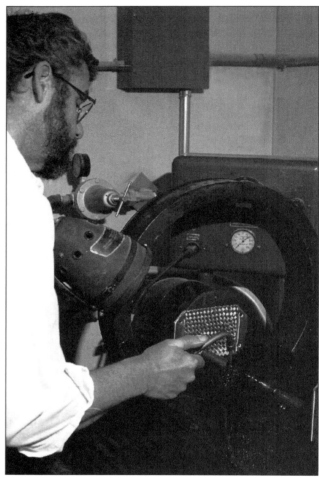

Al Peterson, technician at Omni Flight Helicopters Inc. in Janesville, WI, inspects a gear for cracks using magnetic particle inspection.

to create leakage fields at discontinuities that exist, yet, not so much magnetism that a confusing background is formed. Also, exceeding the recommended amperage can cause burning of the part at the contact points and overheating, which may render it unserviceable.

Magnetization can be accomplished with either AC or DC current. Magnetic particle inspection machines are either AC or DC and in some cases have the capability of both. DC current has the capability of penetrating below the surface of the material, making subsurface discontinuities detectable. AC current creates a skin effect and stays at the surface, providing excellent resolution for fine surface discontinuities. Lycoming specifies that DC current be used.

Continental and Beechcraft publish similar information with the addition of specifications for types of particles to be used and method of particle application. Both manufacturers require the use of wet particles. Wet and dry type magnetic particles are available; however, the wet type is used almost exclusively in the aviation industry.

The particle bath consists of fine particles of iron oxide material, dyed or treated with a visible color or fluorescent pigment, suspended in a petroleum vehicle similar to kerosene. While kerosene has been used for this purpose, commercially available suspensoids, such as Magnaflux Carrier II, are odorless and have flash points over 200°F. The petroleum vehicle allows particle mobility necessary for orientation in the areas of flux leakage.

Continental prefers the use of fluorescent particles while Beechcraft states in its Structural Inspection and Repair Manual that either fluorescent or visible particles may be used. The advantage of using visible particles is that a darkened inspection booth and black light are not needed. However, indications created by fluorescent particles are more readily seen when viewed under a black light.

There are two methods of particle application: continuous and residual method. The continuous method requires that the particles be present while the magnetizing force is being applied to the part.

This method takes advantage of the fact that the magnetic field (and associated leakage fields) is strongest while the magnetizing force is being applied. Therefore, small discontinuities will cause leakage fields strong enough to attract particles while the magnetizing current is flowing. The indication will be formed only while the current is flowing. It's important that the particle bath not be flowed over the part after the current has been removed, or indications of small discontinuities will be washed away.

The residual method relies upon the retentivity (ability to retain magnetism) of the material. The part is magnetized first and then the particles are applied. The leakage fields associated with smaller discontinuities may disappear when the magnetizing current is removed and consequently no particles will gather to form an indication. The continuous provides the greatest test sensitivity.

There are several other factors that affect the sensitivity of a magnetic particle test that may not be mentioned in the manufacturer's publication.

For example, some of these items which shouldn't be overlooked when preparing a magnetic particle test are: concentration and contamination of the particle bath, intensity of the black light and confirmation of proper machine operation.

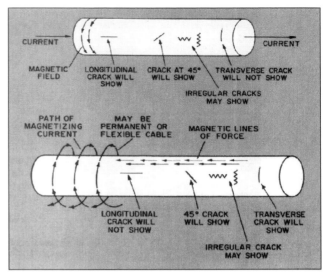

Proper direction of magnetization is essential for successful magnetic particle inspections.

It is important that the concentration of the particles in the particle bath is correct. A weak particle concentration will result in too few particles gathering at a magnetic leakage, possibly making detection impossible. A strong concentration will create a glowing background which reduces the contrast of the indication.

The concentration is measured by turning on the particle bath pump and allowing the bath to circulate through the machine for approximately 30 minutes for agitation. Fill the centrifuge tube, a special beaker graduated in milliliters, with 100 milliliters of particle bath. Let the sample set undisturbed for 30 minutes. The level of the accumulation of particles is read against the scale at the bottom of the tube. Normal concentration for the 14A particles is .2 to .25 ML per 100 ML sample.

The particle concentration changes because of evaporation of the carrier and particles being carried out of the bath on test parts. The concentration should be checked at least once each day that the machine is used and adjusted when necessary.

Before returning the contents of the centrifuge tube to the magnetic particle machine reservoir, examine the material that has settled in the bottom of the tube carefully under the black light. Contaminates will settle on top of the magnetic particles and will not fluoresce. If the volume of the contaminates exceeds 30 percent of the particles, the entire bath should be drained, the equipment cleaned, and the bath replenished using new materials.

The black light typically used in magnetic particle testing is a 100-watt reflector-type mercury vapor lamp equipped with a black light filter. These lamps tend to lose their effectiveness at the desired light wavelength with age and use. Therefore, it's necessary to check the black light intensity on a regular basis.

Airframe Technology

This is done with a black light meter which measures intensity in microwatts per square centimeter. The intensity should be at least 1,000 microwatts per square centimeter at 15 inches from the light source. Allow five minutes from start-up for the mercury vapor lamp to reach full intensity before performing this check.

In fluorescent applications, the inspection will most likely be performed in a darkened booth, so allow five to 10 minutes for eye adaptation. This can make the difference between locating or missing small indications. The inspection area should also be regularly checked with a white light meter to ensure that the level of white light doesn't exceed two foot-candles.

Ensuring that the whole system is functioning properly is simple but sometimes overlooked. Either a part with a defect similar to that which you are looking for, or a test standard with a manufactured defect, should be processed in the same manner as the actual part before proceeding with the actual test.

Ferromagnetic parts that have been inspected will retain a certain amount of magnetism. It's essential that this magnetism be removed before the part is returned to service. Magnetized engine parts, once inside the engine, will attract and accumulate microscopic ferromagnetic wear materials. These particles can get into small clearances and accelerate wear. Magnetized airframe parts can have an influence on the magnetic or electronic compass.

Demagnetizing is accomplished by introducing a magnetized part to an alternating magnetic field and then withdrawing it from the influence of the field. A demagnetizer consists of a large coil energized by AC current. The magnetizing coil of an AC magnetic particle inspection machine can also be used. The part is held in the energized coil and then withdrawn and held at least an arm's length from the coil before the coil is turned off. Demagnetizing equipment is available which electrically reduces the alternating magnetic field to zero, eliminating the need to physically remove the part from the coil.

The part is checked for residual magnetism with a field indicator. If a field indicator isn't available, a small staple and a 12-inch-long thread will accomplish the same end. Suspend the staple from the end of the thread and allow it to contact the vertical side of the part. Attempt to move the staple away from the part. If the staple tends to cling to the part, repeat the demagnetization process. Change the staple from time to time as it may become magnetized itself.

So far, the processes described, are in reference to the wet horizontal stationary unit. This is a relatively large and expensive piece of equipment and may be difficult for smaller shops to justify. Magnetic particle inspections can also be performed with the use of a hand-held yoke. Yokes are available for AC or DC magnetization and can be operated with 110-volt current.

The legs of the yoke become the poles of a very strong magnet when energized. It must be kept in mind that an imaginary line drawn from leg to leg must be aligned perpendicularly with the discontinuity being sought for proper direction of magnetization. The particle bath is also available in aerosol cans, making it handy to use with the yoke. The yoke has limitations, but it is very economical.

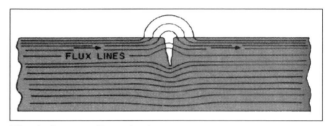

Flux leakage at discontinuity.

Perhaps one of the most important components of any non-destructive test is the inspector. A good working knowledge of the test method is essential. Procedures overlooked or improperly performed can defeat the entire purpose of the test. Overcritical judgment may result in rejection of serviceable parts, or an inexperienced eye may allow a critically flawed part to return to service.

There are many aspects of this test that must be understood by the inspector, but above all, a good inspector must be properly trained and experienced. Magnetic particle inspection, as well as other methods of NDT, when performed properly can provide increased safety and economical operation. However, performed incorrectly, can only serve to create a false sense of security and lead to disaster. *March/April 1992*

The VOM and electrical measurement errors

By Gary Eiff

When it comes to electrical system fault analysis, there is no more universally applied troubleshooting tool than the Volt-Ohm Meter (VOM). This instrument has been used for decades by aviation maintenance technicians to evaluate and solve a wide variety of electrical system failures. Despite its popularity, the proper use of the VOM to make precise measurements remains largely misunderstood.

The secret of making highly accurate electrical measurements lies in an understanding of the various errors which can be made while using the VOM. Once understood, technicians can take appropriate measures to minimize these errors and increase the accuracy of their electrical readings.

Parallax error

According to *Webster's New World Dictionary*, parallax is "the apparent change in the position of an object resulting from the change in the direction or position from which it is viewed." Parallax errors in analog electrical meter readings are caused by the separation of the meter's needle from the face of the scale of the instrument. As shown in Figure 1, if the technician is careless while making a measurement and views the needle from even a slight angle, an incorrect reading is realized. In order to assure that a correct reading is made, technicians must be sure they are reading the instrument from directly over—that is perpendicular to—the meter's scale.

Some analog VOMs have a mirror on the surface of the meter's scale. The purpose of this mirror is to help the technician reduce errors caused by parallax. To ensure that the measurement is taken with no parallax error, technicians must simply move their eyes back and forth while noting the position of the needle and its reflection in the mirror. If the needle is viewed from an angle, both the needle and its reflection will be plainly visible. By carefully moving the eye until the needle's reflection disappears under the meter's needle, parallax error can be eliminated and a true reading achieved.

Scale reading errors

In an effort to make the VOM as versatile as practical, many types of electrical measurements are possible with the instrument. Most VOMs can measure several

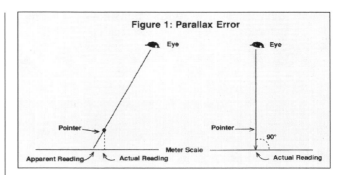

ranges of alternating current (AC) and direct current (DC) voltage, AC and DC current, as well as resistance. In order to ensure greater accuracy, several different scales are provided for each of the different types of electrical measurements possible with the meter. While there are several ways the scales of a meter may be misread, they can normally be grouped into two general categories, incorrect scale errors and scale increment errors.

With analog meters, using the wrong scale for reading the electrical value is among the most common of errors made by technicians when reading VOMs. For example, aircraft technicians most often find it necessary to take measurements of DC voltages. When they take AC voltage measurements, it is common (with some meters) for these technicians to mistakenly read an incorrect value from the DC scale.

Understanding the increment error is a little more difficult. In order to interpret the meter reading correctly, technicians must understand what each division or increment means on each of the ranges of the meter's scale. If, for example, we are making a measurement on the 2.5v AC scale, each division is one-fifth of the major demarcation or 0.1 volt. By comparison, if we choose to make a measurement on the 10v AC range of the meter, each increment is one-tenth of the major demarcation. Each increment therefore represents 0.2 volts.

To further complicate the issue, some of the scales on some meters don't contain a uniform number of divisions between major marks. This lack of uniformity in incremental representations has led more than one technician into making inaccurate readings.

Interpolation error

Technicians often strive to attain greater accuracy in measurement than the fundamental instrument is capable of making. Realistically, measurements are attainable to the nearest increment on the meter. Many technicians try to presume the capability of making more accurate measurements by guessing the position of the needle between the increments and assigning that position a value in the measurement.

For example, if the technician thinks that the needle is three-quarters of the way between the marks, he or she might claim that the reading indicated was 1.175 volts. The difference between this "guess" and the actual value of the measurement is referred to as the interpolation error. When striving for accuracy, the technician should consider that it is generally agreed the addition of an interpolated value into a measurement is limited to one-half of the increment value. In a digital meter, additional accuracy is attainable only through increasing the number of digits the meter is capable of presenting. Thus, a four-digit meter is presumably more accurate than a three-digit one.

Loading errors

Probably the least understood of the errors in using a VOM is that of the loading effect of the meter itself on the circuit under test. In order to take voltage measurements, we must place the meter leads across a source of electrical potential or a component in a circuit. In doing so, we have created, through the meter, an additional path for current and, consequently, affected the total resistance of the circuit under test. This change in resistance will cause the voltage being measured also to change.

Consider the circuit in Figure 2. The voltage measured across resistor R2 is proportional to the ratio of the values of the resistors R1 and R2. If resistor R3 is placed across R2, an additional path for current is provided through R3. Additionally, the total resistance represented by the combination of the two resistors becomes smaller than the value of R2. As a result, the voltage across the combination of R2 and R3 is reduced.

The value of the total resistance of the R2 and R3 combination and the degree to which the measured voltage across the resistors is affected is dependent upon the resistive value of R3. If the value of R3 is very large, there is little effect on the measured value of the voltage across the resistors. As the value of resistance R3 is reduced, the change in the voltage across the resistor combination becomes greater.

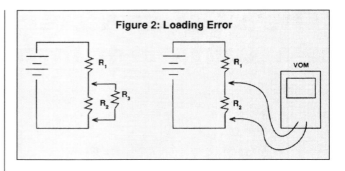

Figure 2: Loading Error

When making voltage measurements with VOMs, it should be remembered that the meter has finite internal resistance. This resistance must be considered when making voltage measurements since, in essence, we are placing the meter into the circuit as we did R3 in the previous example.

When selecting a meter for troubleshooting, pay particular attention to the input impedance of the meter for the range of measurements to be made. The higher the input impedance of the meter the less the voltage under test will be affected and the greater is the measurement accuracy. The meter's input impedance is that mysterious "ohms per volt" value given for the meter. As a general rule, the higher this value the less the meter will affect the reading.

In addition to consideration of the meter's input impedance, technicians should also understand the effects of circuit resistance on measurement accuracy. For example, let's evaluate the circuit in Figure 2 where the voltage source is 12 volts and the voltage across R2 is actually 6 volts. If the value of R2 is 300 ohms and the value of the meter's input impedance is 300,000 ohms, the measured value is not affected very much. If the meter is capable, it will actually read 5.997 volts instead of 6.0 volts because of the meter "loading effect."

By comparison, if the value of R2 had been 100,000 ohms, the meter would have had a much greater effect on the reading. The loading effect of the meter would change the apparent resistance value from the actual 100,000 ohms to a value of 75,000 ohms. This would result in the meter reading of only 5.143 volts. The loading effect of the meter introduced an error in the measurement of almost 1 volt.

As a general rule, as the circuit resistance across which the reading is taken increases, the greater is the error in the measurement due to the loading effect of the meter. For this reason, extra care must be taken when making voltage measurements across high-resistance circuit elements.

Airframe Technology

Calibration error

Probably the most familiar error to most technicians is the calibration error. Meter accuracy is of paramount importance to accurate troubleshooting and superior electrical/electronic maintenance. The ability of the meter to maintain small, uniform errors across its broad range of application is a fundamental attribute of a quality meter. To ensure continued meter accuracy, meters are required to undergo periodic calibrations.

Calibration errors are deviations between the meter readings of the instrument and the actual value of electrical standards. Because of meter circuit uniqueness and non-linearity of the circuit across the meter's range of measurements, these errors may not be the same for all types of electrical measurements or ranges of the meter. When making measurements which require a high degree of accuracy, calibration errors must be taken into account. A properly maintained high-quality VOM will generally have very small calibration errors. The need to reduce calibration errors is another testament for the selection of a high-quality meter for most maintenance applications.

Unlike several of the other errors presented, digital meters are as susceptible as analog meters to calibration errors. Additionally, regular periodic calibration is required to minimize these errors because regular use will affect meter accuracy. Circuit component aging, rough handling, as well as many other factors, affect the need for calibration. Meters used extensively or used in rough-handling environments should be calibrated more frequently than seldom-used or pampered units.

January/February 1992

Gary Eiff is an associate professor at Southern Illinois University in Carbondale, IL.

Airframe Technology

Fuel Quantity Indicating Systems

By Richard L. Floyd

Fuel Quantity Indicating Systems (FQIS) like most aircraft systems have evolved over the years. Some of the earliest systems used a glass tube as a sight gauge; others had a float attached to a graduated rod to give an indication of the level in the tanks.

Other systems consisted of a float arm riding the fuel surface which turned a geared assembly as the fuel level changed. A magnet was attached at the end of a rod, and outside the tank, another magnet in the shape of a pointer would track the movements of the first magnet to give the fuel level indication.

Float-based systems were later enhanced with the addition of an electrical meter movement. In these systems the float is connected to a pivoted assembly to the wiper arm of a variable resistor. DC voltage is supplied to an indicator which has two coils. One coil is connected to a fixed resistor back to ground, while the other is connected to ground through the variable resistor in the fuel tank as pictured in Figure 1.

Capacitance systems

Capacitance-based indicating systems, overcome some of the shortcomings common to float-type systems. They represent the most advanced type of FQIS. Capacitance systems offer greater accuracy, lack of moving parts in the tanks, easy summing of multiple sensors, and true mass or weight indication.

These systems use probes, also called tank units to sense the fuel level changes. The probes act as variable capacitors increasing in value as the fuel level rises.

The first of these capacitance systems was two-terminal or "unguarded." This meant that the system values were directly proportional to the length of the wiring required to complete the circuits. Thus, the probes themselves might make up only a fraction of the overall capacitance of the system.

Later three-terminal or "guarded capacitance systems" were developed. These system capacitance values were now independent of the length of the system wiring. The majority of aircraft have systems of this type. A drawback guarded systems suffer from is the requirement for expensive coaxial cabling and connectors.

A capacitor by definition is two conductors separated by an insulator. Normally the conductors are referred to as plates and the insulator is called a dielectric as pictured in Figure 2.

The fuel probes consist of two concentric hollow cylinders held rigidly in space with respect to each other and form the plates of a capacitor. Since the area of the cyl-

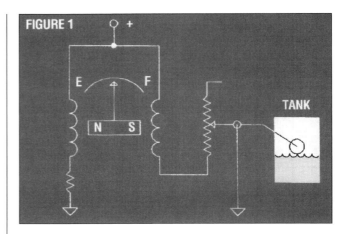

inders and the distance between them is fixed, only a change in the dielectric constant can change the probes' capacitance value.

When the aircraft tanks are empty, air is the dielectric. As the tanks are filled, fuel displaces the air and the capacitance of the probe increases proportionately. For example, if we were to measure the capacitance of an empty tank, then fill completely and remeasure we would find the value to be approximately 2.1 times the empty value as illustrated in Figure 3.

Different fuel types have slightly different dielectric constants and consequently produce different capacitive effects. Many systems use a compensator to correct for such differences. The compensator is a small probe located at the lowest point in any tank, or group of tanks, of a given system. It is usually mounted to and is part of a probe located at this point.

The volume of fuel is also affected somewhat by temperature, so that the weight for a given fuel level changes with temperature. Since the dielectric constant decreases with an increase in fuel temperature, the compensator tends to correct for temperature error also.

The effect of the compensator is directly proportional to the amount of fuel in the tank. That is, the compensator has no effect on the indicator when the tanks are empty, and has a maximum effect when the tanks are full. This can be seen in Figure 4.

So far, all the capacitance systems discussed have been what are termed AC capacitance systems. There are many aircraft which have DC capacitance systems. These terms (AC and DC) are not related to the system power requirements, but rather the type of return current from the fuel probes.

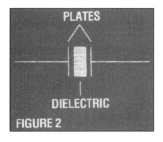
FIGURE 2

In the systems we have looked at the probes were impressed with an AC sinewave excitation and returned an AC current to the indicator or signal conditioner. In a DC system the probes are also impressed with an AC sinewave (some use a triangular wave); however, the return is rectified by diodes on the probes. The negative half wave is returned to the indicator or signal conditioner for processing. DC systems don't require expensive coaxial cabling and special electrical connectors to maintain the guarding shield. The DC types of systems have gained wide acceptance because of this.

System calibration

The most accurate calibration is achieved by removing all fuel from the tanks and leveling the aircraft. The EMPTY adjustment is made at this point. Then the tanks are filled and the FULL adjustment made.

Because it's necessary to empty and fill the tanks and level the aircraft, and because it may be necessary to repeat the operation one or more times to ensure proper calibration, this method ties up the aircraft and manpower for a considerable length of time.

A more convenient method, and nearly as accurate, utilizes precalibrated capacitances to simulate the effect of fuel on the probes and compensator. The preferred method is to empty the tanks (leveling of aircraft is normally unnecessary) and make the EMPTY adjustment. Then, capacitance equivalent to the difference between the EMPTY and FULL values of the aircraft is added in parallel with the tank system and the FULL adjustments made. This equivalent capacitance is the ADDED value referred to in maintenance manuals.

The EMPTY calibration may be rechecked by simply disconnecting or switching out the ADDED capacitance. The operation may be repeated as many times as required to ensure an accurate calibration. Since all of the system's probes and compensator are involved in the calibration, it results in the greatest accuracy.

An alternate method is accomplished with fuel in the tanks. This procedure substitutes equivalent capacitances for both EMPTY and FULL calibration.

First, the capacitance equivalent to an empty tank is inserted into the system with the actual aircraft probes disconnected. If the system is equipped with a compensator, the maintenance manual may specify that the actual compensator itself be used rather than an equivalent capacitance. After making the EMPTY adjustment, the capacitance equivalent to a full tank is inserted and the FULL calibration is made. EMPTY and FULL may be repeated as often as necessary by simply changing the simulated capacitance. The tanks don't have to be drained. If all tank units are in good condition, this method has an accuracy which rivals the preferred method.

Troubleshooting

Some fuel quantity indicating problems can render the system totally inoperative while others cause errors of a lesser degree and may not be readily apparent.

Typical sources of problems may include: poor shielding of the HI-Z (high impedance, shielded coax) lead, insulation breakdown, corrosion in the connections, moisture condensation in the tanks, defective components and open or shorted conditions in the system wiring.

With the proper equipment, testing, troubleshooting and calibration of the system usually become a straightforward, no-guessing-games-required procedure. The idea being to divide and conquer. That is to isolate and test the system elements independently, to determine the source of the problem, and then to repair or replace as necessary.

There are typically three tests performed on the system prior to attempting a calibration: capacitance measurement, insulation measurement and indicator and/or signal conditioner testing. If the probe capacitance and insulation tests prove to be within the allowable tolerances, then the problem is with the indicator or signal conditioner.

Capacitance measurement normally requires that the tanks be drained of all fuel, and the system is disconnected at the indicator interfacing of the test equipment. It's important to note that the indicator must be isolated either through switching arrangements or by leaving it disconnected, and aircraft power (unless specifically advised otherwise) is always removed. Additionally, if the system has a compensator, it (the COMP lead) must be grounded when measuring the tank probes. Conversely, the tank probes (the LO-Z lead) must be grounded to measure the compensator.

Capacitance values vary considerably from aircraft to aircraft. Some systems measure less than 20 pf while others may be greater than 1,000 pf. If the measured value falls within the allowable tolerance, the technician may proceed on to the insulation tests. If not, it'll be necessary to perform additional testing to determine why. Many systems have a bulkhead connector at which the probes may be measured individually. This helps to isolate which probe(s) or section(s) of wiring are faulty.

For those systems not so equipped, it may be that by noting the difference between the measured and prescribed values that the source can be determined. Recalling our earlier example of a tank having four probes with values of 40, 30, 20 and 10 pf, if the tank measured 90 pf instead of the required 100 pf, then the fourth probe becomes suspect and should be checked.

Fuel Indicating System Troubleshooting

SYMPTOM: INDICATOR READS LOWER THAN KNOWN QUANTITY

PROBABLE CAUSE	TEST	REMEDY
Compensator immersed in water	Measure compensator capacitance	Drain and or purge tanks
One or more probes are out of circuit	Measure capacitance Check wiring continuity	Repair/replace wiring or connector
Defective or out of cal indicator	Test/cal indicator	Calibrate or replace
Low insulation between HI-Z and GROUND	Measure insulation	Isolate and repair or replace wiring, connector or probe
Low insulation between LO-Z and GROUND	Measure insulation	Isolate and repair or replace wiring, connector or probe

SYMPTOM: INDICATOR READS HIGHER THAN KNOWN QUANTITY

PROBABLE CAUSE	TEST	REMEDY
Open compensator lead	Measure compensator capacitance	Isolate and repair
Open HI-Z shield	Measure capacitance Check continuity of wiring	Isolate and repair
Defective probe	Measure capacitance	Replace probe
Some water in fuel tanks	Measure capacitance	Drain and/or purge tanks
Defective or out of cal indicator	Test/cal indicator	Calibrate or replace
Low insulation between LO-Z and HI-Z	Measure insulation	Isolate and repair or replace wiring, connector or probe

SYMPTOM: INDICATOR REMAINS AT EMPTY (Regardless of tank level)

PROBABLE CAUSE	TEST	REMEDY
No power to signal conditioner	Measure power	Restore power
Defective indicator	Test indicator	Replace
Open LO-Z	Measure capacitance Check continuity of wiring	Isolate and repair or replace wiring, connector or probe
Open HI-Z	Measure capacitance Check continuity of wiring	Isolate and repair or replace wiring, connector or probe
LO-Z shorted to ground	Measure insulation	Isolate and repair or replace wiring, connector or probe
Various probes are out of circuit	Measure capacitance Check wiring continuity	Repair/replace wiring or connector

SYMPTOM: INDICATOR REMAINS AT FULL SCALE (Regardless of tank level)

PROBABLE CAUSE	TEST	REMEDY
Open HI-Z shield	Measure capacitance Check continuity of wiring	Isolate and repair
Defective indicator	Test indicator	Replace
Large amounts of water in tanks	Measure capacitance	Drain and/or purge tanks
Low insulation between LO-Z and HI-Z	Measure insulation	Isolate and repair or replace wiring, connector or probe

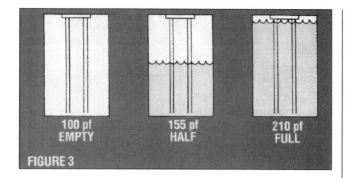

FIGURE 3

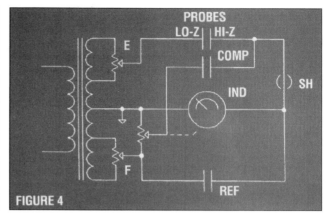

FIGURE 4

The *insulation measurement* also requires fuel drainage of the tanks, disconnection of the indicator and interfacing of the test equipment before the actual test. Again, the indicator must be isolated and aircraft power (unless advised otherwise) removed.

Insulation values will also vary considerably from aircraft to aircraft. Some systems require maximum values of no more than perhaps 1 megohm while other aircraft may have requirements in excess of 1,000 megohms. It's worth mentioning that insulation values always reflect the minimum acceptable reading. Any indication above that is for the better. The procedure usually requires making several actual insulation measurements. This typically involves sequencing of a rotary selector switch. The reason for the multiple tests is that an insulation breakdown could exist between any two or more of the normal four points of the system, that is LO-Z (low impedance tank lead), HI-Z (high impedance coax lead), shield and ground. Systems with a compensator will have additional tests.

If the measured values equal or exceed the listed minimums, the technician may proceed to the next test. If not, and the system is equipped with a connector break, the probes may be measured individually.

For those systems not equipped in this fashion it will be necessary to begin isolating sections of the system by disconnecting at the probes and observing when the insulation values increase. Recalling that, the probes are connected in parallel, LO-Z to LO-Z, and HI-Z to HI-Z from the inner to the outermost probe. All will be affected if there's an insulation breakdown in one probe or section of the wiring.

Again, using our earlier example of a four-probe system, if we were to break the connections between the second and third probe and remeasure with good results, then the problem is in the outer two probes and or associated cabling. By repeating this process of subdividing, we will eventually isolate the cause.

The *indicator test* does not require the aircraft to be drained of fuel. The indicator is disconnected from the indicating system and the test equipment is connected. The test equipment will be preconfigured to simulate the empty capacitance for the system. Power is returned to the system, and the indicator is adjusted if necessary to give an exact EMPTY reading. The capacitance simulators of the test equipment will then be set to simulate the full capacitance requirement, and the indicator is adjusted as necessary to give an exact FULL reading. If either EMPTY or FULL cannot be set, then the indicator must be replaced.

Many indicators also require that a linearity test be performed to ensure proper tracking of the unit and to check for sticking in the movement. This is accomplished by simulating incremental values of capacitance and checking the indicator reading for accuracy. Again, a failure at any point during the testing is grounds for rejection.

It's particularly important to point out that the indicator test doesn't usually constitute an indicator calibration. After the integrity of all points of the system has been ensured through testing, the system should be calibrated to actual aircraft empty conditions.

November/December 1991

Richard L. Floyd, who's been in aviation for 12 years, is a field engineer for Barfield Instrument Corp., headquarters in Miami, FL. He is based at the company's Atlanta, GA, branch.

Airframe Technology

Non-destructive testing
Penetrant processing: Dispelling the myths

By Joseph Hahn

An inspector looks at a turbine blade that has been tested for defects using the fluorescent penetrant method of non-destructive testing. A bright green line jumps out, so the inspector rejects the part. Obviously, it's cracked.

Later, he puts a similar turbine blade back into service because no green indication appears during processing. Obviously, it's not cracked.

The inspector uses penetrant processing to test for flaws in aircraft parts everyday, trusting the test results and acting accordingly. But he wonders from time to time: *Does penetrant processing really work... are the test results really accurate?*

The answer to those questions is yes, but only if the test is performed properly using some simple, fail-safe guidelines.

Penetrant process compared to other NDT methods

Penetrant inspection (PI) is a non-destructive testing (NDT) technique for finding surface cracks and discontinuities on solid, non-porous parts.

The process is very simple. First, a liquid that contains either green fluorescent or red visible dye and possesses penetrating properties, much like familiar household products such as WD-40 and Liquid Wrench, is spread over the surface of a part. Next, excess liquid is removed from the surface, leaving only the liquid that has penetrated into any existing cracks. Finally, by careful observation of these penetrant "indications," you can locate cracks and determine the size and approximate depth of each one.

In general, penetrant processing offers some very important benefits: Penetrant inspection is arguably the best method for finding surface cracks, especially if they are fine, short or shallow. It is also the best method for early detection of fatigue cracks, which usually start at the surface. In many cases, early detection provides a chance to save the part and extend its useful life. However, penetrant processing cannot locate subsurface discontinuities at all. Here are some other important performance comparisons to consider when selecting the right NDT method for your application:

- **Compared to the X-ray method:** PI is faster, more reliable, less expensive and unlimited by the shape or size of the part.

- **Compared to the ultrasonic method:** PI is faster and more effective for overall scanning of parts, better in finding shallow or short cracks, unlimited by the shape or size of the part, and less expensive for volume testing.

- **Compared to the eddy current method:** PI can locate a greater variety of cracks that differ in width, depth and direction, and its use is not limited by the shape or size of the part. However, the eddy current method may be faster for inspecting bars and tubing.

- **Compared to the magnetic particle method:** PI is equally effective in finding open cracks, but not as effective when the cracks are filled with contaminants.

It's a myth that good penetrant inspection results can be achieved with nothing more than a bucket, rag, garden hose and black light.

Important process considerations

It can't be emphasized enough that using the process properly is the key to success. However, there are a number of process variables to consider, each one of which requires a decision that's appropriate for the application:

Selecting the right sensitivity level

Determining the right sensitivity level is a fundamental consideration that depends on the size of the defects that must be located. The smaller the defect, the higher the sensitivity level requirement. On aircraft parts, where very tight defects are critical, a penetrant with level 3 or level 4 sensitivity should be used. (Level 4 is the most sensitive penetrant available.) If there is any doubt, select the higher sensitivity level.

Using a developer

The use of a developer dramatically magnifies the visibility of indications by drawing the penetrant out of cracks; visually, this phenomenon "expands" the apparent size of the cracks. Non-aqueous and wet suspendible developers are made for use with visible penetrants. With fluorescent penetrants, you can also use dry powder and wet soluble developers, in addition to the other types.

Choosing the ideal penetrant method

You can apply penetrant by spraying, brushing or dipping whichever you prefer. Just be sure the entire inspection area gets coated and that the surface is clean and dry before the penetrant is applied. Then, determine which penetrant method best suits your application— water wash, lipophilic (PE), hydrophilic (PR) or solvent-removable. A detailed description of each method follows:

The water wash method involves simple equipment and procedures, and it's ideal for large parts and high production rates. However, it is less controllable than other methods and creates a relatively large volume of emulsified waste water. The proper procedure is:

- Apply penetrant coating
- Allow dwell time (at least 10 minutes)
- Wash penetrant off with 65°F to 75°F water (30 to 120 seconds according to manufacturer's specs)
- Dry the part
- Apply developer
- Inspect for indications (with fluorescent materials, use a darkened room)

The lipophilic method offers greater control than any other penetrant method, but it is more difficult to handle large parts effectively, it creates more waste water than any other method, and it is not as appropriate for high production rates. The proper procedure is:

- Apply penetrant coating
- Allow dwell time (at least 10 minutes)
- Remove excess penetrant by dipping in emulsifier (time duration to comply with manufacturer's specs)
- Wash with 65°F to 75°F water (30 to 120 seconds, according to manufacturer's specs)
- Dry the part
- Apply developer
- Inspect for indications (with fluorescent materials, use a darkened room)

The hydrophilic method is the most common method for aircraft applications because of the control and high production rate it allows. Additionally, it's good for large parts, and waste water is easily separated from prerinse. The proper procedure is:

- Apply penetrant coating
- Allow dwell time (at least 10 minutes)
- Prerinse parts (mechanical cleansing with non-emulsified water)
- Apply penetrant remover via spray or dip (use concentration according to manufacturer's specs)
- Wash penetrant off with 65°F to 75°F water (30 to 120 seconds, according to manufacturer's specs)
- Dry the part
- Apply the developer
- Inspect for indications (with fluorescent materials, use a darkened room)

The solvent-removable method is ideal for spot inspections of in-service equipment because it's portable and creates very little waste. However, it requires the use of solvents that are difficult to apply on large parts. The proper procedure is:

- Apply penetrant coating
- Allow dwell time (at least 10 minutes)
- Remove penetrant with solvent (use a rag with solvent sprayed on)
- Apply developer
- Inspect for indications (with fluorescent materials, use a darkened room)

Airframe Technology

Penetrant method reliability

Penetrant processing is a reliable NDT method as long as you faithfully follow the procedures outlined previously, without skipping steps or taking shortcuts.

To be absolutely certain that your test results are as accurate as possible, here are some things to remember that will eliminate common errors from each processing step.

1. **Thorough precleaning and drying.** If the part is not totally dry, including in the cracks, dirt or water in the cracks can prohibit the entry of penetrant.
2. **Penetrant working condition.** Old penetrant that has been sitting in a tank for a year may not be as sensitive as new penetrant. So, check it regularly and change it when needed. Penetrants can be effective for up to five years, but contaminants often shorten useful life to less than one year.
3. **Dwell time adjustments.** Cold temperatures (less than 40°F) or cold parts will thicken penetrant, so you should allow more dwell time for the penetrant to work its way into cracks.
4. **Emulsifier effectiveness.** Make sure the concentration meets specifications by checking it yourself every shift. But have it professionally evaluated by your supplier once a month.
5. **Wash water temperature.** Water that's too hot (more than 90 F) thins the penetrant and causes over washing. Water that's too cold (less than 60°F) thickens the penetrant and causes too much background. You can use a black light to check for proper wasting.
6. **Air drying blow-off.** If any pools of penetrant remain on a part after washing, air blowing can create a lot of background problems, so be sure to drain or wipe off all puddles before using air blowing.
7. **Drying temperature.** If the temperature is too hot (more than 160°F), it can thin the penetrant and cause it to exit from the crack... thus providing no indication.
8. **Developer condition.** Contaminated developer can cause failure (false indications). And, dry developer that has become wet will probably not provide good coverage.
9. **Test standards.** There is no better way to ensure penetrant processing reliability or to verify accuracy than testing an actual part with known defects. If this is impractical, the next best alternative is using artificial test standards such as cracked aluminum blocks or TAM panels (Tool Aerospace Manufacturer). One side of a TAM panel has five star cracks to test for various sensitivities; another side is sandblasted for removability.
10. **Operator training.** The only way to guarantee reliable interpretation of indications is solid operator training. Almost every part that's processed will have some indications, but they shouldn't all be classified as cracks. To determine whether an "indication" is real, train inspectors to use the bleed-back technique. First, wipe away the "indication" with a cotton swab dabbed in solvent, then reapply some developer. If the "indication" does not return, it was probably just a scratch or a false indication (like a hair). Finally, the best way to make sure that operators don't miss any indications is to avoid operator fatigue; make sure inspectors get frequent, short breaks.

Joseph Hahn is senior chemical engineer and marketing manager for Magnaflux.

When inspecting aircraft turbine blades, the best way to be sure your penetrant processing results are accurate is to test the process against a known standard—preferably an actual turbine blade that has known flaws of the type for which you are inspecting.

Penetrant waste water disposal

Environmental consciousness is rising, so companies using penetrant processing must be more aware of penetrant waste water disposal issues.

It's a big financial issue because disposal by environmentally safe hauling can cost up to $5 per gallon. It's a complicated regulatory issue because local sanitation rules vary widely. And it's a confusing commercial issue because of the advent in recent years of so-called "biodegradable" penetrants.

What does biodegradable mean?

Simply put, a biodegradable material is one that will break down organically, or rot.

Technically, all NDT penetrants currently on the market are biodegradable; they will all decompose in time. Some of them break down faster than others. Some of them can be drain disposed in some parts of the country, but many cannot be poured into the sewer.

Is biodegradable different than drain disposable?

Although all penetrants are biodegradable, they can't always be drain disposed. The ability to drain dispose penetrant rinsings hinges on the rate of biodegradability and on local sanitation authority requirements which vary substantially from one place to another. The safest course of action is to adopt disposal practices that satisfy the most stringent requirements.

What's the best way to dispose of penetrant rinsings?

Pretreatment of penetrant rinsings, prior to disposal, is the best overall solution. Effective pretreatment satisfies the toughest regulations, reduces contract waste hauling and allows for water recycling. It's a win-win-win solution.

The best pretreatment method is ultrafiltration. This type of system removes clear water—which accounts for 90 percent of rinse water volume—from waste water. The equipment is relatively inexpensive, with units available to handle 50 to 4,000 gallons per day. An automatic ultrafilter system can process rinse water for about $0.03/gallon, so it's much more cost efficient than hauling all the waste water away.

Best of all, the water that's removed by ultrafiltration is pure enough to be recycled or drain disposed. And, unlike conventional filtration, or even carbon filtration, an ultrafilter element normally lasts six to 24 months with periodic cleaning. *September/October 1991*

Airframe Technology

The Art of Welding

By Greg Napert

For the average technician, knowing how to weld is becoming less important as time goes on. The move to specialize, combined with a more liability conscious public is resulting in specialty shops throughout the industry that are concentrating on specific repairs. These shops make the repair that they specialize in, tag it serviceable and return the component to the technician for reinstallation.

For those who choose to weld, the components being welded are somewhat limited. Among them, engine mounts, exhaust systems, brackets and some non-structural components. There are, though, some exceptions to this rule. Some of the larger repair stations do keep one or more of their technicians up to speed and certified to do some of the more difficult jobs. In fact, some items require (per maintenance manual instructions) that weld repairs be made by a certified welder.

If you do choose to make occasional weld repairs, it's important to keep skills honed by practicing basic welding techniques.

One repair station that arose as the need for expert welding grew is Kosola and Associates Inc. of Albany, GA. Bob Williams, QC manager for Kosola, says that there are many considerations that have to be taken into account (when welding) that the occasional welder doesn't have the expertise or experience to understand. Kosola specializes in welding engine mounts.

Williams says that among the most persistent problems that he sees regarding the welding of engine mounts include the degradation of the strength of the material, destruction of the metal's corrosion resistance and overstressing of the mounts as the repairs are made.

Keeping the amount of heat used under control and localized, he explains, is very important in assuring a proper weld. In some instances, he says, that can mean just using a wet rag to assure that the area being heated is contained. With excessive heat, he says, the corrosion protection is baked off and corrosion quickly sets up. "Corrosion inside the tubes surrounding the area of a field repair is the biggest problem that we see." That's because this area can't be inspected after the weld repair and cannot be properly protected either. Kosola also minimizes the amount of heat used for the repair by using an inert gas welding process. This process, says Williams, reduces distortion and reduces the amount of inclusions—making for a much better weld.

Williams says that if field repairs are attempted, a coating of some sort should be applied to inhibit corrosion. In many cases that isn't possible. Kosola doesn't even mess with trying to make a patch repair on tubular engine mounts. In all cases, he says, they replace the entire tube if there are any signs of damage.

Another problem with making a field repair, says Williams, especially while the mount is still in the aircraft and engine in place, is the stress that develops in other areas. When technicians replace a piece of tubing on the engine mount, the tube is oftentimes removed and replaced without supporting the rest of the structure. The result is that the engine mount moves slightly and places all of the other weld joints and supports under a highly stressed condition. With the rest of the mount under stress it fails in a relatively short period of time.

Williams says that the only way to correctly make repairs on an engine mount is to entirely remove the mount from the aircraft. It then must be placed in a fixture or jig that will hold it in place with respect to the engine and the firewall.

The only problem with doing this is that a jig must be made for each model and type of aircraft. Something that most shops don't have.

Another item commonly welded and seemingly simple is exhaust system components. Exhaust systems are usually made from stainless material and because of that are not easily repaired.

Inert gas welding equipment is preferable when welding stainless in order to get a weld that's free from weld inclusions. There are many other considerations to take into account when welding exhaust systems, most of which discourage the occasional welder from attempting to repair them.

Determining whether or not to weld

The most important factor in determining whether or not welding is applicable is what type of material is going to be welded. That's why material identification is absolutely critical prior to attempting a weld. In some cases, it may be necessary to call the manufacturer to determine such things as chemical composition, heat treatment and exactly what alloy the material is.

According to Welding Guidelines, an IAP Inc. training manual, items that should not be welded, consist of parts whose proper function depends on strength properties developed by cold working, such as streamlined wires and cables. Also included in the "do not weld" category are aircraft parts such as turnbuckle ends and aircraft bolts which have been heat-treated to improve their mechanical properties. Additionally, brazed or soldered parts should not be welded since the brazing mixture or solder can penetrate the hot steel and weaken it. Most non-heat-treated mild steel (low carbon) parts are weldable.

The weldable aluminum alloys used in aircraft construction include 1100, 3003, 4043 and 5052. Alloys 6053, 6061 and 6151 are also used on aircraft but are heat-treated alloys. If welding is done on these types of metal, they must be heat-treated again.

The welding process

The three most important factors in making good welds, regardless of what type of welding is used, are practice, practice and practice. Enough can't be said of practice because there are so many variations in equipment, setups, types of metal, etc.

Regardless of which type of welding procedure you choose to use, proper preparation of the material is imperative prior to welding. Make sure that bevel angles and the proper fit of the surfaces to be welded are accurate enough to result in a clean weld. Also, arrange fixtures and clamps whenever possible to hold parts in alignment while you're welding. It beats having to start over again after the item you're welding moves out of position.

Cleanliness of the weld joint and surrounding area is also important. Taking the time to remove rust, scale, paint, grease and other foreign substances will result in a stronger weld joint that's less likely to fail.

Generally, it's important to try to minimize the distortion to the item being welded. This can be done in a number of ways. One of them is technique. Distortion can be minimized by making small welding passes with little or no weaving. Whether this technique is appropriate, however, is determined by the type of equipment and type of metal. Other methods of reducing distortion include using fixtures and keeping the area surrounding the weld cool.

After the part is welded, a thorough inspection of the weld and weld area should be conducted. If possible, dye-penetrant the weld and surrounding area or use some other method of non-destructive testing. If the weld contains voids, it's advisable to remove the weld and reweld the affected area.

Protection of the weld area after welding is also critical. The method of protection will depend on the type of material you're dealing with. For instance, if a flux is used on aluminum, the flux should be removed immediately. If it's left on the weld, it will lead to corrosion of the aluminum and premature failure of the joint. Ferrous materials should also be cleaned of the welding byproducts and protected with a primer or paint.

The art of welding requires a considerable investment in equipment, practice, education and experience. But if you decide it's worth it, the investment can save you considerable time and expense. *July/August 1991*

Airframe Technology

Autopilots
How do they know?

By Jim Sparks

Observing today's business aircraft smoothly negotiating an approach and landing can be an incredible sight. What's even more incredible is that there's a good chance that the aircraft is accomplishing the landing itself. Auto flight has come a long way.

Autopilot classification

Classification of auto flight systems is determined by the axis of flight in which they can operate.

For example, some basic autopilots are nothing more than wing leveling devices. More complex systems bring in the pitch and turn coordination. These systems are called three-axis autopilots. Some two-axis models are available that have both pitch and roll circuits and can also interact with a separate yaw damping system.

Most autopilot systems do in fact have some common points. For example, most are not inherently intelligent. This means they depend on outside sources for information. Even when supplied with a wide variety of navigational information they aren't smart enough to sort through and prioritize this available data.

Troubleshooting can be made easier by separating the auto flight systems into three elements: Sensors which provide all the necessary raw data pertaining to flight. A computer, which processes all this data and with the help of a control head or mode selector, can assign priorities. And systems or loads which receive this processed information.

> Sensors ➡ Computers ➡ Load

When troubleshooting, it's helpful to recognize that the sensors supplying the auto flight system also supply other systems on the aircraft. By observing indications elsewhere in the flight deck, a reasonably thorough diagnosis can be made about the validity of raw data.

When attempting to solve autopilot discrepancy, first take into account the big picture and not just basic autopilot components. The big picture includes the flight guidance or flight director system. (Some call it the screening system.) It is, in fact, this system that possesses the knowledge to command the autopilot in its tasks. The flight guidance system observes the same flight data available to the flight crew, then sets priorities on how information will be used depending on flight crew inputs.

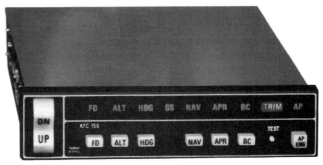

Typical autopilot control panel.

Many of today's aircraft have more than one flight director. This capability can be of great use when troubleshooting as the second system usually has a completely separate set of inputs, and a problem can be corrected in-flight by changing flight guidance inputs to the autopilot. This can often save the technician the time and trouble of analyzing the basic autopilot system.

Most flight director systems also have the capability to supply a pictorial display of summarized flight data. This display is commonly found on the attitude directional indicator (ADI).

Questioning the flight crew about the validity of flight director indication can lead to information regarding the integrity of the flight director sensors and computations.

How the system works

The flight director's primary responsibility is to correct errors. When the flight crew changes an input to the flight director, usually an error is sensed between selected input and actual aircraft position!

Flight guidance can work in two realms. Vertical and lateral.

The "vertical" situation will deal with pitch and speed while the "lateral" mode includes navigation and pertains to aircraft direction such as the roll and yaw axis.

These sensors reflect the SHAPE of the aircraft. Not necessarily geometric shape but, Speed, Heading, Attitude, Position and Elevation.

The *speed* function is considered a vertical mode and depends primarily on the air data system for its reference. Speed can be activated several ways. On many flight guidance selector panels there are function

Airframe Technology

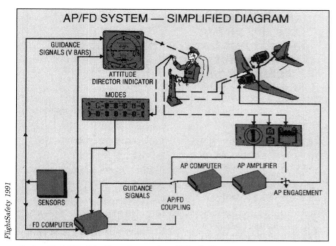

Autopilot/flight director — simplified diagram.

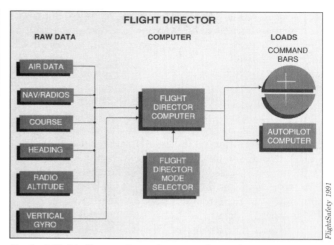

Typical flight director input/output schematic.

switches labeled "Airspeed, Vertical Speed and Mach." In many auto flight systems only one vertical mode can function at any given time. In other words, the system could not hold speed and altitude unless a system of auto throttles was installed to adjust thrust.

Heading data is supplied from a directional sensor usually a directional gyro (DG). This information is then displayed on a horizontal situation indicator (HSI). On most HSIs there is usually an adjustable heading reference indicator. The flight director will compare actual aircraft heading to selected heading. Anytime there is a disagreement, the system will give a command to change aircraft situation. This lateral mode also has preset limits such as max turn angles 3 to 4 degrees again sensed by the DG and roll limits controlled by the vertical gyro (VG). This system will operate when the heading (HDG) selector is engaged.

Attitude information can be used in both vertical and lateral modes and normally requires no additional inputs on the control panel. The normal source for attitude information is the vertical gyro (VG).

Position generally means radio navigation. Most position sensors supply the lateral modes of flight guidance. In addition to receiving a ground-based radio station such as a VOR, the pilot has to assign course information. That is, set the angle of approach to the station. When selecting the NAV function, the system will also observe heading information and, in most cases, will not initiate a turn in reference to a station until it has captured the preset course. Should the aircraft get out of NAV radio range, the flight guidance system will usually maintain present headings. The one vertical mode of navigation is glideslope and is more often associated with approach functions.

Elevation is another vertical mode that is usually supplied by the air data system. When selecting altitude (ALT) on a flight director mode panel, the system should hold present aircraft altitude and display varying pitch commands to return the aircraft to this original position. Altitude Select (ALT SEL) will work with the air data system and some type of altitude selector. In most cases after the flight crew enters a selected altitude, it also needs to tell the aircraft how to get there. That means engaging another vertical mode meaning IAS, VS or MACH; and, unless auto throttles are installed, the pilot will have to adjust engine power. After the system captures selected altitude, the other vertical modes will usually disconnect and the aircraft will continue to hold altitude.

Setting system sensitivity

Automatic flight control system manufacturers also have different ways to regulate system sensitivity. This is accomplished by rate sensors (gyros or accelerometers) aneroid capsules sensing altitude and airspeed information. Radio navigation information can also be used to regulate gains. On many aircraft, when flaps are extended, this supplies an input to the autopilot saying more flight control deflection may be required to accomplish maneuvers. Even during approach to landing, radio altitude can be used as a sensitivity control.

When troubleshooting auto flight systems, the best starting point is a very thorough flight crew debrief. Problems such as autopilot porpoising (oscillation about the pitch axis) can be the result of worn flight control rods, improperly tensioned cables, air data problems, flight director problems, vertical gyro problems or pitch trim problems. A pitch trim problem can usually be recognized by a thorough test of the system including a timed travel check. The vertical gyro can generally be diagnosed by observing the operation of the altitude indicator which can be driven by the same gyro.

Airframe problems can be observed by the flight crew by disconnecting the autopilot and observing the trimmed aircraft response. Also simply turning on the autopilot without any flight guidance input can generally

Airframe Technology

be a good starting point. It is important to find out the aircraft configuration and the result of changing flight director modes or possibly engaging the secondary flight director (if installed).

Most manufacturers have developed operating manuals illustrating the capabilities of the specific system.

Avionic maintenance involves three levels: Level I is removing and replacing components, Level II involves the replacement of printed circuits, and Level III requires extensive electronics knowledge as it deals in the repair of components.

Flightline maintenance of autopilot systems can get down to Level I and in some cases Level II. Most of the time removal of the correct black box will be the extent of the infield repair.

Swapping components can be considered dangerous as a system with a short circuit may cause a failure in one box and after installing a substitute box a similar failure could occur.

Many autopilot manufacturers publish well-detailed manuals on their products. If time permits very thorough checks can be made including voltage tests.

If time is an important factor or if the technician does not have a good operational knowledge of the system, it is usually beneficial to contact a qualified avionic technician to assist in the troubleshooting process.

A good way to become familiar with system operation is to read the operating manual and perhaps spend some time in the flight deck with a knowledgeable pilot and have him describe his normal use of the system.

The operation of auto flight is all based on communications. That is, sensors communicating with computers, then computers performing tasks and communicating with their indicators and follow-up devices.
May/June 1991

Jim Sparks is the director of maintenance training for FlightSafety International and is based out of Houston, TX.

Airframe Technology

Deicer systems
Neglect can lead to early failure

By Greg Napert

Regular inspection and cleaning of deicing systems play a critical role in assuring proper operation. Additionally, early detection of damage or cracks to deicer boots or leaks in the pneumatic lines can make for easy repairs. It's a good idea, therefore, to get the flight crew involved in regular inspections, because letting damage go untended can result in having to replace costly components.

Pneumatic deicers

In order to perform a thorough inspection of pneumatic deicers, shop air must be applied so the system can be cycled. According to BFGoodrich, a pressure regulator that can be regulated to 18 psi +/-2 for high-pressure systems and 10 psi +/-2 for low-pressure systems must be used.

Bart Briggs, shop foreman for Waukesha Flying Services Inc. in Waukesha, WI, who regularly performs inspections and preventive maintenance on deicing systems, says that it's important to eliminate moisture from entering the system. Make sure shop air is dry and that the filter is clean before attaching pressurized air to the deice system.

When pressurizing the system, cycle the boots and observe them for proper operation. Verify that they inflate and deflate according to the cycle specified in the maintenance manual. If the cycle doesn't follow the correct sequence, check the pneumatic tubing for correct routing or the timer for proper wiring. Keep in mind that it's not uncommon for a cycling problem to be caused by both.

All inspections, says Briggs, should include a check for leakage. Use mild soap and water solution to check all hoses and hose connections. Also, apply a mild soap and water solution to the entire surface area of the deicer boots and cycle the boots. As the system cycles, watch for soap bubbles, which indicate a leak. It's important to rinse any soap from the boots and lines with water to prevent it from reacting with the rubber.

To repair any damage, cold patch repair kits are available for the repair of scuffs, cuts, tears and any other damages that result in air leaks. Briggs says that tire-tube patches should not be substituted for original deice boot patches.

BFGoodrich 1991

"Tire-tube patches don't stretch," he says, "because they contain a layer of fabric. And when the boots are inflated, these patches have a tendency to rip off." Patches manufactured specifically for deicer boots, he explains, are made of a material that will stretch as the boot expands. "I've seen guys install a number of tire-tube patches as a temporary repair," says Briggs, "and they only lasted one or two inflation cycles."

When installing patches, be sure to remove the fabric backing material. "I often find patches peeling from the boot because the protective backing wasn't removed when the patch was installed," says Briggs. The backing can be difficult to remove at times and it is often mistaken for an integral part of the patch.

BFGoodrich says that patch kits can generally be considered permanent for the life of the deicer. It's important, however, to make sure that the correct kit is used. For example, estane repair kits cannot be used on neoprene boots and neoprene repair kits cannot be used on estane boots. Estane deicers are rarer than neoprene and can be identified by the brand located on the boot.

A new method of repairing damaged boots, called Bootsaver™ was released in December 1990 by Rapco Inc. of Hartland, WI. Instead of using a patch, the repair is done by working a rubber compound into the cracked or damaged area. The compound, says Rapco, "chemically crosslinks with the base material during applica-

Airframe Technology

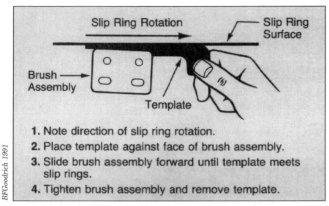

Tool for setting slip ring holder.

tion." The material can be worked to the exact contour of the boot and the surface blended to look like the rest of the boot. This method of repair eliminates having to use patches and goes virtually unnoticed.

Resurfacing kits are available through BFGoodrich to repair scuff damage or severely weathered neoprene pneumatic deicers. The kits aren't intended to repair severely damaged deicers or for repairing deicers with air leaks. Therefore, the boots should always be repaired prior to resurfacing. It's generally recommended, says BFGoodrich, that each deicer not be resurfaced more than twice during its life.

Typically, when it's time to replace one boot, the rest of the boots on the aircraft are ready for replacement. According to BFGoodrich, the boots should be replaced immediately for damage that occurs in the following areas.

Inflatable tube area
- Deicers that have cuts, tears or ruptures that exceed 3/4 inch in length or that are within 1/8 inch of a stitch line (a stitch line is the sewn line that forms the edges of the inflatable tubes) or that cross a stitch line.
- Boots that have broken or cut threads. Such a condition can be detected by inflating the deicer. A broken or cut stitch will cause the deicer to balloon (inflate) between the adjacent tubes where the stitch is cut.
- Separation of the outer neoprene layer in excess of 1 square inch. For smaller areas, BFGoodrich recommends trimming away all loose surface material, then applying a cold patch kit.

Non-inflatable areas
- Cuts, tears or ruptures that exceed 3 inches in length.
- Separation of outer neoprene layer that exceeds 3 square inches.

Additionally, BFGoodrich recommends the following limitations on patches:
- Patches 1 1/4" x 2 1/2"—three per 12-inch square or;
- Patch 2 1/2" x 5"—two per 12-inch square or;
- Patch 5" x 10"—one per 12-inch square or;
- Two patches 1 1/4" x 2 1/2" plus one patch 2 1/2" x 5" per 12-inch square.

These patching limits, says BFGoodrich, are not absolute. The deicer will still function satisfactorily with a higher concentration of patches. However, the aircraft operator should cycle the deicers to assure that inflation has not been impaired.

The process of stripping the boots from the aircraft can be time-consuming. It's done by slowly peeling back the boots from the aircraft and using a solvent to loosen the adhesive. Placing the boots on the aircraft—straight—can also be a trick. Briggs explains that many operators opt to bring their aircraft to a facility that specializes in boot replacements. One such facility is the BFGoodrich Service Center at the Akron/Canton Regional Airport in Ohio.

At this operation, the customer brings the aircraft in the morning, and receives the aircraft back at the end of the day—complete with new deice boots. Although the process appears to be relatively expensive, the excessive downtime and extra man-hours needed to perform the operation at an average FBO is enough incentive to pay extra for the one-day turn-time.

Electrothermal deicers

The first step in inspecting electrothermal deicers is to ensure that all lead strap clamps, terminal clamps, tie straps and retainers are secure and properly installed and that related safety wiring is secure.

The deicing system should then be turned on and checked for electrical shorts and proper heating sequences. If possible, use an auxiliary power unit (APU) to avoid a drain on the batteries. The heating sequence can be checked by simply feeling the boots for warm areas.

"Occasionally," says Briggs, "I find the lead wires swapped and the boots operating in the wrong sequence. It's easy to swap the wires because the system still works, but doesn't sequence properly."

Briggs says that prop boots experience a relatively high failure rate. "Dirt, ice, objects hitting the prop and people moving the aircraft by grasping the boots can lead to broken heating elements. Additionally, because centrifugal forces during operation tend to force the wire against the prop spinner, it frequently cracks or shorts out," he says.

Non-operational boots can be observed on the ammeter in the cockpit, or by feeling for cool boots while they're operating.

Airframe Technology

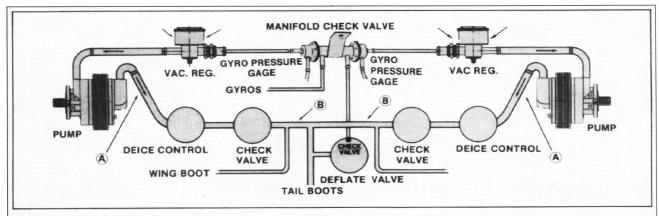

Twin Engine Aircraft — Deice System — Vacuum Gyro Instruments

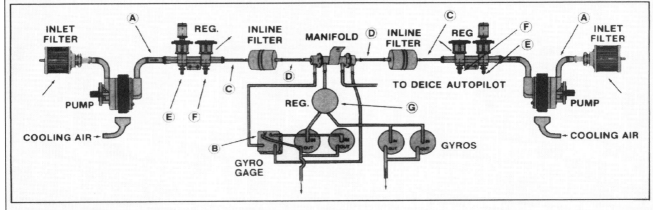

Twin Engine Aircraft — Deice System — Pressure Gyro Instruments

Slip rings should be inspected for excessive wear, roughened surface, cracks, burned or discolored areas and deposits of oil, grease and dirt. The slip rings and all associated components should be cleaned with MEK (methyl ethyl keytone) or equivalent solvent; however, don't use abrasive cleaners on the clip rings. This will result in increased brush and slip ring wear.

Uneven wear or wobble of the slip ring may be repaired if the ring is within proper limits.

Check mounting brackets, brush blocks and modules for cracks, deformation or other physical damage; inspect the brushes for excessive wear, chipping or breakage; and clean all components using MEK or equivalent solvent.

According to BFGoodrich, each brush needs to ride fully on its slip ring throughout the 360 degrees of rotation. Proper alignment between the brush block or module and the slip ring is critical. The proper positioning of the brush block is 1/16 inch from the slip ring assembly and cocked 2 degrees in the direction of the propeller rotation. Correct alignment can easily be obtained using a brush alignment template (see drawing) that is available from BFGoodrich.

Electrothermal deicer boots that contain cracked elements cannot be repaired. Indications that the deicer has an open circuit, the wrong resistance or noticeable damages, such as tears and nicks, that expose the element require immediate replacement. Defective brushes and wire harnesses must be replaced as well.

Extending life

Like any component on the aircraft, the key to keeping boots for any length of time is to keep them clean. Wash deicers regularly with a mild soap and water solution. Oils and contaminants can decompose the rubber and significantly decrease the life of the boots. It's also a good idea to keep the boots coated with some kind of preservative, such as BFGoodrich's Age-Master Number 1 (for neoprene).

Preservatives can help to preserve deicers by inhibiting ozone attack, aging and weathering. For more information, BFGoodrich has established a technical information line for deicing systems. It can be reached by dialing (800) DEICERS (800-334-2377). Customers in Alaska, Hawaii, Ohio and international locations can reach the information line by calling (216) 374-3706.

March/April 1991

Airframe Technology

Problem	Cause	Corrective Action
Boots do not inflate	Open circuit breaker	Reset circuit breaker
	Faulty deflate valve	Check solenoid and valve assembly as follows: Solenoid inoperable: 1) Check for proper voltage 2) Check for clean air vent passage in solenoid 3) Check for proper plunger operation Diaphragm not seated: 1) Check for clear .015 vent orifice located in rivet bottom at center of diaphragm 2) Clean diaphragm seal area 3) If diaphragm ruptured, replace valve
	Two faulty deice control valves	Check solenoid and valve assembly as noted above.
	Relay not functioning	Check wiring or replace relay
	Leak in system	Repair
Slow boot inflation	Lines blocked or disconnected	Check
	Low pump capacity	Replace dry air pump
	One or more deice control valves not functioning properly	Check solenoid and valve assembly as noted above
	Deflate valve not fully closed	Check solenoid and valve assembly as noted above
	Ball check in deflate valve inoperative	Clean check valve or replace deflate valve
	Leaks in system or boots	Repair as needed
System will not cycle	Pressure in system not reaching 17 psi to activate pressure switch	Check deice control valve as noted above Check deflate valve as noted above Repair all leaks in hose Tighten all hose connections
	Pressure switch on deflate valve defective	Replace switch
Slow deflation	Low vacuum	Check for proper vacuum
	Faulty deflate valve (indicated by temporary reduction in suction gage reading)	Check solenoid and valve assembly as noted above
No vacuum for boot holddown	Malfunctioning deflate valve	Check deflate valve as noted above
	Leak in system or boots	Repair as needed
Boots will not deflate during cycle	Faulty deflate valve	Check and replace valve
Boots appear to inflate on aircraft climb	Vacuum source for boot holddown inoperative	Check operation of ball check in deflate valve Check for loose or disconnected vacuum lines and repair

Parker-Hannifin Corp. Airborne Div.

Airframe Technology

Don't fuel with it
Fuel cell maintenance tips

By Greg Napert

Probably one of the least liked jobs in aviation maintenance is changing a fuel cell. Anyone who has done it knows that you can plan on at least two things: Increasing the length of your arms, and discovering a new use for aviation fuel — cologne. The work is tedious, dirty and is usually performed without the benefit of being able to see what you are doing.

The fuel cell is one piece of aviation equipment in which the old saying "If it ain't broke, don't fix it" does apply. The less the fuel cell is disrupted, the longer it is likely to last. Typically, the only time that the fuel cell should be worked on, is when it begins to leak.

Patrick Manning, general manager for Aviation Fuel Cells International in Memphis, TN, says that most damage to a fuel cell is caused by letting the fuel cell sit in the aircraft without any fuel in it. Tripp Volz, vice president of sales and marketing and licensed repairman for Floats and Fuel Cells, also in Memphis, concedes with this fact. He points to fuel cells in aircraft that operate in South America. These cells, says Volz, are probably the most short-lived cells because of the environmental conditions in that part of the world.

Running a close second, is damage that is caused from the removal and installation of the cells. Oftentimes, explains Manning, technicians become frustrated during the removal process, pull on the bladder with excessive force and damage the bladder. This sometimes results in torn fabric, nipples and/or transfer tubes. The customer then ends up with a higher repair bill or possibly, the cell is rendered unrepairable.

Sharp edges on access holes are also a cause of some of the tears, says Manning. It's a good idea to have a roll of duct tape handy to tape up any sharp edges. This will not only eliminate damage to the cell, but damage to your hands as well.

After the cell is removed, it should be tested for leaks. This is done by performing a phenolphthalein (pronounced phee-nol-thalene) test. This test is conducted by inserting ammonia gas into the cell, plugging the holes and filling it with 3/4 psi of air. A white cloth dampened with a chemical solution is placed over the cell. Any leaks appear on the white cloth as pinkish-red spots.

Although removing the cell from the aircraft is a task, installing it is usually much more difficult. It is literally the equivalent of shoving a 4-foot-wide box

Chris Sorensen 1991

through a 6-inch-wide hole. Before installing it make sure that all metal seams and edges are covered with duct tape. This will prevent the cell from ripping while installing it and prevent the cell from wearing on rivets or seams while it is installed in the wing.

One problem that many technicians encounter is that the fuel cell is just too stiff to scrunch through the hole. Manning suggests that the tank should not be unfolded or installed if the temperature of the fabric is below 65°F. If it's below that, says Manning, take the fuel cell into a room and warm it up. He suggests either placing it in a warm, humid room, such as in a room with a hot shower turned on, or leaving it sitting in the sun next to a window until it becomes soft. After it is softened, it should be rolled up immediately to install it into the wing. To assist in getting it through the access hole and into position, some technicians tie the fuel cell in the rolled-up position with rope. Once the cell is roughly in position in the wing, the rope can be removed.

Airframe Technology

This phenolphthalein test, performed at Floats and Fuel Cells, shows an indication of leakage. The bright pink mark on the cloth is a positive indication that the ammonia gas is escaping from inside of the fuel cell.

Some fuel cells are shipped with a warning not to unfold the fabric if temperatures are below 65°F.

According to Volz, the most overlooked reason for improper fit is the sequence of hookup. You should always have your interconnects, he says, loosely attached before attempting to align the clips or snaps on your fuel cell. This allows for proper lineup without distorting the position of the tank relative to the cavity it is placed in. Once clips and snaps have been attached, it is then possible to go back and torque your clamps to proper settings.

Working with the clamps and clips inside of the cell is also a joyous occasion. This is where the fit of the cell is critical. This part of the job seems like it should be easy, but it is actually the most difficult part of the job. Fastening the clamps has to be done using only your sense of feel. You have to feel for where the clamp is located, and feel to make sure that it is in position.

Some cells, because of the fact that they are older, or simply because of how they are constructed, are just going to be more difficult to install. New cells are usually much more pleasant to work with, but according to Volz, at one-quarter to one-half the cost of new, the economic benefits of a rebuilt cell outweigh the installation benefits of a new cell.

Well, you've been working on the cell for a couple of hours, have everything positioned, and guess what? it doesn't fit. There are different reasons for a fuel cell not fitting properly. It could be that the rubber is still too cold. Try placing warm, moist towels inside the cell to try to soften the rubber. You may be able to relax the cell enough to fit it in place.

The hanger clips that hold the top of the fuel cell to the top of the wing are also very difficult to install. They have to be blindly maneuvered into position and pushed into place. Technicians have been known to spend hours just installing one of these clips. Especially when you have to reach far into the tank and push with the tip of your fingers. in this situation, you don't have any leverage and can quickly become fatigued.

One technician, who wishes to remain nameless, suggests that a good way to install a clip in this situation is to use a heavy blunt object such as a bucking bar. Wrap it with tape, he says, so as not to damage the rubber, and use it to gently push the clip into position. Make sure, however, that the clip is, in fact, lined up with the hole that it is being pushed into. This method should only be used if you have located the hole, but cannot muster up enough force to push the clip in.

Repairs can be performed by the individual technician. Patch kits, sold by Goodyear, are available for around $150. However, explains Manning, it doesn't make much sense to spend all the man-hours removing and installing a fuel cell to put on a small patch and disregard the condition of the rest of the cell. He does concede however that in an emergency situation, you may not have any other choice.

Practically all types of damage to a cell are repairable, says Manning. Nipple replacements and repairs to tears and punctures are all typical repairs. "Where we run into a problem, however, is where the fabric shows signs of dry rot." Dry rot can usually be identified by cracking, peeling and a general brittle appearance to the fabric.

"A lot of whether a fuel cell is repairable or not is subjective," says Manning. Don't take a chance and throw a fuel cell away based on your observations unless you have years of experience looking at them. Repair stations are accustomed to looking at fuel cells and can easily determine what they can and, cannot repair. if you think that something is non-repairable, send it to the repair station anyway for a second opinion. They may be able to repair it. *January/February 1991*

Electrical troubleshooting basics

By Gary Eiff
Associate Professor at Southern Illinois University

Many maintenance technicians find understanding and troubleshooting electrical circuits to be somewhat elusive and confusing. Some technicians have even been heard referring to electrical circuit function as "magic" and, therefore, totally beyond their comprehension.

Electrical circuit theory is not nearly as complex as many other knowledge and skill areas already mastered by the successful technician. But first it's necessary to dispel the mystique surrounding electrical troubleshooting and to learn a few fundamental principles and procedures.

The basic circuit

In order for an electrical circuit to function it must have three things: *An applied potential (voltage), a complete electrical path and circuit resistance.*

The basics of effective electrical troubleshooting center around an understanding of these three circuit attributes and an ability to evaluate their presence or absence.

The term "applied potential" refers to the difference in electromotive force (voltage) between the source and a point in the circuit. That is, the difference in electrical pressure, and pressure is what causes electrical energy to flow and work to be done. Without it, electrical energy cannot move through the circuit and the circuit ceases to function. Sources of potential difference include the aircraft's battery, altimeter and other sources of voltage.

Basic troubleshooting

A common first step used by technicians when evaluating an electrical circuit malfunction is to determine if the circuit is "getting power." The loss of applied potential is a frequent cause of circuit failure and is easily evaluated.

The most effective tool for evaluating basic electrical circuit problems is a good quality multimeter. A good meter will measure several ranges of voltage, resistance and current and have a high input impedance. High input impedance means that when the meter leads are placed on the circuit to test voltage or other electrical attributes, the meter represents a high resistance to the circuit under test and does not cause the circuit value being tested to change.

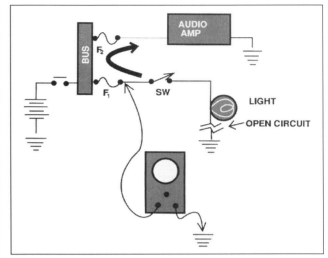

Figure 1. Example of a "sneak" circuit.

For any circuit it's important that the test be performed under normal load conditions. The circuit should be turned "on." Although primary sources may appear to be fine when checked for nominal voltage without loads imposed on them, circuits requiring current may fail to develop the expected nominal voltage once loaded.

The failure of a power source to develop nominal voltage usually occurs for one of two reasons: The power source itself has lost its capacity (the ability to produce its rated current), or there is a high-resistance connection between the point under test and the power source. For example, the screws which secure wire terminations to circuit breakers may loosen slightly causing a high resistance connection to develop between the terminal and the circuit breaker binding post. No current is flowing to the device, but the circuit appears to have the required applied voltage. Once the circuit is activated, however, resistance of the connection drops a significant portion of the power, reducing voltage delivered to the system. This is a dangerous condition since high resistance connections generate excessive amounts of heat and can result in fire.

Since many aircraft have complex electrical distribution systems, it may also be necessary to evaluate the distribution branch specific to the failed system. These branches, called buses, are normally divided into groups

Airframe Technology

of systems with similar functions such as the avionics bus system. These failures are usually easy to identify, since several systems will fail simultaneously.

A common problem which results in a loss of applied potential is the failure of the circuit protection device. Vibration in aircraft may cause fuses to fracture, resulting in a loss of power to the circuit. Although less common, circuit breaker failures can occur.

To determine if power is being delivered to the circuit, measure the voltage of the distribution source side (the aircraft's bus bar) and the circuit side of the protective device. Voltage measured should be that specified in the manual for the bus and for the circuit. Measuring the correct voltage on the bus side of the device and no voltage on the circuit side indicates the device is "open" and has failed or, in the case of a circuit breaker, is "tripped." A voltage on the circuit side of the device which is significantly lower than the bus voltage indicates an abnormally high internal resistance of the protective device, a common circuit breaker failure.

Breaks in the electrical path are perhaps the most common type of failure with electrical circuits. Wires failing, crimp connectors pulling apart, internal components of the circuit fracturing and ground wires corroding are common reasons for incomplete electrical paths.

For an electrical circuit to function, there must be a continuous path through which electrical energy can move from one pole of the applied potential... through the circuit elements... and back to the opposite pole of the applied potential. To ensure completeness of the path, it's often necessary to check the circuit for uninterrupted connections (continuity). Most aircraft technicians use either a continuity tester or the ohms function of the multimeter for this test.

There are several dangers in checking circuit continuity when using these methods. If power is not removed from the system, circuit voltage may damage the continuity tester, ohmmeter or the circuit under test. Both continuity testers and ohmmeters use internal voltage sources to measure continuity. It's possible to damage some circuits by these voltages if the circuits are sensitive or if the device polarity is not observed. Carefully evaluate the circuit before attempting a continuity test. This ensures that the circuit won't be damaged by the tester.

Continuity tests can produce confusing results. It used to be quite common to find continuity testers constructed with electromechanical buzzers. These "buzz-boxes" produce an audible signal when a complete path is detected. The buzzer which produces the signal causes the test voltage to change to a pulsating direct current. However, when checking continuity of shielded wires, the buzzer may not work even if the circuit is complete.

Additionally, make sure that an alternate path (sneak circuit) doesn't exist which would give an erroneous continuity indication. Using the example in Figure 1, it seems possible that the technician could measure the completeness of the entire circuit path by measuring the continuity from the fuse to ground. In this case, however, the technician could misdiagnose the open circuit in the system. The fact that the current used for testing continuity has an alternate path through the system fuse, across the aircraft's bus and through a low resistance path in the audio amplifier to ground, can be confusing. To eliminate "sneak circuits" and confusing results, it's often necessary to disconnect the power or other connections to the system.

There is a lesser-used method for testing continuity which can be simpler and more effective. Use the circuit's applied voltage to determine the completeness of the path. It's often safer, quicker and much more timesaving to use this method for identifying circuit discontinuity.

We can use the system's voltage to determine circuit continuity through a special application of Kirchhoff's Law. This law states that in an open circuit, the source voltage for the circuit will be apparent across the point of discontinuity. Using the same circuit as before, Figure 2 illustrates how this technique works. The technician first references a voltmeter to the aircraft's ground. Using the positive probe of the meter, check various points throughout the circuit for the applied voltage. Measurements at points A, B and C show that the source voltage is present. When a reading is taken at Point D, no voltage is indicated. This tells us that we have passed the "open" point in the circuit. The point in the circuit where the voltage disappears is the point of discontinuity.

Technicians who have a good understanding of electricity would have known immediately that a voltage reading on the ground side of the lamp (Point C) was incorrect, and that an open circuit existed in the ground leg of the circuit. In a functioning circuit, the light, being the only load in the circuit, should drop all of the applied voltage. Having a voltage at Point C indicates a complete path between that point and the source voltage. The only remaining alternative as to why the circuit is not working is that the ground or "return path" for the electrical energy in the circuit is open.

Open ground circuits are a common cause of failures in aircraft systems. Because the ground circuit is the aircraft's structure, it's not perceived as part of the circuit and is often taken for granted. Broken ground wires and "bonding straps" for radio racks, instrument panels and engines mounted on rubber "shock absorbers" are a common cause of circuit failure. Corrosion between wire terminal and structural surfaces also cause high resistance or open ground circuits.

Every functioning circuit has resistance. Lights, motors, radios and all other system elements exhibit resistance to the flow of electrical energy. Malfunctions of circuit elements will cause changes in their normal resistance values. Some malfunctions cause resistance exhibited by a circuit to increase, while others cause it to decrease. Normally, changes in circuit resistance are caused by internal malfunctions within circuit components. Occasionally, however, high resistance connections or shorts in wires and connectors may cause changes in circuit resistance.

Most technicians easily recognize the two extremes of change in circuit resistance; that is, the open circuit and the shorted circuit. However, more subtle changes in circuit resistance may go undetected. If a malfunction causes the apparent resistance of a circuit component to change from its normal value to some higher resistance value, the current in the circuit will decrease. Conversely, if the malfunction causes the component resistance to appear less than its normal value, current in the circuit will increase. Any substantial deviation from the manufacturer's specifications for the current required by a circuit component indicates that the circuit has a malfunction.

To accurately check circuit current, it's most often necessary to open the circuit and place a current meter in the path of the electrical energy. The need to disconnect circuit elements and disrupt wires and connections discourages many technicians from using current measurement as an electrical troubleshooting technique. But it can provide fault identification and isolation. Consider it when having difficulty isolating electrical troubles.

While the road to effective troubleshooting may seem arduous, the rewards are many. Starting with basic concepts and continuing to collect, develop and refine electrical troubleshooting skills, will in time allow you to identify, isolate and repair even the most complex electrical problem. *Amt* *November/December 1990*

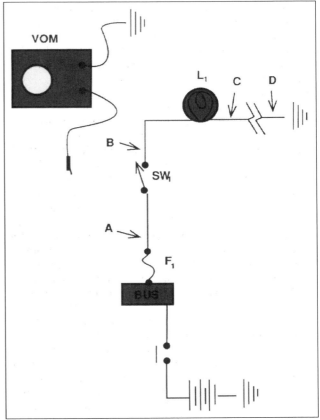

Figure 2. Measuring the source voltage at Point A means two things: There is an applied potential presented to the circuit, and continuity exists between Point A and the voltage source.

Measuring the source voltage at Point B indicates that a continuous path exists between Point B and the source. The switch is therefore working.

Measuring source voltage at Point C indicates that a complete path exists between the point and the source voltage. Voltage should not be present on the ground side of the load in the circuit. The technician knows, therefore, that an open circuit exists between Point C and ground.

Taking measurements along the ground path, the point at which the source voltage is no longer present, is the point of discontinuity. Since we do not have voltage at Point C, we have passed the "open circuit" point.

Airframe Technology

The pressurized window
More than something to look through

By Greg Napert

TORRANCE, CA—"We go through life looking through windows, not at them," comments Dave Crettol. Crettol, general manager of Aircraft Window Repairs Company, explains that there is a tendency for the technician to overlook the obvious and take windows for granted.

Inspecting pressurized aircraft windows takes a trained eye, and attention to detail is critical. "It's a matter of safety," says Crettol.

There is much more abuse than necessary when handling and caring for aircraft windows, says Crettol. "People just don't realize that it is a structural part of the aircraft and as critical to the pressure vessel as the pressure bulkhead itself. From a maintenance perspective, windows are not just there to look through."

Steve Baumann, quality control manager at Aircraft Window Repairs, explains that technicians need more information regarding the handling and care of aircraft windows. And most, says Baumann, are really not equipped with the proper information to make repairs. When dealing with pressurized windows, there is more to taking out a small nick or scratch than simply polishing away the damage. Attention must be paid to dimensional tolerances, depth of repair and distortion limits, he says.

Baumann says that approximately 15 percent of windows that are sent in for repair need to be scrapped out. "That's a shame," he says, "especially at a cost of roughly $3,000 to $4,000 and more per window. And especially since some of the most common types of damage are the most preventable."

Razor cuts

Razor cuts are often found on outside edges and radiuses of the windows where razor blades were used to cut sealants or used to cut masks that were used during painting. Baumann explains that these razor cuts have lead to failure of the window because the cuts set up stress risers where stresses concentrate. Some razor cuts are repairable, says Baumann, but it all depends on the thickness of the window and the depth of the cut. Cuts that are located in the edge of the window are more likely to lead to cracks or panel loss, he explains.

Crazing

Crazing is a chemical- or stress-related phenomenon that causes the molecular structure of the plastics to break down or self-relieve. Crazing can be caused by unusually high concentrations of stress, paint strippers, adhesives, excessive heat (such as from shorted or

Measurement of the window's thickness is critical after performing any type of repair.

improperly used heating elements), oils and even fumes from chemicals used on or around the aircraft. Telltale signs of crazing are usually a cloudy or scratched appearance that is visible but cannot usually be felt on the surface.

Any type of crazing, says Baumann, should be removed as soon as possible and should be removed completely. The nature of crazing damage is such that it will continue to progress if not removed. Chemicals actually become trapped beneath the surface layers of the window and will continue to deteriorate the window's structural integrity. If the crazing is due to stress concentrations, the reason for these concentrations should be determined and the window repaired immediately. The ability to repair crazing damage, however, depends on the depth of the crazing and the thickness of the window.

A common problem is damage from strippers that are used prior to painting. The stripper will seep into fine sealant voids around the window and begin working on the plastics. This usually can be prevented by properly masking the outside of the window assembly so that strippers cannot penetrate the voids or come into contact with the window in any way.

Cleaning agents can also be a problem, says Baumann. Always refer to the manufacturer's maintenance manual for acceptable cleaning agents. Certain non-approved cleaning agents will cause a window to craze. "Crazing is potentially more damaging to the window than razor cuts. Even though you cannot feel it, crazing that appears to be close to the surface oftentimes actually penetrates the window up to 30 thousandths deep. It is one of the primary reasons for scrapping out a window," says Baumann.

Time spent protecting and properly masking windows while working around aircraft with any chemicals or cleaning agents is time well spent.

Sometimes, however, crazing damage is not that easy to prevent. There was one occasion, says Baumann, where an aircraft was running up its engines in front of another and happened to be leaking synthetic oil. The aircraft leaking oil sprayed a fine mist of oil onto the other aircraft's windshield and caused it to craze.

Another common type of crazing is something called volcanic crazing. Volcanic crazing, according to Baumann, is identified as fine hairline-like diagonal scratches that can usually be seen in bright sunlight or under a powerful light. They are usually found on aircraft that fly at high altitudes. This type of crazing is distinguishable from scratches in that if you run your fingernail over it, you cannot feel any grooves. The crazing is actually beneath the surface layer of the plastic and continues to penetrate further with each pressurization cycle. It's theorized, says Baumann, that these crazing marks result from chemical attack that is present at high altitude. The chemicals, most heavily concentrated over the North Atlantic, are reportedly from volcanic eruptions that have taken place over the years. Volcanic crazing is not considered to be preventable; however, a thorough cleaning after each flight will help slow the process.

Removal and installation

Damage from removing and installing windows is another common problem. Damage caused by improper fit, prying and scraping can easily lead to window failure.

Baumann points to one instance where a technician filed a small area off of the edge of a window to make it fit. This particular window had a metal inner layer around the edge for mounting purposes that was originally corrosion protected. The modified edge, with no corrosion protection, began to corrode. And without the support of the metal edge, the outer ply of the window began to crack.

Baumann also points to numerous windows that were perfectly repairable except for cracks that resulted from improper removal. He suggests that you do not use a screwdriver or similar device to pry the window from the aircraft. If there is sealant holding the windshield in place, he says, use a plastic scraper to remove as much of the sealant as possible. Baumann suggests that if the window will still not come loose, use a light gauge safety wire or fishing line to cut the sealant from around the window. With one person inside and the other outside the aircraft, feed the wire through the sealant. With a technician on each end of the wire or line, move it back and forth—cutting the sealant as you move along, says Baumann.

The repair process

Although there are repair kits available on the market, Baumann recommends not using any of them on pressurized windows. "Repair kits that are sold commercially should be for use on non-pressurized aircraft only," he says. "Such kits only allow you to remove damage in one specific area and result in distortion and/or windows that are out of limits with respect to thickness. We recommend that only qualified repair shops repair damage to pressurized windows," says Baumann.

The repair process for removing scratches, razor cuts and other types of identifiable damage utilizes a progressively coarse-to-fine grit abrasive sanding process. The most important criteria of this process is the removal of all stress risers. With regard to damage caused by various types of chemicals, it is important to go beyond the obvious damage (i.e. crazing or discoloration) to ensure the chemical contaminant is completely removed from the transparency. Otherwise, the deterioration process will begin again and eventually require replacement of the window.

Repair of a window usually can be more efficiently accomplished by removing the window from the aircraft. It's just not practical to attempt to repair most types of

Airframe Technology

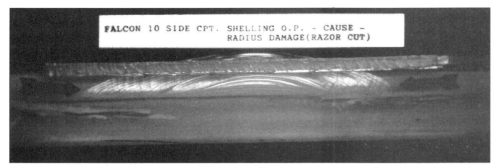

Damage resulting from razor cut.

windows while they are installed on the aircraft, says Baumann. However, windshields are a different story. Because of the level of difficulty and man-hours involved in removing a windshield, it is usually more cost-effective to leave the windshield on the aircraft to repair it. "We do send technicians on the road to accomplish these types of repairs," says Baumann.

Even if damage is only confined to one tiny location, material must be removed over the entire surface of the window or windshield so that the repair limits the amount of distortion. Although there are a number of tests to assure straight vision, the normally accepted practice is to place the window in front of a grid board. A double exposure photograph is taken and then measured for "minutes of arc" tolerance. In the case of windshields this method cannot be used. Instead, the window is viewed through from the inside looking outward and the amount of distortion is determined by viewing straight lines in the hangar.

Measuring thickness

A thorough repair means not only completely removing the damage—and doing so without distorting the optical clarity but using equipment to measure the thickness of the window after the repair. Baumann explains that measuring window thickness is not as easy at it appears. He suggests that anyone using an ultrasonic measuring device should be certified to do so.

Equipment must be calibrated properly, says Baumann. This involves using a known thickness of material and a known material density. One common error occurs when a material of one density is used to calibrate the instrument; then a window of a different density is measured. This causes false readings and possibly results in acceptance of a window that should be rejected or rejection of windows that are acceptable, he explains.

Debonding

Debonding is another potential problem. Cabin windows usually consist of two panels. The primary and secondary window layers are normally bonded together with an adhesive. Over time, the adhesive will break down and allow moisture to seep between the surfaces—fogging up the window, says Baumann. The only solution to this problem is to remove the window and have it overhauled, he says. The overhaul process includes rebonding of the panels.

Some sealants or adhesives are not compatible with acrylics and will cause crazing. Aircraft manufacturers must be consulted for recommended materials prior to bonding or installing any window. This is critical because any interaction between the sealant and acrylic is hidden from view and could cause premature failure of the window before problems are detected.

Direct vision (DV) windows, located on either side of the cockpit that can be opened by the flight crew on the ground, can create even larger problems if debonded. Debonded DVs allow excessive moisture to enter between the inner and outer plies—resulting in fogged windows. Although desiccant systems are installed to absorb moisture that normally is present, they usually are not sufficient to handle the excess moisture resulting from cracked or debonded windows.

Before attempting to repair DVs, however, don't assume that the window has debonded just because there are fogging problems. The desiccant may need to be changed or the tube connecting the desiccant to the window may be plugged or broken.

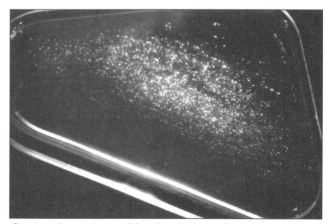

Crazing damage caused by chemical attack.

Airframe Technology

Care and cleaning

It is important, says Baumann, to use only tested cleaning agents on windows. One aircraft operator, he explains, wiped down all of his cabin windows with a glass cleaner that was on the market and left the aircraft in the hot sun. Upon returning the owner found that the combination of the cleaning agent and the heat caused all of the windows to craze. The result — all of the windows had to be removed and sent in for repair. "He wasn't very happy about it," relates Baumann.

The first step to cleaning windows should be to thoroughly flush the surface with water. This removes any dirt or abrasives that can scratch the windshield. After flushing the window, use an approved cleaner and gently wash the window with your bare hands so that you can feel for and dislodge any abrasives or dirt that remain.

After cleaning, use a soft cloth and pad the moisture from the window surface. Do not rub the window dry because this will build up a static charge which will tend to collect dust and dirt particles. The use of an approved polish that has an anti-static agent also helps to reduce static.

Shipping damage

"It's unfortunate," says Baumann, "but some of the damage that we see here is the result of improperly packing the windows for shipment." Use a container that is appropriate for the window, he explains. If necessary, build a container that will protect the window from damage, and inspect the container for protruding nails or other objects that can damage the window. Also, package the windows in plastic trash can liners, as plastic won't scratch plastic. When boxing, be sure to use enough soft packing material to keep the windows from coming into contact with the box or with other windows. "There's no need to put more damage in the window than is already in it," he says.

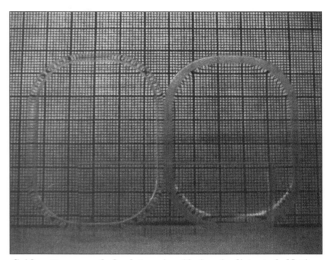

Grid pattern can help determine if view is distorted. Notice how the edges of the window on right appear distorted.

Who's qualified to repair your windows?

Robert Cupery, president of Aircraft Window Repairs Company recommends using this checklist of "musts" to determine who's qualified to repair your windows.

- Must be NDT certified and must meet ore exceed the recommendations of ASNT recommended practice SNT-TC-1A in accordance with military standard 410D. (Note: Merely having an ultrasound unit or machine does not make that person qualified.)
- Must be FAA certified and approved.
- Must have aircraft manufacturer's approval or certification.
- Must have capability, knowledge and technical data sufficient to inspect the entire window assembly.

September/October 1990

Airframe Technology

Corrosion
How to eliminate the eliminator

By Greg Napert

LAKELAND, FL—Corrosion a fairly straightforward natural process, whereby metal is converted into a by-product of that metal. Theoretically, the process is simple to halt. So why, then, is this age-old phenomenon claiming so many aircraft in this day and age?

Harry Shannon, service manager for Lakecraft in Lakeland, FL, believes that corrosion is controllable. Understanding corrosion is the key to preventing it, says Shannon. The amphibious aircraft that are cared for at this facility are regularly exposed to one of the most corrosive environments possible—saltwater.

Prevention begins with quality construction and quality materials, says Shannon. Most aircraft are manufactured without painting the internal structure of the aircraft. The aluminum is manufactured with a small protective layer of aluminum Alclad. The Alclad very slowly oxidizes and this oxidation creates an airtight layer that protects the base metal. But in reality, says Shannon, the Alclad continues to deteriorate. Normal wear and scratches also contribute to exposing the base metal to the environment, allowing it to corrode.

Lake treats all metal used on the aircraft with a five-step process before the aircraft is even fabricated. This prevents metal-to-metal contact between lap joints, and contact between aluminum skin and rivets, fasteners and other structural components. It also preserves the Alclad layer and substantially reduces the amount of exposure of the base metal.

Taking this extra step to prevent the initial corrosion of the aircraft by preventing metal-to-metal contact and reducing the exposure to the environment goes a long way toward extending the life of these aircraft.

But even with this extra step, unless preventive maintenance practices are employed to reduce exposure to the environment or corrosive materials, corrosion will eventually set in and destroy the aircraft.

What is corrosion?

Simply stated, corrosion is the breakdown of metal by either chemical or electrochemical attack.

According to the Airframe and Powerplant Mechanics General Handbook (AC65-9A), the two basic classifications of corrosion which cover most of the specific forms are direct chemical attack and electrochemical attack. Both types of corrosion work on the metal in the same manner, converting metal into a compound such as an oxide, hydroxide or sulfate.

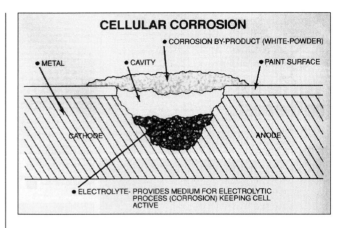

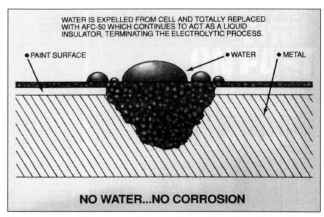

These illustrations, provided by Corrosion Block® of Allison, TX, explain how cellular corrosion is formed and how ACF-50 works to prevent it.

The basic process, in either direct chemical or electrochemical attack, involves the flow of electrons from one area of the metal (called the anode) to another portion of the metal (called the cathode), through a liquid or contaminant (called the electrolite). In direct chemical attack, anodic and cathodic changes occur simultaneously at the same point. In electrochemical attack, the changes usually occur a measurable distance apart.

Some of the more common causes of direct chemical attack are spilled battery acid, flux deposits from welding and entrapped caustic cleaning solutions. It's apparent that the way to avoid direct chemical attack is to keep the aircraft clean and free from any corrosive chemicals.

Electrochemical attack, however, is more complex. It involves the exchange of electrons due to dissimilar metals, stress and fretting. This form of corrosion is not always readily apparent and can go unnoticed until the structure fails. Electrochemical attack often cannot be halted and is usually beyond the control of the technician.

Causes

Of all factors that affect corrosion, the one that is the most predominant is the operating environment of the aircraft. Marine environments that include exposure to saltwater or other contaminants can rapidly accelerate the corrosion process. Temperature also influences the rate of electrochemical attack. This process is accelerated in a hot, moist climate. Although the technician often has no control over the operating environment of the aircraft, it should be a consideration when determining frequency and extent of cleaning and inspections.

Foreign material such as soil, dust, oil, grease, exhaust residues, acids and caustic materials can also rapidly accelerate corrosion. Keeping the aircraft clean is a major factor in eliminating the onset of corrosion due to these factors.

Removal and prevention

Once corrosion has set in, removing the corroded material or replacing parts can be extremely labor intensive and expensive.

Complete corrosion treatment, according to AC65-9A, is accomplished by completing the following steps:

- Clean and strip the corroded area.
- Remove as much of the corrosion by-product as possible.
- Neutralize any residual materials remaining in pits and crevices.
- Restore protective surface films.
- Apply temporary or permanent coatings, treatments or paint finishes.

Theoretically, this seems quite simple, but in reality, the extent of corrosion usually dictates repair procedures. And in many cases, the item that is corroded requires replacement.

Treatments

Proper preparation of metal prior to applying treatments, coatings or paint is probably the most important part of assuring that the aircraft does not continue to corrode, says Shannon. Make sure that all of the corrosion is thoroughly removed.

Shannon also recommends that if the aircraft is going to be painted, paint it immediately after preparing the surface. Many paint jobs that peel prematurely, says Shannon, are often the result of waiting too long to paint the aircraft after preparation of the surface. "You don't wash your car two weeks before the wedding," says Shannon, "so why wait so long after cleaning the aircraft to paint it?" Shannon says that dirt and dust build up rapidly, and that unless it is painted promptly the paint will not adhere properly.

In addition to traditional methods of surface treatment, rust inhibitors are also quite popular in the aviation industry. Two products, LPS 3 (LPS Laboratories Inc.) and ACF-50 (Corrosion Block) are touted by most in the industry to be among the most effective.

LPS 3 is a product that, according to Joseph Tarpley, technical service manager for LPS Laboratories Inc., "is a water displacing corrosion inhibitor. It penetrates into seams, lap joints and around rivet heads, displacing water and leaving a long-term protective film." This product has good adhesion characteristics, and because of its waxy consistency, says Shannon, is best for use in exposed areas. According to Shannon, "It will not wash away as readily as ACF-50."

The consistency of ACF-50 is that of 10 weight oil. "Because of its unique capillary action," says Jim Van Gilder, chairman of Corrosion Block, the Addison, TX-based distributor of ACF-50, "this product has a powerful attraction for metal, causing it to spread over wide areas, between the tightest skin laps, around rivets and screws, and cover blind spots not directly touched by a sprayer." It disputes moisture and, according to Van Gilder, actually has the ability to stop existing corrosion without having to remove any of the corroded materials. Van Gilder says that ACF-50 has proven itself in the harshest of environments and if retreated as directed (every 18 months), is most effective in preventing corrosion.

Despite all of the technical advances made in corrosion prevention and treatments, however, the best and most inexpensive method for preventing corrosion is to simply keep the aircraft clean.

Airframe Technology

ACF-50
What the doctor ordered?

Due to an aggressive marketing campaign and widespread acceptance of ACF-50, a revolution in corrosion prevention is slowly taking place. What is ACF-50? And why is it so different from other products?

Jim Van Gilder, chairman of Corrosion Block, distributor of ACF-50, says that ACF-50 is a clear, clean chemical which effectively stops or prevents corrosion from forming. The product is a thin compound (the consistency of 10 weight oil) with a strong affinity for metal. This thin consistency allows the product to be spread, through capillary action, into minute cracks and crevices.

"When moisture is present for prolonged periods of time," says Van Gilder, "corrosion cells are formed. Through electrolysis, electrons are moved from the skin or fitting to the surface and a white crusty residue becomes apparent. These corrosion cells act as miniature batteries, feeding the transfer of electrons as long as oxygen-providing moisture is present. This reaction is more complete when salt is present because the salt holds moisture longer and acts as a catalyst to feed the abuse." According to Van Gilder, ACF-50 isolates the moisture from the surface, stopping the process.

What sets ACF-50 apart from similar products, says Van Gilder, is that it provides a thin, water-repelling chemical barrier to the metal substrate, after it has removed moisture from the surface. "ACF-50 then actually searches out moisture in the smallest seams and lap joints. If it can't push the moisture off the surface, it has the unique ability to absorb the moisture, causing an oil-in-water emulsion to form." LPS 3, says Van Gilder, and similar products are strictly one dimensional, passively repel water, and once the solvents evaporate, they leave behind a static wax film.

ACF-50 is a "nearly perfect insulator," says Van Gilder, with a dielectric of 38,400 volts. Because of this, and the fact that it is also a lubricant, it can be used in a number of other applications as well. ACF-50 works well as a penetrant and lubricant, says Van Gilder. According to Van Gilder, it can also be used safely to spray electrical equipment, including avionics and circuit boards.

ACF-50 is a sacrificial coating that is "chemically absorbed" and must be reapplied every 18 months.

Types of corrosion

Many forms of corrosion exist. But according to AC65-9A, the most common forms are as follows:

Surface corrosion—
usually appears as a roughening, etching or pitting on the surface of a metal. This form of corrosion is accompanied by a powdery deposit of corrosion by-product and may be caused by either direct chemical or electrochemical attack. Surface corrosion will sometimes spread under painted or plated surfaces and can go undetected until the paint or plating begins to lift and bubble.

Dissimilar metal corrosion—
results from contact between dissimilar metals in the presence of a conductor. This corrosion usually occurs where the insulation between two dissimilar metals has broken down. Metals have varying electrical potentials and when placed in contact with one another, transfer electrons. According to Shannon, dissimilar metal corrosion can also exist when two apparently similar metals are in contact as well. For example, corrosion can take place between rivets and the skin. They are both aluminum alloys, but the alloys may be slightly different allowing for a small amount of electron transfer.

Intergranular corrosion—
is an attack along the grain boundaries of an alloy. Lack or uniformity in the alloy structure caused by internal changes during heating and cooling is usually the cause for this form of corrosion. Aluminum alloys and some stainless steels are especially susceptible. Intergranular corrosion is often undetectable; however, ultrasonic and eddy current inspection methods have been successfully used to identify it. According to AC65-9A, severe intergranular corrosion may cause the surface to exfoliate. This flaking of the surface is due to delamination of the grain boundaries.

Stress corrosion—
is the result of the combination of sustained tensile stresses and a corrosive environment. This type of corrosion is particularly characteristic of aluminum, copper, certain stainless steels and high-strength alloy steels and usually occurs along areas where the metal has been cold worked. Some examples of components that are susceptible to stress corrosion are aluminum alloy bellcranks with pressed-in bushings, landing gear shock struts with pipe-thread-type grease fittings, clevis pin joints, shrink fits and overstressed tubing B-nuts.

Fretting corrosion—
occurs when two mating surfaces are subjected to rubbing. Surface pitting is usually accompanied by finely ground debris. This debris remains entrapped between surfaces and often results in an extremely localized abrasion. Small sharp grooves resembling Brinell markings may be worn into the rubbing surfaces. Bearing surface corrosion of this type is often referred to as false Brinelling. *July/August 1990*

Airframe Technology

Sheet metal repairs
Canadair Challenger Inc. stresses importance of integrity

By Greg Napert

WINDSOR LOCKS, CT—The word "integrity" resonates from within the walls of Canadair Challenger's service facility in Windsor Locks, CT. These technicians, who are working on some of the most sophisticated aircraft in the industry, have a great respect for their work, and more importantly, they practice what they preach.

Lead technician Stephen Drziak, who has 43.13-1A/2A and a few other technical manuals permanently imbedded in his memory banks, serves as a mentor for the other technicians who enthusiastically follow his lead.

With a great deal of exposure to sheet metal repair, modifications and alterations, the technicians at Canadair have become craftsman with a unique perspective on their trade.

Most technicians use approved techniques when making sheet metal repairs, says technician John Mozonski, but because they are often rushed, their only goal is to get the aircraft back in the air. Therefore, the repairs are usually temporary and serve to keep the aircraft flying until it is scheduled into the Canadair facility.

Some of the areas that the technicians at Canadair say they see abused most often are as follows:

Bucking a rivet

One of the major problems that technicians have, says Drziak, is properly bucking a rivet. The bucking bar not being perfectly parallel to the rivet is the cause of the problem. And this is quite easy to correct, without much practice. With the rivet set placed squarely against the head of the rivet and the bucking bar against the tail, apply pressure to the bucking bar so that the bar is resting on the sheet metal. This will align the bucking bar so that it is perfectly square to the surface. After the bar is square to the surface, the man on the rivet gun should push the rivet in and begin riveting. This technique can be used on flat and curved surfaces.

Tooling marks

Tooling marks are another common problem. There are ways to reduce the amount of tooling marks, says Mozonski. The use of improper tools is probably the most common cause, followed by carelessness.

Make sure that you are using the right size rivet set for the rivets with which you're working.

Drziak suggests applying nylon tape on the riveting set to help, eliminate some of the "smiles" that are caused by not having the rivet set perfectly parallel to your work. He also suggests placing nylon tape on the edges and corners of the bucking bar, especially when the bar is in close proximity to other surfaces.

"Scratches can be a real problem," says Challenger technician Eric Bennett. Removable panels is one area where he sees many. When the panels are removed, the sealant is often cut and scraped with a razor blade. The blade scratches the surface of the metal and sets up stress marks that can lead to cracks. Use a plastic scraper to remove the adhesive. Plastic will not harm the surface of the metal.

If there are scratches anywhere on the aircraft, they must be dressed out. If the scratches are located on a machined surface or on a surface that has previously been shot peened, the area must be shot peened to relieve stress. Burnishing, says Drziak, is really not sufficient. "All you are doing is pushing the clad layer of material back into the scratched area. This does not remove the stressed area." You have to remove the damaged area in order for the repair to be effective.

When sanding any aluminum parts, whether dressing a scratch or preparing the surface for painting, use an aluminum oxide based sandpaper. Silicone carbide is not recommended as it will set up corrosion in the metal.

Blind fasteners

Blind fasteners also present certain problems. Mozonski says that the proper use of blind rivets takes a little practice. It's important not to pull the rivet too quickly. This often results in the skin not being pulled tightly against the surface. The rivet then has to be removed. Press the trigger of the rivet gun very lightly and pull the rivet to the point that the material is together, then finish peg the rivet.

Also, it's important that the hole be prepared properly. Use a guide to align the drill perpendicular to the surface. The blind rivet is not as forgiving as the standard rivet. If the hole is drilled at an angle, the rivet head will not seat properly. Oversized holes will cause the rivet to pull right through the sheet metal. There is nothing more frustrating than ending up with a hole that is larger than the largest available oversized rivet.

If it becomes necessary to remove a blind rivet, take care not to damage the skin surrounding the rivet. Thick materials are easier to work with. Simply use a center punch to knock the pin out and remove the lock. The head should then be easy to remove. Thin materials, however, require that you drill the rivet from the aircraft. Because of the way that the rivet is designed it's difficult to keep the drill from drifting off of center.

Hole finding methods

Technician Doug Roper suggests that hole finding, especially on compound curved surfaces, is not always easy. A strap duplicator helps, but one is not always available or the best thing to use. He suggests that there are a couple of methods that work quite well. One is to use plexiglass, or similar product, as a template. The fact that it is clear allows you to locate the holes and drill the plexiglass for use as a template.

Another method that he suggests is to fabricate metal straps approximately 1 inch wide by 12 inches long. Drill a hole in one end of the metal strap and

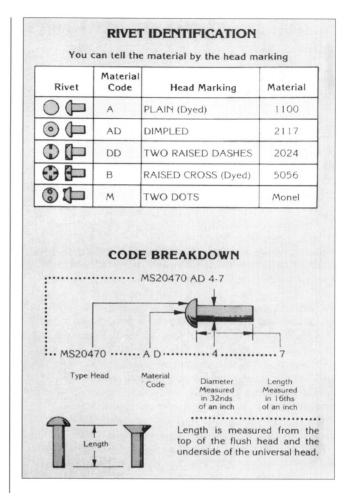

position the strap over the hole with the non-drilled end of the strap over the surface of the aircraft. Place a strap at each hole. Place a screw through the hole and into the nut plate and secure the non-drilled end of the strap to the aircraft with aluminum tape or duct tape so that the strap doesn't move. When the new piece of metal is inserted beneath the strap, the hole in the strap will provide an exact location for the hole to be drilled.

Drziak claims that most sheet metal repairs don't take a special talent. Instead, complete knowledge of the task, and everything associated with it, is the key to successful repairs.

Blind riveting guidelines
By Roger Nikkel

Since its introduction in the 1930s, the blind rivet has evolved from a simple, pull-through, non-structural tacking fastener into a widely used structural fastener. Using either a wire draw or compressed column concept of installation in conjunction with a mechanical locking component, these newer structural blind rivets offer performance and labor-saving features that truly benefit general aviation maintenance operations. However, their proper use is crucial if performance comparable to solid riveting is to be achieved.

Over 95 percent of the problems associated with blind riveting stem from the lack of adequate attention to hole preparation and/or awareness of tooling requirements. Typical problems that cause rivet failures include over or undersizing the hole due to worn drill bits, drill bit wobble, improper bit selection (i.e., reaming the hole out with an undersized drill), improper sheet thickness calculation caused by blind side burr protrusion, metal chips in between the sheets or deformed sheet stack components.

Sheet stack misalignment due to sheet creep (particularly critical when working around radiused surfaces) and lack of perpendicularity when drilling also frequently produce rivet failures. It is recommended that the installer stay within 1.5 degrees from centerline while drilling.

Using a liberal number of reusable mechanical fasteners (clecos) to tightly clamp the sheets together while drilling and riveting will greatly reduce sheet misalignment. Using a drill guide with universal headed rivets and/or a microstop drill cage in conjunction with a piloted countersink for countersunk rivets will aid in preventing countersink depth and hole angle problems. Destacking and deburring after the hole drilling operation will allow for removal of any chips or foreign matter that may affect the riveting process.

Be sure that you are using the proper type tooling with the rivet you have selected (i.e., some rivets require shifting or double action tooling, whereas many require non-shifting or single action tooling). Be sure the rivet tooling manufacturer's recommended air pressure is provided in the case of power tools. Too fast or too slow installation cycle can often cause rivet malfunction. It is also essential that the tool, hand or air operated, be kept perpendicular to the work surface. Side loading from improper tool angle can easily cause premature rivet stem breaks and/or head gaps. In most cases, maintaining firm steady pressure on the head of the rivet during the entire installation cycle is recommended.

It's important to try and save the existing hole if a rivet malfunctions. If it is a non-structural fastener, use the original drill size and simply drill down to the radius under the head until the head is removed; then punch out the rivet sleeve. If a structural, mechanically locked fastener is used, it is essential that the locking device be extracted before punching out the rivet stem. Drill down to the locking device until it is free, pick it out, and then punch out the stem. Next, using the original drill size, continue drilling until the head comes free; then punch out rivet sleeve. Check the hole with a hole gauge to confirm hole size. If it has become oversized or eccentric due to the removal process, go to the next available rivet diameter. *May/June 1990*

Roger Nikkel is president of Fastening Systems International Inc., Sonoma, CA.

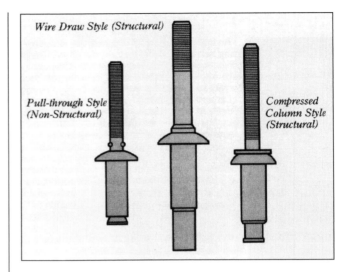

Airframe Technology

Oxygen systems
Don't hold your breath wondering if they're going to work

By Greg Napert

Oxygen systems normally don't require an extensive amount of attention, and problems with oxygen systems actually occur quite infrequently. But keeping up with servicing requirements and regularly verifying operation greatly improve the safety margin for those involved in flying the aircraft.

Cylinder savvy

Jack Coloras, director of sales and service for Tec-Air/Comprogas Services, E. Northport, NY, says that for the most part, technicians are very dependable and conscious of servicing requirements as they relate to oxygen systems. But regulations aren't clear enough when it comes to explaining how items should be overhauled, explains Coloras.

Coloras points to the case of retesting cylinders. "Most manufacturers," says Coloras, "recommend a complete overhaul of the cylinder assembly at each cylinder retest interval. A complete overhaul would consist of a cylinder hydro-test and an overhaul of the valve or regulator. This work must be performed by an authorized FAA facility for the cylinder assembly to become airworthy once again."

In many cases, says Coloras, the cylinder is sent to a DOT-approved hydro-test facility where the valve or regulator assembly is removed, the tank is hydro-tested and the valve or regulator assembly reinstalled, without overhaul.

Coloras recommends that the entire assembly (cylinder, valve and regulator) be sent to a facility that can handle all components and return the assembly completely overhauled. According to Coloras, overhauling the regulator assembly is not something that should be done unless properly equipped. There are too many special tools and various types of testing equipment that are needed to overhaul these components. "Safety concerns and potential problems resulting from attempting to overhaul the cylinder assembly yourself outweigh any benefits that can be viewed as advantageous," says Coloras.

Types of cylinders

The three most commonly used types of cylinders today are thin-wall steel (3HT), thick-wall steel (3AA) and composite. Composite cylinders have been just recently introduced, and because of their light weight, are becoming quite popular. Wire wound cylinders are primarily

Maintaining a clean environment free of any oils or grease is imperative while servicing oxygen systems.

used in military applications and in some older aircraft. Low-pressure portable cylinders are rare, but still used in some military applications.

Keeping track of when cylinders are due for hydro-test is important, says Coloras. The rating of the cylinder is the key to the frequency of overhaul. If the cylinder has a DOT3HT1850 rating, the retest period is three years. The DOT3HT1850, which is rated for 1,850 psi, is currently the most commonly used cylinder for aviation.

If the cylinder has a DOT3AA1800 or 2100, the retest period is five years. Additionally, DOT3HT oxygen cylinders have a 24-year life limit and the DOT3AA oxygen cylinders have an unlimited life. Composite cylinders must be reinspected and hydrostatically retested in

The Best of Aircraft Maintenance Technology Magazine

accordance with exemption DOT-E8162-1850. The service life on composites shall not exceed 15 years with the retest period being three years.

Think safety

The most important thing to remember when servicing oxygen systems is to think safety. AC65-15A points out a few good practices:

- Tag all repairable cylinders that have leaky valves or plugs.
- Don't use gaseous oxygen to dust off clothing, etc.
- Keep oil and grease away from oxygen equipment.
- Don't service oxygen systems in a hangar because of the increased chances for fire.
- Valves of an oxygen system or cylinder should not be opened when a flame, electrical arc or any other source of ignition is in the immediate area.
- Properly secure all oxygen cylinders when they are in use.

Other precautions such as never attempting maintenance unless the oxygen supply is turned off, purging the connecting hose before connecting to the aircraft filler valve, filling the system slowly to avoid overheating, opening any valves slowly and checking pressures frequently during servicing should also be kept in mind.

Servicing

Servicing the oxygen system with oxygen is simple and straightforward, but it is easy to become complacent. It is important, therefore, to stay familiar with and regularly use the pressure/temperature correction chart supplied with the aircraft. Ken Krawczyk, technician for Van Dusen Services in Milwaukee, WI, says that pressure differences when filling in cold weather compared to warm weather can vary as much as 200 psi. Krawczyk says that it is especially deceiving in cold weather. Because the cylinder warms up while servicing and then cools down after servicing, you are left with less pressure than you thought you had serviced it with. "In cold weather, I normally service the system with slightly more oxygen than what the recommended pressure is," says Krawczyk. "That way, after the cylinder has cooled down, I end up with my desired pressure."

Maintenance

Any time that maintenance is performed on the oxygen system, the supply cylinder should be turned off and the system drained of any pressure. Pressure should be drained slowly so that condensation doesn't form in the lines.

Three types of oxygen cylinders. From top to bottom: steel, composite and wire wound.

If any fittings or lines are removed, they should be removed slowly to allow any residual pressure to drain. Cap any open lines or fittings immediately to prevent contamination of the system. When reinstalling fittings and lines be sure to keep them a minimum of 2 inches from oil, fuel, hydraulic or other fluid lines, electrical wiring, and kept clear of all hot ducts, conduit or equipment.

Leak testing should be performed periodically as recommended by the manufacturer or every time that the system is opened for maintenance. A leak test is performed by completely filling the oxygen system, and checking the pressure a specified period of time later to make sure that the pressure has not leaked off.

Corrections for temperature and pressure must be accounted for by using a chart that is supplied by the aircraft manufacturer. Keep in mind that at least one hour should pass to allow the cylinder to cool before taking pressure or temperature readings after the system is initially filled.

Purging of the oxygen system is required if it has been depleted and not recharged within two hours, or if it is suspected that the system is contaminated.

To purge the oxygen system connect an oxygen, dry nitrogen or dry air source to the fitting that the system oxygen cylinder is normally attached to and fill and drain the system at least three times. According to AC65-15A, systems that have a line connected to each end of the cylinder (one for filling and one for distribution) should be purged by opening all of the regulator emergency valves, connecting a source of oxygen at the filler valve and passing oxygen at a pressure of 50 psi through the system for at least 30 minutes. This job should be performed in a well-ventilated area.

Airframe Technology

Keep it clean

The cleanliness of the environment around any source of oxygen is of utmost importance, according to AC65-15A, any parts that are exposed to oxygen or tools that are used to work on oxygen systems should be cleaned with anhydrous (waterless) ethyl alcohol, isopropyl alcohol or any other approved cleaner. If masks to regulator hoses are contaminated with oil or grease, the hoses should be replaced.

The basic concept of an oxygen system of any type, says Coloras, be it fixed or portable, whether continuous flow, demand or pressure breathing system is to reduce a highly pressurized source of oxygen to a breathable pressure for human consumption at various altitudes. Without periodic inspection of regulators and verifying the operation of the oxygen system, there is no way to determine if it is actually working. Operate the system whenever you get the chance. By doing so, you won't have to hold your breath wondering if it is going to work when needed. *March/April 1990*

Airframe Technology

Composite rotor blade inspection and repair
Getting over the fear

By Greg Napert

The latest in high-tech manufacturing materials and composites offers new challenges to today's technician. Learning the new technology and developing skills, much like the men who crafted the wood, dope and fabric airplanes of the past, will be a requirement for those who wish to keep up with today's changing technology.

Helicopter rotor blades offer an exceptional challenge to those wishing to make repairs because of the critical role that the rotor plays in flight. Most technicians today choose not to make any type of repairs to the rotor in fear that an incorrect repair might result in disaster.

While these fears are understandable, taking the time to know composites and learn basic preventive maintenance and minor repairs can actually result in the prevention of catastrophic failures.

Practice makes perfect

Prior to attempting any types of repairs on any composite component, it is imperative that two things be done. First, there must be a thorough understanding of the type of material that is going to be repaired or inspected. And second, the technician's skill level when working with the material should be developed by practicing the repair on practice parts. It is highly advisable to attend a composite school where practical hands-on experience can be gained and repair technique, that can contribute to an airworthy repair, can be learned.

What's a composite?

According to FlightSafety, "A composite is a combination of two materials: a mass of fibers, the principal load-carrying material and a matrix (commonly epoxy) which bonds the fibers together and gives them lateral support. The combination of the two leads to a very high strength-to-weight ratio material." Today's advanced composites are also referred to as "fiber-reinforced plastics."

Materials that are used in today's composites include fiberglass, Kevlar, carbon, boron and ceramic fibers.

Composite rotor inspection

Many sophisticated methods of inspection are available, but the most valuable and economical inspection is done with no special tools. The eyes, ears and hands can give you valuable indications that there are defects in the rotor. It is important, however, to refer to manufacturers' recommendations for proper inspection techniques.

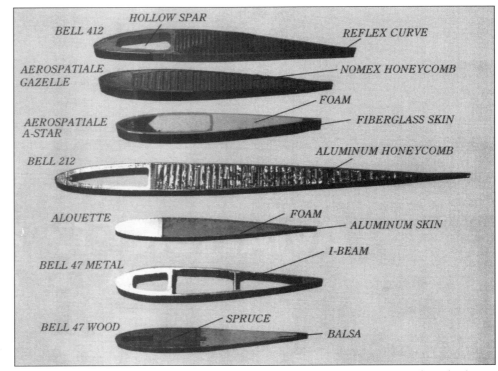

Cross sections of rotor blades showing old technology (bottom) vs. new technology (top).

"I like to get a mechanic in the habit of touching and feeling along the trailing edge of the rotor blade for damage all the way to the tip," says Dana Kerrick of Composite Technics Inc., Dallas, TX. "Many people don't realize how critical a nick on the trailing edge of a blade is. It is extremely important to repair it right away." Kerrick explains that the trailing edge can ultimately determine the life of the blade. A small nick can set up stresses that can propagate to the leading edge and cause blade failures. Kerrick suggests polishing out any small nicks as soon as they are discovered to relieve stress in that area.

Kerrick explains that many technicians are reluctant to remove the rotor from the helicopter for a good thorough inspection, but it is important to place the rotor in a position where the entire blade can be inspected and there is suitable lighting.

According to FlightSafety, a good visual inspection with the aid of a magnifying glass should be the first step in inspecting a blade.

An acoustic impact (coin tap) should follow to verify visual indications of damage. The coin tap test is conducted with a special lightweight (1-ounce) hammer, coin or other object (depending on the manufacturer's recommendation). By simply tapping on suspected areas of the blade, a change in the sound can indicate delamination, bond separation or other damage. Becoming proficient at coin tapping takes practice. Different materials, complex curves and varying thickness in materials can affect the sound of the tap, thus, the importance of knowing the material and composition of the blade that you're inspecting.

Other methods of inspection that may be spelled out by the manufacturer include hardness tests, moisture tests, ultrasonics, X-ray and thermography, to name a few.

"If technicians have questions regarding damage," Kerrick says, "call us—info is free—one of the only things that's free anymore. Send a photo. We'll give a reasonable assessment of the damage."

Preventive maintenance

"Practically all damage," says Kerrick, "is due to one of four things: corrosion, erosion, trees or wires." Of these four, corrosion and erosion are the only two that can be addressed by the technician.

To prevent corrosion, Kerrick recommends a daily washing of the blades with whatever solution is recommended by the manufacturer. "You can't use just any soap or cleaner," Kerrick says. "There are certain clean-

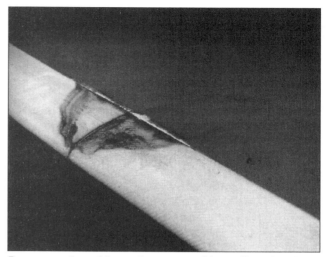

Damage such as this can be prevented by sealing the joints where the erosion strips meet on the leading edge of the rotor blade.

ers that are approved and others that are forbidden." Some solvents or cleaners may contain chemicals that will deteriorate the blade.

"Rotor blades should not be waxed," says Kerrick. "Many technicians wax blades assuming that the blades will be protected, but the wax can imbed itself in the composite and prevent paint or adhesives from sticking. This can make repairs more difficult and possibly more expensive."

Erosion is an ever present problem and should be addressed frequently, explains Kerrick. Flying in rain or in harsh environments, such as near the ocean or in dusty areas, can quickly erode the blade to the point of putting it out of service. There are steps that can be taken to reduce erosion and significantly reduce operating costs.

Exposed areas where paint has been worn off the blade should be repainted immediately. It is much cheaper to wear out paint than to wear the actual blade.

Kerrick also suggests sealing joints where leading edge erosion strips meet. If this joint is neglected, explains Kerrick, the original sealant will wear away. Moisture will then enter into the joint and propagate from that area, causing the erosion strip to separate from the rotor. Kerrick says that the manufacturer usually provides kits to perform this repair and that it can be done in a couple of hours.

"Technicians don't have to perform these tasks," says Kerrick, "but they'll realize a much greater blade life if they do."

Airframe Technology

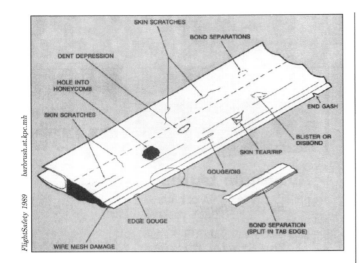

Repairs

There are actually many repairs that a technician can comfortably perform in the shop, but the principal factor that limits the repairs is the manufacturer. Procedures are clearly spelled out that specify repair limits, techniques and methods. It is important, however, to remember that if a repair cannot be accomplished based on manufacturer's specifications, it may still be repairable by an approved overhaul shop or the manufacturer.

Minor damage such as nicks, scratches and minor delamination can usually be repaired if the damage is in a non-critical area.

Kits are available from the manufacturer for specific repairs and are typically quite easy to use. There are usually no special tools that are required, but to accomplish professional looking repairs, it may be necessary to purchase special equipment such as vacuum and heating devices. *AMT January/February 1990*

Airframe Technology

Avionics removal and replacement
Avionics tips for aircraft technicians

By Peter S. Lert

It has been said that the job of the aircraft technician consists of "doing the impossible to the inaccessible." Nowhere is this more evident than when it comes time to get at something behind the instrument panel, or buried under the center control pedestal, or, in larger aircraft, somewhere in a nose or aft avionics compartment.

This article is aimed at the Airframe and Powerplant, rather than avionics, technician. It's a guide to the safe and easy removal and replacement of avionics components in order to get at general aircraft work that would otherwise be inaccessible.

Panel-mounted systems

It's usually the panel-mounted systems that cause the aircraft technician the most trouble. This is partly because the mass of cabling behind the panel-mounted radios is often more extensive than that behind the control heads of remote systems, but also because panel-mounted systems tend to be found in lighter aircraft, where there's less working room.

With newer installations, radio manufacturers provide Markwidth mounting trays for permanent installation in the panel. The radios themselves are then slid into the trays and retained by various locking arrangements—usually screws.

In older installations, the mounting tray is, in fact, nothing more than that: the actual electrical connectors are on the radio itself, and come out with it. Antenna coax connectors and multipin cable connectors must be removed.

We'll look first at the removal of the radios themselves. Unfortunately, this often isn't enough; if you have to get at something behind the radio stack, merely pulling the radio still leaves the tray and all its cabling in place, so we'll address that next.

NOTE: If you're removing more than one radio, or even a single one with several cables, make sure you know which connector goes where for replacement. Many older nav-comm sets, for example, use identical coax connectors for the nav and comm antennas; swap them, and while neither nav nor comm will fail outright, neither one will work properly (comm antennas are vertically polarized, while nav antennas are horizontal).

Locking arrangements

In older radios, locking arrangements tend to be fairly simple, ranging from a slotted or hex-head screw that runs the full length of the radio (starting at an inconspicuous hole in the front panel) to a simple Dzus fastener at the back of the mounting tray.

If you don't have access to the radio manufacturer's manuals, examine the front and back of the radio carefully, using a light and mirror as necessary.

Security locks

With the rise in radio thefts, some systems may have more holding them into the panel than just the normal front-panel locking screw—for example, tabs that protrude through the back of the mounting tray to accept a locking bar or even a small padlock. It doesn't hurt to ask the aircraft operator if such a provision is installed.

Out it comes

Once all the connectors are removed and the lock released, there's nothing holding the older radios in their trays but friction, so they should slide out easily. If they don't, look carefully to see that you haven't forgotten one last connector—some of the older sets may have four or more.

Radio removal

Once the radio is unlocked, a reasonable pull should slide it right out. Unfortunately, this isn't always the case, particularly if the connectors are stiff and the radio has been in place for a long time.

The basic rule is NEVER to pull on the knobs. At best, if the radio is stiff, you'll pull the knob off the shaft; at worst, you can do all sorts of internal damage. Ideally, the place to pull is by the side edges of the front panel.

If it is difficult to grasp the edges of the radio, remove the adjacent radios as well; it's often easiest to pull everything in the stack at once. Be sure to note which radio goes back where.

Another tip is to have a look at the back of the stack; often, even though the connectors are part of the mounting tray, there will be a hole through which you can push against the back of the radio to at least start it out of the stack.

Airframe Technology

At worst, you may have to gently pry the radio out from the front. This is where the most cosmetic, if not structural, damage occurs; front panels are generally made out of plastic, and a screwdriver prying at them doesn't help at all.

Moreover, prying from only one side tends to jam the radio tighter into its tray. Far better to make a couple of radio-pulling tools by bending a little right angle lip in some thin metal or shim stock an inch or so wide, with a finger hole to pull by (see illustration). Work one behind each side of the radio you need to pull (starting the tool at the top or bottom of the stack if necessary), and apply a gentle pull from both sides at once.

Older rack removal

In a good installation, mounting trays will be supported from the rear, as well as by the mounting channels at the front. In any case, expect lots of screws (often countersunk to allow the radios to slide in and out). In general, there will either be nuts with lockwashers, or metallic or fiber locknuts.

Typical instrument panels have a punched aluminum angle running down each side of the stack opening in the panel. Normally, during reinstallation, the punched holes never quite match the ones in the radio trays! A good practice, upon removing the trays and retrieving all the nuts and washers, is to note the location of the active holes, discard the nuts and washers, and install Rivnuts.

Removal of modern panel-mounted radios

About 10 years ago, manufacturers started installing panel-mounted radios in more elaborate mounting trays in which the tray itself included all the connectors, with mating connectors on the back of the radio itself.

Theoretically, these radios should have been easier to remove—just unlock them and out they come. In the real world, however, things are never quite so simple—now, in addition to mere friction, you have to overcome the holding forces of the connectors, which can vary from non-existent to pretty stiff.

Modern rack removal

It can be particularly tough to remove factory-installed radio packages, since these were often put together before the rest of the panel and have little or no slack to work with. Bear in mind the many connections on the cable bundles that are held in place by not much more than their solder joints.

If you really have to pull the trays, you might consider leaving the connectors themselves in place. The actual connectors are held into the backs of the trays by screws or Rivnuts, installed from the back or the front. A long-bladed screwdriver will help remove the connectors, allowing you to leave all the wiring in place and remove only the mounting tray itself for behind-the-panel access through the radio stack opening.

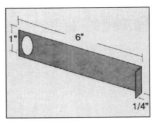

Radio removal tool (make two from shim stock)

Note that in some radio trays, the connectors are mounted with "floating" washers or spacers allowing them to move slightly and align themselves with the connectors on the radios as they are inserted. Failing to "float mount" such connectors when you reinstall them can make the radio impossible to insert—or, worse, hard to insert and even harder to remove next time!

Modern solid-state radios often require forced-air cooling, and this can complicate rack removal and replacement. In some installations, each radio has a cooling hose connection at the back of its mounting tray; in others, one or more perforated "piccolo tubes" run up the sides of the radio stack.

Installation of modern panel-mounted radios

Slide the radio into its rack, using GENTLE pressure if necessary to seat the connectors, and then lock it in place. Use FINGER TIGHT FORCE ONLY when tightening the locking screws. (The reason Narco went to its spring-lock arrangement was to avoid radios locked forever in place by hex or Phillips screws with stripped-out heads!)

One nice touch your customers will come to appreciate is to give the contacts on the mounting tray and radio connectors a light squirt of a good lubricating-type contact cleaner—the sort sold in electronics stores as "TV tuner cleaner" works well. Many avionics problems can ultimately be traced to loose or corroded connections, and the light lubricant action makes the radios easier to remove and replace later on.

Removal and replacement of remote radios

Remote radios and their control heads are generally much easier to remove. The radios sit in racks in the avionics bay of the airplane (usually, although not always, in the nose), and their control heads are usually mounted in the panel or center pedestal, either by simple screws or the even more convenient standard size "Dzus rail."

Some remote radio boxes—particularly the larger airline-standard ATI units—have mating connectors similar to those of panel-mounted units at the rear of the radio box. Others, such as the more recent Collins ProLine, have cable connectors at the front of the box. In either case, a simple locking mechanism—sometimes safety-wired—must be released before the box can be removed, usually by a convenient handle.

November/December 1989

Accessory Technology

Accessory Technology

Troubleshooting aircraft alternator systems

By Robby Starr

Most aircraft today utilize an alternator system in combination with a battery to provide electrical power. There are still some planes flying around with generator systems, but they seem to be few in number. The alternator and associated components that make up the "alternator system" is fairly simple to understand as long as it's working properly. However, when the system has failed or not quite operating up to par, it can be something of a nightmare. Several steps can be taken to help eliminate some confusion when dealing with alternator or charging problems, but first let's define some basics about alternator systems.

Type I and II systems

Whether on a single-engine or twin-engine aircraft, there are just two basic types of alternator systems: Type I and Type II. The major difference between these two types is how the wiring between the voltage regulator and the alternator is configured. Also the voltage regulator for the two systems doesn't operate in the same manner.

In a Type I system the current controlling element of the voltage regulator is located in series between the aircraft bus (DC) and the alternator field (F1). In a Type II alternator system the current controlling element of the voltage regulator is located in series between the alternator field (F2) and ground. If an alternator has a single lead brush rack, then it has to be wired as a Type I system with the appropriate type voltage regulator. If an alternator has a two lead brush rack, then it can be used with either type of voltage regulator and can be wired for either type of system. The alternator and voltage regulator are by no means the only components of the charging system, but when it comes to maintaining or troubleshooting, most of the attention seems to be focused on these two components.

Identifying the system

When beginning to troubleshoot a discrepancy in the alternator system, the first thing to do is to determine which type of system the aircraft has. The easiest way to do this is to take a peek at the rear of the alternator. The only thing that is to be determined at this point is whether or not one side of the field coil is grounded. If the alternator only has one field terminal available, then the other is internally grounded and we know that it is a Type I wired system. If the alternator has two field connections available, look to see if one of those is externally grounded. If so, it is also a Type I system. If neither one is grounded, then the system is a Type II. There are a few more basic things to check that will help get closer to the real problem.

Basic checks

Now that we know which system we are working on, let us check the voltage at the input to the voltage regulator with the master switch and/or the alternator switch on. The engine can be running but it isn't necessary. Most systems incorporate an overvoltage device that is in series between the alternator switch and the voltage regulator. An exception is the newer generation voltage regulators which have the overvolt protection built into the regulator itself. We now want to measure the voltage at the input of the voltage regulator on either type of system, and if we read bus voltage or battery voltage at this point, we are in good shape. If the system is a Type II system, the voltage at one of the alternator field terminals should be the same as the input to the voltage regulator.

In either case, if no voltage is read at the input to the voltage regulator, go upstream from the voltage regulator and measure the voltage at the input to the overvolt relay with the same conditions. If no voltage can be read at this point, there may be a loose wire, or more likely, a faulty field switch, alternator switch, or master switch. Once this is corrected and the proper voltage is present at the input of the voltage regulator, the system should work. If not, you have at least eliminated this part of the charging circuit so a more specific problem can be analyzed.

Again, these are just basic steps to take in the interest of time and money. Under a really tight time table, it's common to point the finger at the regulator or the alternator and replace them, when in reality, it may be a simple problem that you are overlooking. It pays to look at the simple things first.

Common problems

Some specific types of problems are encountered with the alternator system on various aircraft. The explanations to the alternator problems that we are discussing are the most common problems I field on a regular basis.

Accessory Technology

When replacing a voltage regulator, a common error is to fail to adjust the regulator at the prescribed bus voltage. Some regulators on the market cannot be adjusted, but the ones that can, need to be. Several technicians have indicated they think voltage regulators are preset from the factory. This is not true. When the regulator is tested at the factory for certification, it is impossible to represent all the different parameters of an aircraft system in the test. So if the regulator has an adjustment, always set it when checking the system.

Another problem that often occurs when replacing a voltage regulator is the shorting of the field output to ground—especially on a Type I system. When this happens and the alternator switch is on, the regulator tries to drive full field current to the airframe ground and the regulator gets fried.

When hooking up the regulator, all the switches should be left in the "OFF" position, and the field connection should be made first and isolated before wiring the rest of the system together. Some of the newer voltage regulators or ACUs available have built-in ground fault protection for the field circuit and will eliminate the effects of accidental grounding of the field wire. Ground fault protection also will eliminate a damaged regulator due to a defective rotor coil in the alternator.

Let's discuss the subject of field grounding problems. Many times pilots report that when flying, the alternator drops off line, and after waiting awhile they can recycle the system by turning off the alternator switch and then back on. Everything seems to be OK they say, until the system drops off again. However, when the technician is testing and adjusting the system on the ground, the problem doesn't occur. The culprit in this situation is often the alternator rotor. Under certain conditions, usually at altitude, it can develop a short in the windings.

Sometimes it is a short to ground and other times it is a coil-to-coil short. It seems to only happen under certain engine speeds and thermal conditions which are present during flight and not on the ground. We call this a "flying short." If it's a short to ground, then either the field circuit breaker will pop or the regulator will get toasted. If it's a coil-to-coil short, the output voltage will rise and the overvolt relay will trip the system off.

This type of alternator problem almost never manifests itself on the test bench. So when replacing a voltage regulator that has been shorted or toasted, or when fixing a repeatedly overvoltage tripping problem, always check the rest of the field circuit for shorts or low resistance. Never replace just the regulator and assume all is well. If all checks fine and the problem is still present, replace the alternator or at least the rotor in the alternator.

One specific case that can mean trouble if not recognized relates to the Prestolite 100-amp alternator, and this is common to many twin-engine aircraft. This particular alternator has an isolated ground—which means that its case is not common with the ground terminal. Therefore, if with this alternator a regulator is shorted or the field fuse is blown, check for a short between the field terminal and part of the case that the rotor shaft is in continuity with. If the continuity is checked to the output ground terminal, a short may not reveal itself. Always keep in mind to compensate for resistance that is present due to brushes and dirty slip rings in the alternator.

Another often reported problem is severe fluctuations in the ammeter movement. Assuming that the meter is good and if it is used with a shunt, the shunt should be checked for tight and clean connections. If all of this has been checked and the problem remains, then check to see if the rate of movement is consistent with engine rpm.

Also, check to see if the amplitude of the movement is reduced if more load is added to the system. If the answer to these is yes, the problem is more than likely a diode problem in the alternator. Under normal operation, there's always some ripple voltage at the output of the alternator and is normally dampened by the capacitance in the battery. But if one or more of the diodes in the alternator is shorted or not switching properly, the output will basically be an AC voltage wave riding on top of a DC signal. If the AC portion is strong enough, the load current will follow the voltage (partially sine wave), hence, the ammeter will follow also.

Sometimes the alternator checks good under a load. If this is the case, then adding some filter caps to the alternator positive output terminal may help. If not, replace the diodes in the alternator. This type of problem will also add noise to the system and can sometimes be eliminated by adding some capacitance to the alternator output.

Noise can also come from a voltage regulator if it is a Type II regulator. The main trade-off between the Type I and Type II regulators is heat vs. noise. Many Type II regulators are switching regulators, and the switching action introduces noise into the system. This is not a problem with Type I regulators, but switching regulators will operate much cooler and require less space and ventilation for heat dissipation. This is something to keep in mind when dealing with a noise problem or when selecting a regulator for a new aircraft design.

Accessory Technology

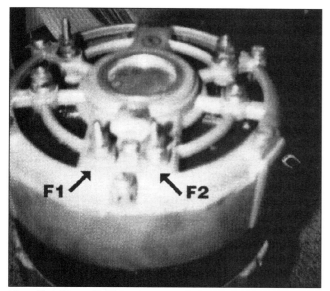

Type I or Type II. This alternator is a Prestolite 100 amp with isolated ground.

When checking through an electrical charging system, it's important to keep in mind the requirements of the components that make up the system. For example, the voltage regulator only does what it is told to do. For the regulator to do its job properly, it must have access to the voltage that is present at the alternator output. If there is a great difference between these two voltages (anything greater than half of a volt), the regulator will deliver incorrect field voltage to the alternator. Most of the time this will result in an overvolt situation.

There is a path for current between the alternator and the input (bus) to the regulator that we call the regulator feedback circuit. There are several components in series making up this circuit: components such as shunts, main output circuit breakers, master switches, alternator switches, field switches, overvolt relays. These will vary depending on the type of aircraft. If any of these components offer more resistance than normal, they will also contribute a voltage drop across that particular component, robbing the voltage regulator of the signal it needs to operate properly.

This happens more often when the aircraft is older and the panel switches become worn and pitted. Therefore, always check the voltage at the alternator and compare it with the voltage at the input to the regulator. If they differ, measure voltages starting at the alternator. Work toward the regulator and you will find, along the way, the component giving you the problem.

Finally, one of the most frustrating problems to technicians is the task of paralleling alternators on twin-engine aircraft that utilize the paralleling type voltage regulators.

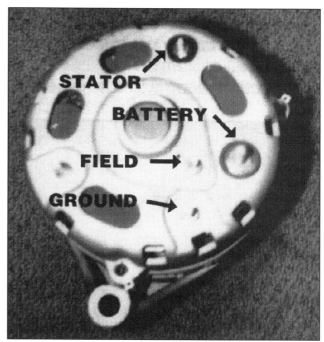

Type I or Type II. A Prestolite 100-amp alternator with isolated ground.

These are Type I systems and are used extensively on aircraft made by Cessna, Beech, Piper, Aerostar, Grumman, Rockwell, and a few others. On these systems, there are two voltage regulators (one for each alternator) tied together by a single wire for balancing purposes. These types of regulators are designed to balance the load current on the two alternators in the system. They are designed based on the idea that all things are equal between the two separate sides of the electrical system (i.e., alternator internal characteristics, wiring, engine rpm, etc.).

The regulators balance by means of field voltage balancing. This is very similar to a water pump that has one output pressure and is pumping water through two hoses of equal diameter and length. Under these conditions, the water flow at the end of the two hoses will be the same. If anything in one of the hoses changes so that it is no longer exactly like the other hose, the flow at the end will no longer be the same. All the while there is nothing wrong with the pump. The problem the technician faces occurs when changing a component in a twin-engine system and only replaces the component on one side; the system will no longer be balanced. This is most apparent when the component changed is the alternator.

Accessory Technology

The left and right alternators will no longer have equal output. This is not really a problem but rather a fact that can be dealt with in a limited manner. One of the first things to do is to make sure that each of the alternators and its associated components for that side of the aircraft work fine by themselves as far as voltage adjustment and load carrying capabilities. Once this has been established then they should be paralleled as close as can be.

This will more than likely not be ideal but it is OK. If one alternator will not come on line at a small load, check to see at what load demand on the system will cause it to come on line and share part of the load. The rule of thumb is 30 to 40 percent of the system load capabilities. This may not be acceptable for some people. If not, replace the brushes and clean the slip ring on the old alternator.

If this does not quite satisfy the operator, let the new alternator gather some time on it so that the brushes will properly seat. The imbalance in the system will probably improve with time.

All components of the system, such as switches, connections, circuit breakers, relays, etc., can affect the balance of the system and should be checked to alleviate some of these problems. *November/December 1995*

Robby Starr is an electrical engineer for Aero Electric in Wichita, KS. He has been with the company since 1980.

Accessory Technology

Battery care
Are you aware?

By Joseph F. Mibelli

Let's face it—a battery—even on aircraft, does not rank high on most people's list of important items to take care of—until it fails that is. Then, all of a sudden the failure generates an instant awareness and a dire need to be educated on the subject of battery care—this, after first recuperating from the minor shock caused by the cost of battery overhaul or replacement.

The batteries in an aircraft are part of an overall system that includes emergency backup for various items, some of them very critical. The main battery in a small aircraft is used for engine starting and for main emergency backup. In larger aircraft, the main battery is used to start the APU and also for main emergency backup. Smaller battery packs are also used to provide emergency illumination and emergency power backup for avionics.

Battery maintenance, then, has to be considered as important (more in some cases) as the maintenance for engines, structure and other vital parts. Improper battery care can result in problems that range from nuisance to deadly.

Even if battery problems don't generate life-threatening situations, the cost of overhauls or outright replacements can be very expensive. Unlike the typical car battery, the aircraft battery is a precision device and must be serviced accordingly.

The main battery in the majority of larger aircraft is a nickel-cadmium type, with sealed lead-acid gradually gaining acceptance, particularly in the smaller aircraft. But, regardless of the technology, all batteries are required to perform the same task: supply current when required.

This being the case, how do we know that the battery will deliver the required amount of power when required?

Bench testing, under specific conditions, as set forth by the battery manufacturer, is the only reliable way to determine the condition of a battery.

There are no simple direct measurements, such as placing a voltmeter across the terminals, to determine the condition of the battery. The voltmeter reading may tell us something about the state of charge (with an enormous margin of error), but it cannot tell us how well the battery will deliver current when demanded. This is particularly true for nickel-cadmium batteries that have a very shallow discharge curve, but it is also true for lead-acid batteries, even though they have a more pronounced discharge profile.

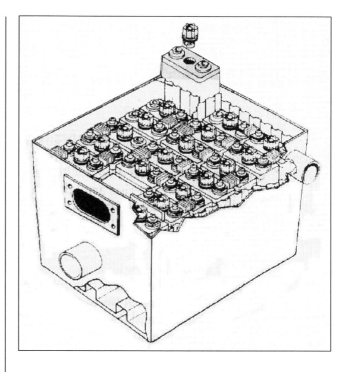

The most basic and crucial bench test is the capacity test. This test determines if the battery will deliver the rated current in the minimum time interval, while maintaining a terminal voltage above some minimum.

A typical 40A-Hr battery is required to deliver 34A (85 percent), for one hour with a terminal voltage of no less than 20 volts.

The currents under real conditions such as engine starting are many times higher, or they can be many times lower, such as in backup-power situations.

The bench test, even though not a realistic condition, gives us a reliable and uniform method to determine the condition of the battery.

As with any kind of a test, where the resulting numbers will lead to a pass or fail decision, the equipment used to test the batteries has to be of instrumentation quality. The equipment must allow the operator to program and monitor the test parameters as set by the manufacturer.

Why do batteries have to be tested in the first place?

Accessory Technology

Just as with any other part of the aircraft, normal use wears out the various elements of the battery system (plates, separators and electrolyte), which slowly reduces the capacity of the battery.

In addition, normal use under extreme conditions, abuse and improper maintenance will significantly shorten the life of the battery. Of these, inadequate maintenance can become the major cause of short battery life.

The proper approach for long battery life begins with periodic maintenance, most of the times at intervals much shorter than those established by the aircraft manufacturer.

Test intervals, as given by aircraft manufacturers, are for reference only. The actual time between tests has to be established by usage (number of starts) and by the results of bench tests (water consumption).

Battery abuse

Reducing usage under extreme or abusive conditions may not be as easy for the pilot as it sounds. The following are examples of battery abuse and some ways to limit the abuse of the battery:

First of all, what constitutes usage under extreme or abusive conditions?

The starting of engines is the mechanism that rapidly wears out the main batteries. So you need to communicate to the pilot this point.

The peak discharge current, often exceeds 1,000A and lasts for several seconds. It is followed by a drain of several hundred amps, lasting for several tens of seconds. Finally, as the engine fires up, the battery is hit with a high recharge current as the starter motors become generators.

Furthermore, usage of the batteries under extreme temperature conditions further contributes to the wearout mechanism. Multiple short flights (less than 30 minutes) where there is not enough time to replenish the charge used to start the engines also contribute to the wear of the battery.

Remember also that the batteries are used to power other electrical devices, such as air conditioning, while the aircraft is on the ground. When the engines are at idle, the rpms are not high enough to allow the generators to replenish the current consumed by the equipment.

If the batteries are not properly recharged, they will have less and less charge reserve every time they are asked to deliver starting current. If this practice is continued to the point where one or more of the cells reach full discharge, subsequent starts will cause those depleted cells to reverse, severely cutting the overall available capacity and also reducing their useful life.

The engines will also suffer with batteries in poor conditions. With a battery that cannot deliver proper power, the result is a hot start, which results in costly engine overhauls (well in excess of the cost of maintenance or replacement of the batteries).

The simplest and best solution to this problem is to use ground power equipment for engine starting. The next option is to shorten significantly the period for testing of the batteries. If the manufacturer of the aircraft calls for battery testing every 100 hours, reduce the interval to 75 hours.

Better yet, alternate electrolyte check and deep cycle every 50 hours for best performance. This, at least, will ensure that the batteries are in proper conditions to meet the heavy demand.

Another condition that can shorten the life of the battery is overcharging while in flight. Batteries can be literally cooked when subjected to overcharge by an improperly set or malfunctioning voltage regulator. This condition can also lead to catastrophic results. The battery can go into thermal runaway with heavy release of toxic fumes, and it can also explode if the charge is not terminated.

Emergency batteries, such as used to power the artificial horizon, lights and other types of essential equipment, are normally not in use. They remain in standby under continuous charge while the aircraft is active, and they do not deliver current unless called for in emergencies or routine equipment tests.

Batteries that are subject to such a state of inactivity will gradually degrade in capacity and will fail miserably when called upon to deliver power (such as in case of a main power failure).

This type of failure is known as capacity fading, and can be prevented by testing all emergency batteries (capacity test, deep cycle) as often as the main batteries are tested.

Failure to do so will result in little or no backup power under emergency situations (the consequences of which need no additional reminders).

When such batteries are taken for service after prolonged periods of inactivity, the cells may be so degraded that no amount of deep cycling will restore the required capacity. Thus resulting in costly overhauls.

Inactive batteries

When an aircraft is brought in for routine service, the main battery and the many additional batteries on the aircraft may also be serviced. No problems here.

What constitutes a problem for batteries occurs when the aircraft is going to be inactive for a prolonged period of time. Depending on climatic conditions (usually high temperature), the batteries will begin to self-discharge, or worse, experience a severe loss (evaporation) of electrolyte. The time periods could be several months to as short as a few weeks.

If the time period is short, all that is needed is to top charge the battery (which includes verification of electrolyte level).

Batteries that have been "abandoned," however, will require a deep cycle to bring them back to proper levels. If the evaporation of the electrolyte has exposed the cell separators, the affected cells may be ruined and thus require replacement.

It's important, then, to remove and to service the batteries immediately of any aircraft that are expected to be inactive for a prolonged period of time.

If this is done, the batteries will be ready for service when the work on the aircraft is completed. Batteries that are serviced and are not immediately placed on the aircraft to be used will require a top charge prior to installation if they have remained on the shelf for more than two weeks.

One way to keep batteries in a ready-to-go state is to connect them to a trickle charger. Batteries can then be kept on standby for several months without any detrimental effects (the low charge current is controlled to simply offset the loss due to self-discharge).

Some final points

- Test batteries at the intervals given by the aircraft's manufacturer (the main batteries and the emergency batteries).
- Reduce the time periods if there is a heavy engine starting demand on the battery, or if prior test records deem it necessary (i.e., high water consumption and loss of capacity).
- Reduce the engine starting load on the battery by the use of ground support equipment (battery carts).
- Service and properly store batteries that are not immediately needed.

Joseph F. Mibelli is president of JFM Engineering Inc. in Miami, FL.

A new approach to battery maintenance
State-of-the-art documentation

By Scott Marvel

An aircraft technician must keep accurate battery maintenance records in order to support a battery's release to service. But who has the time or desire to babysit batteries?

Does this sound familiar? After making the visual inspection, you attach the charger/analyzer to the battery, set the controls and the egg timer, and then wait until it's time to conduct the cell voltage measurements. Now, grab the clipboard (where did my pen go?), probe 20 cells with the multimeter, record each cell's voltage, reset the egg timer and repeat. Is this the best use of your time?

Or how about this—you're inside the aircraft. Back in the battery shop, a timer goes off to tell you it's time to read the cell voltages. Trouble is, you are so far inside the plane that you can't hear it go off! The result—lost time while you reset the charger/analyzer, and lost efficiency while the battery sits on the bench.

This can be a repetitive and frustrating job. Wouldn't it be great if it could be automated?

The challenge

The challenge aircraft maintenance technicians face is that the most common way to measure and record cell voltages in a nickel-cadmium battery is by hand. With a multimeter, the technician measures voltage levels cell by cell during battery servicing.

The technician must also interpret the voltage measurements to determine if each cell is serviceable, which is fundamental to flight safety—and vulnerable to human error.

Don't forget lead-acid batteries. Although the individual cells are not accessible, documenting the battery service data is just as important in determining the battery's condition.

Manufacturers recommend that you measure and log cell voltages at the following points during battery servicing: 1) start of topping charge, 2) end of topping charge, 3) end of capacity check, and 4) end of final recharge. Voltages on lead-acid batteries should be measured at the end of the charge and cap check.

But, these measurements must be taken within a specific, critical time frame, which requires constant vigilance that's not always possible in a busy aircraft maintenance shop.

Furthermore, after the measurements are taken, the aircraft maintenance technician must evaluate them to identify faults and take corrective action. If the measurements are not taken within the specified time frame, the technician's ability to identify possible critical faults is impaired—he just doesn't have the right information at hand.

Accessory Technology

Capacity and the four faults

What are these cell voltage measurements used for? First, they are used to determine the ampere-hour capacity of the battery, and second, to detect the four most common fault conditions—gas barrier fault, cell reversal danger, cell balance fault, and cell overvoltage fault.

For a nickel-cadmium battery to be certified as airworthy, or "serviceable," the battery must yield at least 85 percent of its rated ampere-hour capacity during the cap check. Each individual cell must measure at least 1 volt when the battery has delivered 85 percent of its rated ampere-hour capacity.

For example: All 20 cells in a 40 ampere-hour battery must measure at least 1 volt after being discharged at 40 amperes for 51 minutes (85 percent of one hour).

What are the fault conditions, exactly?

Gas barrier fault (negative slope)

During charge, any cell's voltage that peaks and then drops by 0.10 volt may have a damaged gas barrier membrane. This is a nonrepairable condition and the cell must be identified for replacement.

This condition is particularly important to detect because it could lead to thermal destruction of adjacent cells, or even catastrophic failure of the entire battery if charging is allowed to continue. Early detection of gas barrier damage can significantly reduce battery maintenance costs.

Cell reversal danger

During the capacity check, if any cell drops below 1 volt before the battery has delivered 85 percent of rated capacity, then that cell has failed the cap check. If no action is taken, and the voltage is allowed to drop below zero, the cell can be driven into a reverse-polarity condition that can cause permanent damage. When this fault condition is identified, the cell or cells should be clipped off with resistor clips, or shorted out with shorting straps to prevent cell reversal.

Cell balance fault

Any cell that is not within 0.05 volt of the average shows that the cell is out of balance. This indicates the battery may need to be deep-cycled to restore cell balance. Manufacturers recommend all of the cells be within 0.05 volt.

Cell overvoltage fault

Any cell that exceeds 1.70 volts during charge may have a low electrolyte level, or it may be shorted. Continued charging of this cell will cause permanent damage.

When a cell overvoltage fault is detected during the initial charge, a small amount of distilled water should be added, and the cell voltage should be measured again. If the voltage then does not fall below 1.70 volts, that cell should be identified for replacement.

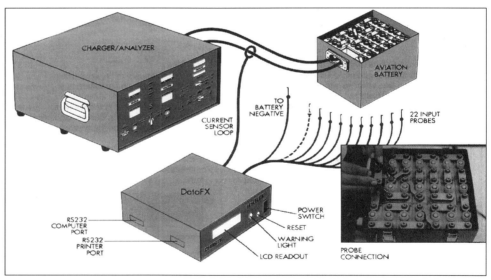

An example of a currently available battery monitoring system is Christie's DataFX™ aircraft battery measurement and documentation system.

Benefits to automation

Monitoring and recording cell voltages, detecting fault conditions, and measuring cell voltages at fixed intervals are naturals for automation. And no matter how attentive the operator may be, battery measurement and documentation is subject to human error. Automating the process would not only increase accuracy and safety, but it would also reduce labor costs.

With an automated battery monitoring system, a consistent maintenance method can be established. When this type of system is used, the battery servicing, data collection and documentation processes all become standardized, and the variations otherwise inevitable between different technicians are eliminated.

With an automated battery monitoring system, charge/discharge rates are documented electronically, faults identified, ampere-hours calculated and cell voltages recorded for evaluation. This increases credibility with the FAA, your boss, your customer and the battery manufacturer when submitting warranty claims.

The question remains, what is the present state-of-the-art in automated battery monitoring systems?

State-of-the-art battery monitoring

An example of existing state-of-the-art battery monitoring is installed at the Aerospace Energy Systems Laboratory (AESL) at NASA's Dryden Research Facility, Edwards Air Force Base, California.

This facility supports a variety of battery types for Dryden's flight research aircraft, which range from an SR-71 "Blackbird" to a Boeing 747.

The AESL is a computerized aircraft battery servicing facility that utilizes a distributed digital system which consists of a central computer and battery servicing stations, connected by a high-speed serial data bus. Each battery station contains a digital processor, data acquisition capability, digital data storage and operator interfaces.

The central computer provides data archives, manages the data bus, band provides a time share interface for multiple users.

While this setup is appropriate for flight test work, it's not commercially available.

Commercially available systems

So what's practical for the commercial market? Remember, an automated battery monitoring system should measure ampere-hour capacity, detect gas barrier damage, cell reversal danger, cell balance and cell overvoltage faults and be compatible with all battery types and charger/analyzers. Ideally, the system should print out a report tabulating all of the important data about the battery.

A data interface is also desirable, which opens up various possibilities for data logging and multiple-unit operation. Several charger/analyzer stations could be integrated through a PC with software that permits the operator to view, print or store battery servicing reports.

The next step

So what's in store for the battery shops of tomorrow? Development of a true integrated battery monitoring system. This system would consist of state-of-the-art charger/analyzers, real-time battery monitoring, data gathering, analysis and archiving, and specialized reporting capabilities. It's only a matter of time before we see systems like this hit the market. Stay tuned!

Scott Marvel is the director of marketing and sales at Christie Electric, (310) 715-1402.

Accessory Technology

A closer look at inspecting aircraft windows

By Andrew E. Geist

With a recent increase of crazing and other damage to acrylic aircraft windows, many technicians express concern over what preventive maintenance can be completed to help their windows last longer, and hopefully avoid premature costly replacements.

The common question asked is, "What can be done to increase the serviceable life of our aircraft windows?" The key is being educated about what to look for, looking for the warning signs of trouble, and finally following some simple window maintenance tips:

What to look for

There are many different elements and causes of window damage. By far, the most common is crazing. There are numerous types and causes of crazing which include volcanic, chemical, and age crazing. The most prevalent type seen at the corporate and airline level is caused by volcanic crazing. Crazing can be defined as fine micro-cracks on or extending beneath the surface of the acrylic. When the combination of the outside applied forces to the window surface exceeds the tensile strength of the plastic, the surface cracks and crazing begin. Any additional forces applied to the window just accelerate the rate at which the window crazes. Crazing is easily identifiable as a glaring reflection of light from the surface of the window.

Possibly the best place to begin this education on how to inspect for damage is by reading Chapter 56 of the manufacturer's maintenance manual. Many of the manufacturers have revised this section recently with new data regarding what to look for and an explanation of the damage. Most of the information contained within this section will describe acceptable damage criteria and immediate attention items.

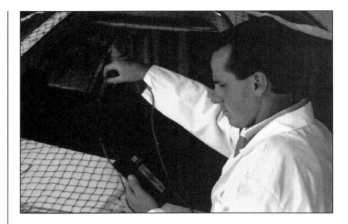

Several inspection methods including ultrasonic (above) and optical micrometer (below) are used to check window thickness before, during, and after repair.

With the manufacturer's description of damage and a note pad in hand, it is now time to inspect each window of the aircraft. Remember, the key is to take your time and not be rushed. The following procedure for inspecting for damage works very well:

1. Remove from your person all articles that could scratch during your inspection (i.e., rings, watches, buttons, etc.).
2. Use a high intensity light. (500 watt is suggested.)
3. Hold the light at a 45-degree angle to the exterior window surface, or have an assistant do the same with the inspector sitting on the inside of the aircraft
4. Slowly move your head around to change your vision perspective and angle.
5. Take your time and look for any reflections of light, scratches, pits, cuts.
6. Document any damage noting the date and aircraft time; then consult Chapter 56.

Once you have started a maintenance tracking log for your windows, you can check them on periodic intervals throughout the year to note any changes. An industry suggestion is to inspect windows at least annually and/or every 250 hours. The maintenance manual, of course, takes precedence.

Window maintenance tips

Following these simple guidelines will help save your organization expensive window replacements and downtime:

- Clean your windows often using water and your hand to clean off any debris, as this will not scratch.
- Polish your windows often using approved products outlined in Chapter 56.
- Finally, repair or have damage repaired as soon as detected. Waiting will only greatly decrease your chances of saving a window.

September/October 1995

Andrew E. Geist is general manager of AeroScope Inc. in Broomfield, CO.

Accessory Technology

Certified vs. qualified welders
If you're welding aircraft parts — you need to know the difference!

By Greg Napert

Aviation maintenance technicians are faced with new products and newer technology every day. With this comes increased responsibility.

The question many face today on the subject of welding is whether or not they should use their knowledge and sometimes limited skills to make repairs on aircraft. As in any profession, there are some talented technicians who have taken it upon themselves to learn the skill of welding to a degree that is suitable for the application.

However, for every one of those skilled individuals, there are dozens of technicians who never have the opportunity to become proficient at welding. The basic skills that are taught in A & P school are not meant to make the technician a professional welder, only to assist the A & P in identifying welding processes and make the technician aware of acceptable vs. unacceptable practices.

There are also regulatory considerations that need to be taken into account when determining whether or not to take on the challenge of performing a weld repair. For instance, if you're performing major repairs, you may need access to approved data, which may prove difficult to obtain.

This combined with the skill question often leaves a technician with two choices: Send the part(s) back to the manufacturer or manufacturer/FAA-approved facility to be repaired, or in the case of minor repairs or preapproved field approvals, bring in a welder who is qualified to perform the repair.

Qualified vs. certified welders

Certified welder is a loosely used term according to Ron Scott, director of skill training for the Hobart Institute of Welding Technology.

Normally, certifications are only good where the welder is employed. That is because certification is not only based on skill, but on materials and process. In other words, you are certifying the results of a welding process as employed by an individual. Someone who is certified on titanium is not certified for cobalt or stainless steel, and a welder certified with one process may not be certified with another process.

That's why most welding certifications, as a general rule, are only good at the facility where in use since it also includes the welder's skill, process and type of material being welded.

Exceptions are the individuals who have a mobile company, such as welding equipment mounted on a truck. Typically these companies are incorporated and have certification(s) for the type of welding done in conjunction with the processes and materials used. Rather than looking for certified welders, Scott suggests companies seek qualified welders who are capable of being certified and then get those individuals certified at their facility.

To become certified, says Jim Collins, certified welding inspector for HIWT, "A welder must undergo recurrent training and be part of a program where they are welding every day to maintain skill levels. Most welders in the industry are certified to a general specification called MIL STD-1595A which was designed for aircraft. This certification requires the welder use the process at least once every three months in order to keep current. An eye examination is also required and 20/30 vision necessary to pass. Reexamination of the eyes every five years is a requisite. The standard also eliminates the three-month requirement if retesting for certification is performed every two years."

Scott says short of qualifying an individual at your facility you should be sure of the following items before contracting someone to perform a weld repair:

- The individual is proficient with the specific type of material being welded and the welding process being utilized.
- The individual has documentation to show they qualify for the material and process.
- The individual has a track record and is currently using the applicable materials and process.

Scott suggests sending weld samples to a lab for examination whenever you contract with someone who has not been certified at your facility. There are several labs located around the country which will examine weld samples and determine whether they conform to code.

"The bottom line is if you are using a contractor to perform a critical weld, you must question their credentials and possibly take the time to have them qualify to your company standard or other code," explains Scott.

Welding processes

Contrary to what you may have heard, the size and/or cost of a welding machine does not make welding any easier. Large or small size of machines will not eliminate the need for excellent welding skills. Scott tells us that the same basic welding principles and talents are still required whether using a $3,000 machine or one that costs $500.

He explains that the primary and preferred process used in the aviation industry today is gas tungsten arc welding, more commonly referred to as TIG welding. (TIG stands for tungsten inert gas.) Other brand names encountered are Heli-arc and Heli-weld. These names were all derived from the early TIG welding being done with helium shielding gas. Today argon shielding gas is used in the majority of applications due to it being less expensive.

There are many different materials which may be involved in aircraft welding, several of them considered exotic. Titanium, inconel, stainless steels, monels and aluminum all require special skills and knowledge compared to welding mild steel.

"You need basic skills to weld any type of material; however, additional knowledge is necessary when specialty materials are encountered because of unique requirements. For instance, if you are working with titanium, there are special gas purging procedures that must be used to shield the back side of the item being welded to prevent oxidation. There are also trailer gas nozzles that keep the heated titanium shrouded in shielding gas so it does not oxidize. The shielding of titanium is critical during the cooling phase because it will oxidize at relatively low temperatures," says Scott.

Aluminum, stainless steel and inconel are particularly susceptible to oxidation and must be protected by a gas shield. If the weld penetrates 100 percent through the parts being welded, then the backside of the joint should be protected by gas shielding.

"There are certain skills welders doing aircraft repair welding must have that are entirely different from those required in other industries," states Scott.

"I cannot emphasize enough how important it is to know what you are doing. There are so many persons who proclaim to be professional welders and perform inferior welds by not utilizing back purging or the right process," he explains.

Another good example of knowing what you are doing involves the use of wire-feed welders which are quite popular in the industry. Collins says, "The problem with inexperienced welders using wire feed welders is the possibility of cold lapping. When cold lapping occurs, welds may look good, but in reality what happens is inadequate penetration into the base metal. The puddle of molten metal just lies on the surface of the base metal and lacks strength and integrity.

"These gas metal arc welding machines utilize a continuous wire feed electrode which is deposited as filler metal in the joint being welded. Because of the reasons stated above, these machines require highly skilled welders who are able to maintain correct travel speed, 'stick-out' distance, and gun angle and remain on the leading edge of the molten puddle to prevent insufficient penetration and cold lap," he says.

Gas tungsten arc on the other hand uses a nonconsumable tungsten electrode, and filler metal is added by hand. This type of welding requires better hand-eye coordination since both hands are used, one for holding the torch, and the other for feeding filler metal into the puddle. Usually much better welding results are obtained with gas tungsten arc welding (GTAW) due to the increased control by the welder.

Collins says that "welding titanium with GTAW process may often require welding with the entire part enclosed in an inert environment, or with a fixture that feeds inert gas shielding to the back side of the joint. Stainless steel and inconel present the same problem and must be protected by a back purge. If protection is not provided to the backside of the joint on these materials, carbide precipitation or sugaring will occur causing the weld to be defective.

"We would also suggest purging of the atmosphere and other contaminants from the backside of any thin materials that will have 100 percent penetration. An example is thin wall tubing that is completely penetrated. Purge the interior of the tubing during welding with a shielding gas to assure that oxidation does not take place," comes as a recommendation from Collins.

Chrome molybdenum (chrome-moly) is a commonly used metal tubing for fuselage trusses. Scott says that chrome-moly tubing is sensitive to cracking and should be preheated prior to welding. In the past oxyacetylene welding has been the most common method of joining this material. However, GTAW does an excellent job and provides more control with less contamination.

"It's not that you can't do an adequate job with oxyacetylene," says Scott. "I am a firm believer that oxyacetylene is the foundation for all welding processes. Regardless of the type of welding you do, the important thing is being able to read and control a puddle of molten metal and oxyacetylene teaches a person to do that," states Scott.

Scott is very adamant in stating, "Welding is a science combined with art. To do it properly takes education, training, and practice. Unqualified welders perform a disservice to the ones who are highly skilled and qualified in the industry. In addition to that, they endanger the lives of people who place their trust in aviation maintenance technicians."

Accessory Technology

Causes and cures of common welding troubles

Poor penetration
Why
1. Travel speed too fast
2. Welding current too low
3. Poor joint design and/or preparation
4. Electrode diameter too large
5. Wrong type of electrode
6. Excessively long arc length

What to do
1. Decrease travel speed
2. Increase welding current
3. Increase root opening or decrease root face
4. Use smaller electrode
5. Use electrode with deeper penetration characteristics
6. Reduce arc length

Magnetic arc blow
Why
1. Unbalanced magnetic field during welding
2. Excessive magnetism in parts or fixture

What to do
1. Use alternating current
2. Reduce welding current and arc length
3. Change the location of the work connection on the workpiece

Inclusion
Why
1. Incomplete slag removal between passes
2. Erratic travel speed
3. Too wide a weaving motion
4. Too large an electrode
5. Letting slag run ahead of arc
6. Tungsten splitting or sticking

What to do
1. Completely remove slag between passes
2. Use a uniform travel speed
3. Reduce width of weaving technique

Porous welds
Why
1. Excessively long or short arc length
2. Welding current too high
3. Insufficient or damp shielding gas
4. Too fast travel speed
5. Base metal surface covered with oil, grease, moisture, rust, mill scale, etc.
6. Wet, unclean, or damaged electrode

What to do
1. Maintain proper arc length
2. Use proper welding current
3. Increase gas flow rate and check gas purity
4. Reduce travel speed
5. Properly clean base metal prior to welding
6. Properly maintain and store electrode

Cracked welds
Why
1. Insufficient weld size
2. Excessive joint restraint
3. Poor joint design and/or preparation
4. Filler metal does not match base metal
5. Rapid cooling rate
6. Base metal surface covered with oil, grease, moisture, rust, dirt, or mill scale

What to do
1. Adjust weld size to part thickness
2. Reduce joint restraint through proper design
3. Select the proper joint design
4. Use more ductile filler
5. Reduce cooling rate through preheat
6. Properly clean base metal prior to welding

Undercutting

Why
1. Faulty electrode manipulation
2. Welding current too high
3. Too long an arc length
4. Too fast travel speed
5. Arc blow

What to do
1. Pause at each side of the weld bead when using a weaving technique
2. Use proper electrode angles
3. Use proper welding current for electrode size and welding position
4. Reduce arc length
5. Reduce travel speed
6. Reduce effects of arc blow

Distortion

Why
1. Improper tack welding and/or faulty joint preparation
2. Improper bead sequence
3. Improper setup and fixturing
4. Excessive weld size

What to do
1. Tack weld parts with allowance for distortion
2. Use proper bead sequencing
3. Tack or clamp parts securely
4. Make welds to specified size

Spatter

Why
1. Arc blow
2. Welding current too high
3. Too long an arc length
4. Wet, unclean, or damaged electrode

What to do
1. Attempt to reduce the effect of arc blow
2. Reduce welding current
3. Reduce arc length
4. Properly maintain and store electrodes

Lack of fusion

Why
1. Improper travel speed
2. Welding current too low
3. Faulty joint preparation
4. Too large an electrode diameter
5. Magnetic arc blow
6. Wrong electrode angle

What to do
1. Reduce travel speed
2. Increase welding current
3. Weld design should allow electrode accessibility to all surfaces within the joint
4. Reduce electrode diameter
5. Reduce effects of magnetic arc blow
6. Use proper electrode angles

Overlapping

Why
1. Too slow travel speed
2. Incorrect electrode angle
3. Too large an electrode

What to do
1. Increase travel speed
2. Use proper electrode angles
3. Use a smaller electrode size

July/August 1995

Accessory Technology

Starter-generator overhaul

By Greg Napert

The starter-generator has a tough job to perform. It must produce no load speeds of 3,500 to 4,500 rpm and produce up to 300 amps during the entire operating cycle of the engine. In addition, it is expected to last nearly 1,000 hours of operation, as some turbine engines are running with TBOs (time before overhaul) of over 4,000 hours.

Joe Megna, maintenance manager for Jet Air Corp. in Green Bay, WI, a business jet and general aviation maintenance facility that has decided to overhaul starter-generators to service its customers, says that even though the overhaul intervals run around 1,000 hours, the brushes won't make a thousand hours. "The brushes average, depending on what model starter-generator they're used on, somewhere in the neighborhood of 400 hours—higher if it's turning a free-spooling turbine such as the Pratt & Whitney PT6. On engines such as the Garrett 331 engine, however, you're turning the prop, gearbox, and accessories. This is much harder on the brushes and on the starter."

Megna says the starting cycle is where most of the wear and stress is placed upon the starter. And it's here where technicians should warn pilots to adhere strictly to manufacturer's start and cooling cycles.

The overhaul process

The primary starter/generator that Jet Air Corp. overhauls, says Megna, is the Lear/Siegler/Lucas Aerospace.

"Basically," says Megna, "when we receive a starter-generator for overhaul and/or repair, we make sure everything is there and observe the condition for warranty claims. We completely tear down the unit and go through the air inlet, the brushes under the cover, drive shaft, dampner hub., etc. Then we'll look at the stator subassembly and (if applicable) the rpm pickup; then we clean, inspect, and test them, prepare all good components, assemble them, and test and undercut the commutator. We then balance each and every armature. Elimination of vibration is critical to the longevity of the starter.

"Once we're done reassembling the starter, we have run sheets designed by the manufacturer that we must follow to run the starter/generator," says Megna.

A typical starter-generator run-in requires a variable speed test stand capable of driving the starter-generator at speeds of 6,000 to 12,000 rpm at rated load, and 13,000 rpm at no load. It also needs to be equipped with suitable instrumentation to measure torque, speed, voltage, current and temperature, and the power supply must have a capacity of not less than 700 amperes at 10 volts.

Additionally, the ambient temperature must be controllable, and a means for providing cooling air must be provided.

Lead technician Jon Perlberg at Jet Air Corp. says, "Essentially, you're trying to simulate conditions of the starter-generator when it's in the aircraft. When the starter-generator is in the aircraft, a cooling duct is attached and ram air provides cooling. So that's exactly what we do on the test bench. We also perform tests with the ducting detached and allow it to self-cool. This simulates idling conditions where the aircraft is not being provided with ram air and has to pump air with its own fan. We have a checklist that we go by and one of the requirements is that we run it self-cooled."

Megna says one area where there's a lot of wear and tear is in the starter drive. Garrett went to a new style with a Torlon® insert that doesn't require lubrication. That way, after so many hours, you just replace the insert. You don't have to replace the entire starter drive or the gear on the gearbox of the engine, which sometimes requires tearing down the gearbox. Megna says sometimes you can get the starter-generator drive out of the engine without tearing down the gearbox, but not always. There's a bearing into which you install a puller, but if it locks up at all, you end up tearing down the gearbox.

Megna says, "Although the company performs the complete teardown and assembly, there are a couple of starter-generator components we've found that are more beneficial to send out for overhaul. We typically send out the field assemblies to be overhauled individually. There are places that have the equipment to overhaul them much more efficiently than we ever could.

"We also send out the armature for overhaul. The armature is always balanced as part of the overhaul process, but when we get it back, we always take the time to rebalance it to ensure that it's perfect. We feel this is a really critical stage in overhauling the starter-generator. Balance is too important to make assumptions."

Replacing brushes the right way

Megna says, "There are many people who don't have the resources to disassemble the unit to replace the brushes, and they just replace the brushes on the aircraft. And you don't necessarily want to overhaul the unit every 400 hours when the brushes wear out.

Accessory Technology

"But my feeling on replacing brushes is you just can't do it properly without turning the armature, which requires disassembly of the starter-generator. Putting a set of new brushes on a grooved armature just doesn't make sense. And as long as you are disassembling the unit to turn the armature, you've got to replace the bearings. Because using a bearing puller can damage the bearings, you don't want to take the chance of using the old ones and have it fail.
It's just worth the little extra time and a couple hundred dollars to do it right. The starter may not seem flight critical to many people, but if you're up there flying and you have an engine failure, you're going to have to rely on it as much as anything else."

He says that the two bearings in the unit basically support the whole armature which is spinning at a speed in excess of 12,000 rpm so they take a lot of abuse. "Many manufacturers recommend that you overhaul at 1,000 hours. Everytime our customers need brushes, we change the bearings. For relatively small expense, it's a good move.

"We had one like that on a Citation. The technician put a new set of brushes in but couldn't get it started. Well, it was because the brushes weren't making full contact. Even if the commutator appears to be smooth, the brushes might lie at a different angle and possibly don't make full contact," says Megna.

"Also if you're arcing excessively, you end up burning the commutator segments and may end up needing an armature overhaul," says Megna.

The acceptance run

The following is a list of typical acceptance run tests that are performed before an overhauled starter-generator is returned to service:

Minimum speed for regulation

The generator is operated at 13,000 rpm, 30 volts, no load, and the field current and voltage are measured to determine field resistance.

Typical starter-generator troubleshooting chart

Trouble	Possible cause	Isolation/correction
Excessive sparking at brushes	Shorted or grounded stator windings.	Check stator for shorts and/or grounds. Replace stator and housing assembly.
	Overload.	Check KW load against a generator rating and reduce load as necessary.
	Shorted or grounded armature winding.	Check armature for shorts and grounds. Replace defective armature.
	Damaged, worn, eccentric, or dirty commutator.	Visually inspect commutator and check total and bar-to-bar concentricity. Refinish commutator.
	Brush bounce.	Check bearings for excessive wear and/or damage. Replace bearings. Check armature balance. Correct balance.
	Improper brush seating.	Check brush spring tension. Replace brush springs. Check brushes for proper installation. Check for brush hang-up or stacking due to damaged brush holders. Replace damaged brush holders.
No power output	Sheared drive shaft.	Rotate/manipulate shaft by hand to verify. Inspect for internal debris, broken winding retaining band, seized bearing, or other damage that may have caused armature to jam.
	Defective stator windings.	Check stator for opens, shorts, or grounds. Replace defective stator and housing subassembly.
	Defective control unit in external circuit.	Verify acceptability of generator control unit. Replace defective control unit.
Overheating	Excessive load.	Check KW load against generator rating and reduce load as necessary. Check external circuits for shorts and grounds.
	Insufficient cooling.	Check for and eliminate any obstructions in air inlet and outlet areas. Check blast air supply and adjust if required.
Noisy operation and/or vibration	Interference between fan and air inlet adapter or cover.	Check for damaged or deformed air inlet adapter or cover. Check for deformed fan blades. Repair or replace as necessary.
	Loose components.	Check all accessible attaching screws and bolts and tighten as necessary.
	Worn or damaged bearings.	Replace bearings.
	Unbalanced armature.	Check and correct armature balance.
	Worn or damaged drive splines.	Replace drive shaft.
	Misalignment between starter and drive unit.	Check and correct alignment.

Continuous operating speed and equalizing voltage

The unit is operated at 12,000 rpm, 30 volts, and 300 amps until stabilized. Once the temperature is stabilized the output voltage must fall within specified limits. This is the normal speed and maximum output that the starter-generator runs on the aircraft. Before taking any measurements, you've got to stabilize the temperature of the starter-generator. That usually takes about 20 minutes of operation. At this point we should have no more than a 2°F rise in for each five-minute period of time. Also, we've go to make sure that the brushes are seated with 95 to 99 percent contact, depending on the model. We do this by pulling the brushes after a short run and observing the surface. If the brushes are shiny over a portion of the brushes, that's the portion that's making contact. Where they are dull looking, it is not contacting.

We can usually achieve full contact just through sanding and properly seating the brushes. But sometimes we run across a generator that requires an extended run-in before performing the acceptance test. So, we just run it till the brushes are seated fully. If it's so critical here, then you could imagine that it's just as critical you get full brush contact in the field also. That's why we don't recommend just replacing the brushes without anything else. You put them in, it goes a few turns and arcs and sparks, and you burn it up and end up paying for the full price of an overhaul. It just doesn't make any sense.

Minimum speed for regulation

The generator is operated at 6,550 rpm, 29.5 volts, and 150 amps. The field current and voltage should be within specified limits.

Minimum speed

The generator is operated at 6,100 rpm, 27.0 volts, 150 amps for 15 minutes, and the current and voltage must fall within limits.

Compounding

The generator is operated at 12,000 rpm, 30 volts, with loads of zero, 75, 150, 225, and 300 amps applied. The field current must rise accordingly with each load applied.

Commutation

Operate the generator at 12,000 rpm, 30 volts, and 300 amps. Commutation shall not exceed pinpoint sparking when observed.

Speed pickup test

The generator is operated at 6,000 rpm with field switch open. A 20K ohm ±10 percent load an oscilloscope is used to observe the out-put Voltage. The peak-to-peak voltage shall not be less than 2.5 volts nor more than 4.5 volts.

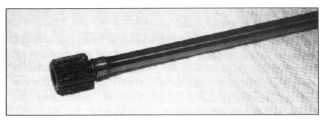

The starter shaft is under heavy load and the drive splines are often damaged. Inspect the splines carefully for wear, such as exhibited by the steps on these splines.

Dielectric test

The terminal block is removed and leads from the brush box filters are disconnected. With the starter-generator still hot from the previous operation, a test voltage of 220v AC, 60Hz is applied between the stator lead and the starter-generator frame for one minute. No evidence of insulation breakdown shall occur as a result of this test.

Locked rotor test

A jumper is installed as specified, and the shaft is restrained by a suitable torque measuring device. Increased voltage is applied until output torque is 20 pound-feet. The terminal voltage and line current are recorded and must be within limits.

Commutator runout

The armature is rotated on its own bearings and the bar-to-bar runout of the commutator is measured. It shall be within specified limits. *May/June 1995*

Accessory Technology

Phosphate ester-based hydraulic fluids
Demystifying Skydrol and HyJet

Although technicians are exposed on practically a daily basis to phosphate ester-based hydraulic fluids, many technicians never really understand the makeup, the reason for, or the consequences of exposure to this fluid. Skydrol®, manufactured by Monsanto, and HyJet-IV-A^{plus}, manufactured by Chevron, are the two most popular synthetic hydraulic fluids. They are essentially the same in makeup and chemistry, and the same general precautions should be followed when using either. The two products, in many cases, can even be intermixed and/or substituted, one for the other, on most aircraft.

The following material, obtained from the Skydrol Hydraulic Fluid Service Manual, is presented in the interest of giving a full understanding of the reasons behind using these hydraulic fluids, proper handling, servicing methods, proper safety precautions and effects of exposure:

Brief history of aircraft hydraulics

The first aviation hydraulic systems were used to apply brake pressure. These systems used a vegetable- or castor oil-based material and natural rubber for seals and hoses. As aircraft design produced larger and faster aircraft, greater use of hydraulics was necessary to operate landing gear, wing flaps and cowl flaps. This required higher pressures (1,000 psi) and improved fluids.

The industry turned to petroleum-based fluids such as Mil-O-3580 to meet these needs. Since the petroleum oil caused the natural rubber system materials to swell, the industry changed to synthetic rubbers such as neoprene and buna N. The newer and faster transport aircraft required hydraulic systems to work at even higher pressures (3,000 psi). And the location of hydraulic equipment exposed the fluid to many points of ignition within the aircraft. Both airborne and ground fires were traced to the use of this flammable hydraulic fluid.

Shortly after World War II, a growing number of aircraft hydraulic fluid fires drew the collective concern of the aviation industry and public. Operators of commercial transport aircraft had particular cause for alarm. The loss of life and equipment due to the flammable Mil-H-5606 fluid could not be tolerated. In 1948, Douglas Aircraft Co. requested that Monsanto Co. help develop a fire-resistant hydraulic fluid. The new resistant hydraulic fluid developed was based on phosphate ester chemistry—and named Monsanto Skydrol 7000 fluid.

Skydrol-compatible safety equipment is a must when working on or around hydraulic fluid and when servicing equipment.

As the transport industry moved toward jets, Skydrol 500A fluid was developed to meet the needs of the new aircraft. The development of more advanced aircraft required modification to the formulation of Skydrol hydraulic fluids. The changes required by the aircraft manufacturers were known as modifications to the fluid specification—or simply as Type I, II, III, and now IV fluids.

Monsanto Skydrol LD-4 and Skydrol 500B-4 hydraulic fluids are Type IV fluids formulated to the rigid specifications of the aircraft manufacturers.

Likelihood of encountering problems while mixing fluids

All Type IV fluids are miscible and compatible and may be used with each other in any and all proportions.

When mixing Type IV fluids with other type fluids (i.e., Type I, Type II or Type III), it's recommended that a minimum of 20 percent or more of the fluid be drained and replaced by the Type IV fluids. As only Type IV fluids have been produced since 1981, this situation is rare today.

Mixing of fluids

Mixing Type IV with:	Recommended Type IV fluids fluid concentration:
Types I, II, III (no longer produced)	20 percent minimum
Type IV	Mixable in all portions

Shelf life

If Skydrol fluids are protected from environmental effects, they do not have a limited shelf life. However, only the quart cans of Skydrol fluids are hermetically sealed and not subject to contact with the environment. Therefore, if there is a doubt about the condition of the fluid after three years of storage in containers other than the sealed quart cans, we suggest that an analysis be performed before the fluid is introduced into the aircraft. All fluids should be analyzed when in storage for up to five years.

Proper storage methods

The basestocks of Skydrol fluids are blends of phosphate ester materials which are mildly hygroscopic. Most problems associated with storage occur because of poor environmental conditions such as:

- Storage out of doors—exposure to weather
- Storage in buildings with high humidity levels

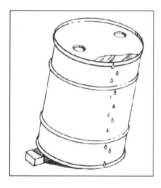

Storage recommendations

Skydrol fluid drum should be stored horizontally whenever possible. If the drums cannot be stored on their sides, they should be blocked up and tilted so that any water which may collect on the top does not cover the bungs.

Factors affecting compatibility with rubber-like materials

Swelling and softening of rubber compounds immersed in Skydrol fluids are greatly influenced by factors other than the type of rubber. Though a particular type of rubber may swell excessively, the compounded material may be entirely suitable for a given application.

Swelling and softening generally decreases as:

- percent of compounding material (sulfur, carbon black, zinc oxide, etc.) increases.
- amount of cut surface decreases (molded surfaces are more resistant than cut surfaces or buffed surfaces).
- hardness increases.
- mechanical restraint increases. (Restraint from surrounding structures or from fabric incorporated in the rubber will prevent or minimize swelling and softening.)
- temperature decreases. (Increased temperature greatly accelerates the effects of immersion.)
- mechanical working during immersion decreases.

These factors make it necessary to test borderline cases, under simulated use conditions.

Effect of fluids on tires

All hydraulic fluids, engine oils and cleaning fluids have a softening effect on rubber materials used in aircraft tires. Petroleum-based fluids have a more severe and permanent effect than Skydrol hydraulic fluids on natural rubber tread stock, commonly used on new tires. The reverse is true of most retread stock on which Skydrol hydraulic fluids have a greater softening effect. Both petroleum oils and synthetic fluids may cause soft spots in tires which will shorten their life. Neither should be permitted to remain on tires.

Protection of aircraft tires

Tires may be covered with polyethylene sheets to protect them from jet turbine oils or hydraulic fluids.

In the event fluid gets on the tire, it should be immediately removed by wiping the excess fluid with dry rags. If the tire has soaked in the fluid, it should be cleaned with soap and water and rinsed with clear water. Petroleum solvents should not be used for cleaning tires because the solvents will deteriorate the natural rubber. To remove tackiness, an alcohol may be used for cleaning.

Component cleaning solvents

Skydrol fluid components can be cleaned with nonhalogenated cleaning solvents. These nonhalogenated cleaning solvents do not contain chlorine, a serious hydraulic fluid contaminant. These solvents should be used in the cleaning of all critical hydraulic fluid components.

Noncritical hydraulic fluid components can be cleaned with halogenated solvents. The use of such solvents can produce a significant hazardous waste disposal problem. The use of such solvents is discouraged due to the waste disposal problems and the possibility of hydraulic system contamination.

Effect of Skydrol fluids on lubricants

Long experience with Skydrol fluids reveals no problems in using standard greases and lubricants in wheel bearings or other areas. If a leak occurs, dilution of the grease or pressure sealant with Skydrol will take place much the same as if mineral oil were used. The effect of diluting lubricant with Skydrol is comparable to diluting lubricant with mineral oil.

Testing and preservation of hydraulic units

Components for use with Skydrol fluids should be checked in a test stand equipped with Skydrol fluid. Either Skydrol LD-4 or 500B-4 fluid may be used in the test stand. If the test fluid is not the same as used in the aircraft, the unit should be drained and filled with the proper Skydrol hydraulic fluid before storing or installing. Units tested on other fluids may shorten the seal life and bring about expensive system contamination. It is best to test a unit in the same fluid as used in the actual hydraulic system.

Nonhalogenated cleaning solvents

NTA-L (NSN 6850-00-264-9039)	Texaco Co.
PD-680	Chevron Intl. Oil Co.
Safety Kleen Solvent	Safety Kleen Corp.
Stoddard Solvent**	Ashland Chemical Co.
Odorless Mineral Spirits	Chevron Intl. Oil Co.
White Spirits	—
Safety Solvent	—
Varnoline	—
Isopropyl Alcohol	—

**A mixture of 85 percent nonane and 15 percent trimethylbenzene.*

Halogenated cleaning solvents
Freon TF
Freon 113
Trichloroethylene

O-ring specifications

The following specifications serve as a guide in obtaining materials for use with Skydrol fluids. National Aerospace Standards (NAS) on seals compatible with phosphate ester fluids helped to establish guidelines for the right seal in the right hydraulic system.

NAS-1611—This is the O-ring specification covering sizes equivalent to MS-28775.
NAS-1612—This is the boss seal gasket specification covering sizes equivalent to MS-28778.
NAS-1613—This is the procurement specification for O-rings.

Each package shall be legibly marked in accordance with procurement specifications.

Note: NAS-1611 and 1612 parts should be checked to ensure that they are of the new EPR material rather than of the older butyl rubber.

The need for taking fluid samples

The performance of the entire aircraft hydraulic system can be affected by the condition of the hydraulic fluid. If the hydraulic fluid additives are damaged or depleted, protection provided by the hydraulic fluid is lessened.

Only after sampling the hydraulic fluid for the purpose of fluid analysis can the condition of the additives be determined and damage to the hydraulic system avoided.

Safe handling of Skydrol fluids

Skydrol fluids should be handled the same way any aviation fluid or lubricant is handled. Care should be taken in handling Skydrol fluids to keep them from spilling on certain plastic materials and paints which might tend to soften. If a small amount of Skydrol is spilled during handling, it should be wiped up immediately with a dry cloth. When large pools form, an absorbent sweeping compound is recommended.

The use of Stoddard solvent, denatured alcohol or MEK to remove traces of the fluid would then follow. Finally, the area should be washed with water and a detergent.

Tools should not be allowed to soak in Skydrol fluids if they have painted areas or vinyl chloride plastic handles. Many nonmetallic materials are resistant to Skydrol fluids and will not be adversely affected by it.

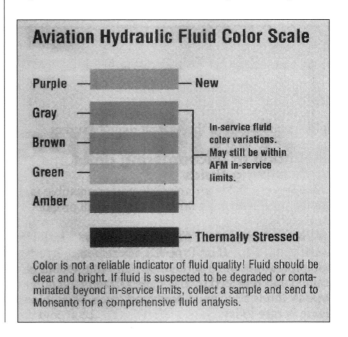

Accessory Technology

Skydrol compatibility chart

RATINGS OF COMPATIBILITY
- **Excellent Resistance** = Material may be used in constant contact with the fluid.
- **Good Resistance** = Withstands exposure to the fluid with minimum swell (for plastics and rubber) or loss of integrity.
- **Poor Resistance** = Should not be used near the fluid.
- **No Resistance** = Disintegrates in the fluid.

Resistance to Attack by Aviation Phosphate Ester Fluids

Material	Excellent	Good	Poor	No
Fabrics				
Acrylic[1]			•	
Cotton, Wool, Rayon		•		
Fiberglass, Nylon, Polyester[2]		•		
Carbon (Graphite)	•			
Coated Fabrics				
Buna N-coated Cotton		•		
Buna N-coated Nylon		•		
Butyl-coated Nylon	•			
Ethylene Propylene Nylon	•			
Polyethylene Nylon		•		
Neoprene-coated Nylon, Cotton, Polyester		•		
Silicone-coated Fiberglass	•			
Silicone-coated Polyester	•			
Vinyl-coated Cotton, Nylon, Polyester				•
Vinyl-coated Fiberglass				•
Fluoroelastomer-coated Nylon		•		
Metals				
Aluminum	•			
Brass		•		
Bronze		•		
Cadmium		•		
Chromium	•			
Copper[3]			•	
Ferrous[4]		•		
Lead		•		
Magnesium[3]		•		
Noble (Gold, Silver)	•			
Stainless Steel	•			
Zinc[5]		•		
Titanium[6]		•		
Exotic (Hastelloy, etc.)	•			
Beryllium Copper	•			
Conversion Coatings				
Anodizing (Aluminum)	•			
Dow[7] and[1] (Magnesium)	•			
Paint Finishes				
Alkyd[7]		•		
Acrylic			•	
Asphaltic			•	
Cellulosic Lacquer			•	
Epoxy	•			
Epoxy-Amide	•			
Heat-resistant Aluminized		•		
Latex			•	
Polyurethane		•		
Linseed Oil		•		
Shellac	•			
Silicone		•		
Urethane		•		
Varnish		•		
Vinyl			•	

Skydrol compatibility chart Continued

Material	Excellent	Good	Poor	No
Thermoplastics				
ABS			•	
Acetal		•		
Acrylic		•		
Cellulosic		•		
FEP (Fluorocarbon)	•			
Nylon	•			
Polycarbonate[8]		•		
Polyethylene	•			
Polypropylene	•			
Polyvinyl Chloride				•
Polyvinylidene Chloride		•		
Reinforced TFE	•			
Styrene				•
TFE (Fluorocarbon)	•			
Polyphenylene Sulfide (PPS)	•			
Polyetherketone (PEK)	•			
Polyetheretherketone (PEEK)	•			
Thermosets				
Melamine	•			
Polyester	•			
Phenolic	•			
Polyamide		•		
Polyimide (Vespel)	•			
Elastomers				
Butadiene Acrylonitrile (Buna N)				•
Chlorosulfonated Polyethylene[9]		•		
Epichlorohydrin[10]	•			
Ethylene Propylene[11]	•			
Fluoronated Hydrocarbon[12]			•	
Polyacrylic			•	
Polybutadiene			•	
Polychloroprene (Neoprene)			•	
Polyisoprene (Natural & Synthetic Rubber)			•	
Polysulfide (Thiokol)			•	
Polyurethane				•
Isobutylene Isoprene (Butyl)		•		
Silicone		•		
Styrene Butadiene (Buna S)			•	
Perfluorohydrocarbon[13]	•			
Fluorcethylena (TFE, FEP)	•			
Misc. Materials				
Cork[14]			•	
Leather[14]			•	
Vinyl Asbestos Tile				•

*Based on material from *Machine Design*, January 21, 1971 Copyright, 1971 by Penton IPC, Inc., Cleveland, Ohio

Footnotes
1. Includes Acrilan, Cresian, Orlon, Zefran acrylic fibers, and Dynel and Verel modified acrylic fibers.
2. Includes Dacron, Fortel, Kodel and Vycron polyester fabrics.
3. Copper and magnesium are not recommended for use in hydraulic systems due to the corrosion which occurs as a result of moisture and system temperatures.
4. Ferrous alloys should be plated (except stainless steel).
5. Lead and zinc are not recommended for use in a hydraulic system. They tend to form oxidation products which tend to form soaps and cause emulsions.
6. Titanium is not recommended for service at elevated temperatures (>3,250 F/1,630 C). Hydrogen embrittlement will occur from the phosphate ester fluid.
7. Includes alkyd-phenolic, alkyd-silicone and alkyd-uren finishes.
8. Includes Lexan & Merlon.
9. Hypalon.
10. Includes polyepichlorohydrin (Hydrin 100) and epichlorohydrinethylene oxide (Hydrin 200).
11. Nordel, Royalene, Vistaton, Epsyn, etc.
12. Viton, Fluorel, Kel-F, etc.
13. Kalrez & Trefzel.
14. Leather and most similar materials act as wicks, absorbing and holding the fluid.

Since it is difficult to visually distinguish between materials that are resistant and those that are not resistant, it is highly recommended that all materials wet with Skydrol fluids be wiped off and cleaned as soon as possible.

Skydrol fluids will not harm clothing with the exception of a few materials such as rayon acetate. Some types of rubber-soled shoes may soften and deteriorate when exposed to Skydrol fluids.

Material Safety Data Sheets (MSDS) available from the manufacturer should be consulted for all health and safety concerns.

Accessory Technology

Chemicals in Skydrol fluids which may cause irritation

Skydrol fluids are phosphate ester-based fluids blended with performance additives. Phosphate ester is a good solvent, and as such will dissolve away some of the fatty materials of the skin.

Repeated or prolonged exposure often causes drying of the skin, and if unattended could result in complications such as dermatitis or even secondary infection from bacteria.

Skydrol fluids are not known to cause allergic-type skin rashes.

Hand protection

To avoid exposure, a worker should wear gloves that are impervious to Skydrol hydraulic fluids.

Some manufacturers claim "Nitrile" gloves are suitable for this purpose. Nylon and polyethylene "throw-aways" are also acceptable.

Monsanto does not recommend the use of protective barrier creams as they do not provide complete protection.

Proper first aid treatment for eye exposure

Monsanto is not aware of any case of eye damage from Skydrol fluid exposure. When the fluid gets into the eyes, it can cause severe pain, but it has not been known to cause any damage to the eye. First aid should consist of washing the fluid from the eye with potable tap water or a standard eye irrigation fluid (such as dacriose solution).

We advise copious flushing. Afterward, the installation of sterile mineral oil or petrolatum is preferred over unsterile products (such as milk). Any additional treatment would not be considered first aid and should be administered under the supervision of a physician.

Where splashing or spraying is possible, chemical-type goggles should be worn.

Inhalation of fluids

Upper respiratory irritation, including nose and throat irritation and tracheitis and/or bronchitis, can occur from inhalation of a Skydrol fluid mist. People with asthma may have a more marked reaction. If through some accident liquid Skydrol fluid is aspirated directly into the lungs, such as by swallowing a large amount and breathing it at the same time, or by breathing in while vomiting, it is possible that chemical pneumonitis could occur. This occurs following deep aspiration of any foreign material into the lungs. There is no record of this happening with Skydrol fluids, and the possibility of this occurring under normal industrial conditions does not appear to be likely.

When mist or vapor is possible because of high pressure leaks, or any leak hitting a hot surface, a respirator capable of removing organic vapors and mists should be worn.

Ingestion

Ingestion of small amounts of Skydrol fluid does not appear to be highly hazardous. Should ingestion of a full swallow or more occur, we suggest the immediate ingestion of milk or water, followed with hospital supervised stomach treatment, including several rinses of saline solution and milk.

Protective materials

Gloves

Where continuous exposure of the hands to Skydrol hydraulic fluids is required, protective gloves should be worn. Monsanto does not recommend the use of protective barrier creams as they do not provide complete protection.

Shoes

A synthetic sole material resistant to Skydrol should be worn. Natural rubber compounds will offer no resistance to Skydrol fluids.

Eye glasses and goggles

Eye protection should be worn when splashing or spraying of Skydrol hydraulic fluid is possible. Chemical-type goggles are recommended.

Disposable clothing

Disposable clothing materials resistant to Skydrol hydraulic fluids should always be used. One example of resistant material is Durafab, made by E.I. dupont de Nemours.

Clothing manufactured from resistant materials can be obtained from most safety equipment supply houses.

Protective clothing manufactured from resistant materials such as cotton, wool, nylon or polyester is resistant to Skydrol hydraulic fluids.

Respirators

When Skydrol fluid mists or vapors are possible because of high pressure leaks or any leak hitting a hot surface, a respirator capable of removing organic vapors and mists should be worn.

Accessory Technology

Skydrol fluid safe handling summary

- Avoid direct exposure to Skydrol fluids.
- Impervious gloves and chemical-type goggles should be worm.
- Eye baths should be available when there is potential for eye contact.
- When mist or vapor is possible, a respirator capable of removing organic vapors and mists should be worm.
- Consult the Material Safety Data Sheets (MSDS) or contact Monsanto Chemical Co. for health/safety information.

March/April 1995

Accessory Technology

Helicopter rotor track and balance
Advancements offer an interesting selection of equipment and techniques

According to John Beach, president of Dynamic Solutions Systems, since helicopters were first produced, a variety of methods and techniques have been used to reduce vibrations produced by their rotor systems. "Rotor smoothing," he says, "is a term describing the use of dynamic track and balance to achieve vibration-free flight."

Main rotor tracking methods

"Theoretically, main rotor blades should all fly in the same plane and maintain equidistant angular spacing during flight. Pitch links and trim tabs are adjusted to compensate for blade differences to keep the blades in line at all forward speeds. Rotor tracking systems," he says, "have focused on providing information that can be used by the mechanic to adjust pitch links and trim tabs to coax the blades to fly 'perfectly' in track."

Flag tracking

Beach explains that the earliest technique for rotor tracking was flag tracking, where the tip of each blade was marked with colored chalk, crayon or grease pencil and a white strip of cloth mounted to a pole was pushed into the edge of the operating rotor's blade path. The marks on the cloth gave a measure of the blade track. In fact, some helicopter manufacturers still use the color convention to identify blades (i.e., blue blade, yellow blade, etc.).

Electro-optical tracking

In the 1960s, an opti-electronic method of measuring rotor track height and lead lag was developed and patented by Chicago Aerial. Chicago Aerial built a small single lens system that could be mounted to the aircraft and measure track in flight, but achieved only very limited success with it. The system they sold the most was a large, heavy ground-based dual lens/sensor system that was very accurate but could only measure track on the ground. This system of tracking was superseded with the development of strobe tracking.

Strobe light tracking

Then, in the early 1970s Chadwick-Helmuth adapted a strobe light and retro-reflective tip targets to allow blade track and lead lag to be measured on the ground and in flight. This technique requires significant operator skill and training.

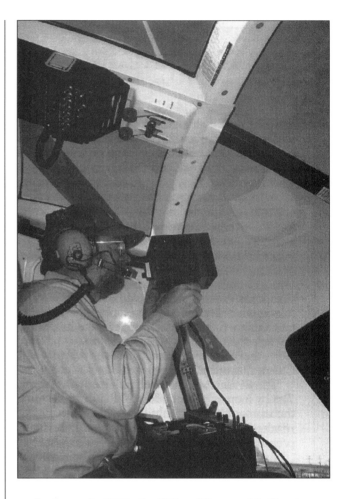

In the early 1980s the U.S. military and helicopter manufacturers made it clear they were looking for a system that could measure blade track consistently and accurately without highly skilled human operators.

Electro-optical tracking revisited

Beach says that shortly thereafter, Stewart Hughes of the UK revived the method originally developed by Chicago Aerial and applied up-to-date electronics and packaging methods to make this technique practical for in-flight tracking. "This system is built under license by Scientific Atlanta in the United States," he says. "Chadwick-Helmuth also developed a similar tracker which is currently being offered for sale. Helitune Ltd. of the UK developed a different tracking method using a line scan video camera."

Tracking using vibration sensors

Beach says that "theoretically, vertical vibrations in level flight result from asymmetric lift vectors produced by uneven loading on the rotor blades. Users found the track conditions of the rotor directly related to these vibrations in the airframe." He continues, "Experimentally it was found that vibration information could be used to adjust pitch links and trim tabs to produce minimum vibrations at all forward speeds. After this process was complete, the blade track could be measured and surprisingly the blades were most frequently not in perfect track!" Beach says that this lead to a quandary. Do we want perfect track or minimum vibrations?"

What value is tracking?

He explains that in the process of using these tracking methods and measuring the vibrations resulted, users found "perfect track" rarely produced minimum vibrations. He says that "various theories have been proposed to explain this effect. One theory is that each blade has a slightly different shape, twist, flexibility, etc., and only by putting them slightly out of track can these variations in lift be compensated. Another theory is that each blade produces a 'turbulent wash' the trailing blade must fly through. If alternating blades are set to fly high and then low, each blade will have 'calm air' to fly through resulting in smoother flight. This effect is more pronounced on aircraft with four or more blades on the main rotor."

This fact has lead some manufacturers to conclude blade tracking is of little value and the only purpose of rotor smoothing should be to minimize vibrations without regard to blade track. This approach has a few drawbacks however. First, a "bad" blade must be put way out of track to minimize vibrations and can only be detected if a tracking system is used. Second, blade track and lag information may make finding some problems with the rotor much easier. For example a bad damper may produce a subtle transient vibration effect during turns, but a lead lag measurement will show the damper problem as a blade obviously unstable in angular position.

Jim Fenton, engineer at Scientific Atlanta, says that, in fact, blade tracking is very important. "Without acceptable blade tracking, if you do any type of maneuver that places stress on the blades, you can get severe blade split. The distance the blades are apart will be exaggerated in adverse maneuvers, and this will load the blades unevenly. Because of this, it's critical to make sure the track is flat. Also, the smaller the track split, the less energy it takes to run the helicopter. It puts more stress on the gearbox, engine, and airframe, and will burn more fuel. It's correct to want to reduce vibra-

Vibration and optical sensor placement depend on helicopter make and model, and type of balancing equipment used.

tion, but if you ignore track, you do it at your peril. You need to reduce vibrations, but you've got to be sure you're not doing something detrimental and producing a big track split."

Rotor balancing methods

Beach explains that early methods of balancing helicopter rotors were limited to static "bubble" balancing of rotor heads and "weighing" blades to be sure the rotor was symmetrically loaded. In the 1970s maintenance personnel began attaching vibration sensors and using spin balancing techniques common in industry on large industrial blowers. Using a strobe flash to establish the phase of the vibration and a tuned filter to establish the amplitude of the vibration, charts (nomograms) could be used to determine where to add weight and how much.

Early complex algorithms

With helicopter rotors, however, there is interaction between mass imbalance and blade track at hover. If one blade is flying high and producing more lift, it will also lag its normal position due to higher drag force. This induces an effective mass imbalance. Because of this and other interactions, users developed procedures or "algorithms" that allowed the rotor to be smoothed by performing steps in a particular order. For example, first, track blades with pitch links on the ground; second, track the blades with pitch links in hover; third, spin balance the main rotor at hover; fourth, adjust tabs based on track data and/or vertical vibration data in forward flight These procedures required a good deal of skill and accuracy from the maintenance personnel. This has led to the industry acknowledging a need for automated computerized methods that reduce the workload and skill level needed to accomplish satisfactory rotor smoothing.

Accessory Technology

Computer-based algorithms

In the 1980s, hand-held computers were programmed to perform the rotor smoothing algorithms. In some cases, the engineers programming the new computer systems wanted to simplify the algorithms currently in use. A popular concept was that all helicopters of the same type were sufficiently similar to allow a single computer math model to be used.

Track and balance equipment based upon this concept was subsequently developed and introduced to the market by Chadwick-Helmuth with its Model 8500 and Scientific Atlanta with its RADS AT (built under license to Stewart Hughes of the UK).

These units "use customized programs for specific model aircraft and utilize a single flight method of balancing. Single flight methods are quick, but a downside to the single flight method is it does not lead to any method of verifying the changes recommended were executed properly. The single flight method requires the user to make several adjustments at once, and due to the interaction of these adjustments it is difficult to determine if adjustments are done correctly."

However, Fenton at Scientific Atlanta says single flight equipment offers the option of performing the track and balance in a single flight, but it is not limited to that. He says, "The technician always has the option of verifying the solution." In fact, he explains that in practice, more than one run is usually made. Additionally, he says the more runs that are made, the more you can reduce vibration levels on the helicopter. It's all up to how much time the operator and technician has to spend on balancing and how far they want to go beyond maintenance manual specifications.

"Single move" method

Beach refers to the rotor smoothing algorithm developed by Dynamic Solutions Systems (DSS) as the "single move method." Using this method, each flight only results in instruction to make a single adjustment. The next flight is used to verify that the change matches the math model; if it does not match but is within a tolerance, the math model is corrected to match the current aircraft

One advantage of this method will be the first flight may only consist of pulling the aircraft up into hover for 30 seconds. Forward flight will not be attempted until hover vertical and lateral vibration is low. This will avoid the safety hazard of trying to fly an aircraft that is badly out track or balance in forward flight conditions as could be the case with the single flight method.

Tail rotor balance

"Helicopter tail rotors can be surprisingly difficult to balance," says Beach. He explains that this is partially because their bearings and pivots allow the rotor to tilt on its axis dynamically. Another problem is tail rotors are often very sensitive to minute changes in tip weight.

Small changes are normal due to small errors in weight measurement and location. But larger changes are indications of large errors in weight placement or measurement, or they are indications of a defective component in the rotor hinge mechanism.

Basic selection criteria

"A very serious and fundamental problem with most modern helicopter track and balance systems on the market is in their limited basic analysis capabilities. Some systems have become so highly applications specialized they have left out many basic features required of any good general purpose vibration analyzer/balancer," he says.

"It's also been my experience the first few things people look for when they're buying equipment are: what the equipment capabilities are, if the equipment interfaces with a computer, how much the equipment costs, and what the price includes," says Beach.

But he stresses that probably one of the most important things you must often take into account, is even with the most sophisticated equipment, it's essential the concepts of balancing be fully understood in order to accomplish good balancing. And because of this, the companies supplying this equipment should also offer computer training support. With the extremely high costs of helicopter and crew time, computer balance simulation training for the mechanic is almost a necessity," says Beach. *January/February 1995*

Accessory Technology

Hydromatic propeller governors
Basic operation and troubleshooting

By Greg Napert

Although the reciprocating engine propeller governor performs a fairly complex task, it's a relatively simple device that works on basic principles of motion and mechanics. And although there are four primary manufacturers of these governors, McCauley, Hartzell, Hamilton Standard, and Woodward, they are all very similar in operation and design.

Basically most governors consist of a gear pump which takes oil from the engine lubricating system and boosts it in pressure to actuate the propeller pitch change mechanism; a pilot valve actuated by spring-balanced flyweights that controls the flow of the high-pressure oil to the prop; a pressure operated transfer valve which on some feathering installations allows high-pressure oil from an auxiliary source to feather and unfeather the prop; and a relief valve system which controls the output pressure of the gear pump within a specific range.

The operating principle of these components is quite simple. Essentially, a set of flyweights is connected to a pilot valve. The repositioning of the pilot valve, via a rack assembly, affects the tension on a spring that opposes the force that the flyweights generate.

The more force that opposes the flyweights, the more engine rpm it takes to position the pilot valve so that it is in the desired position.

In practice, the action of the pilot via cockpit controls, that raises the rack, decreases the compression of the speeder spring and thus reduces the speed at which the flyweights must rotate to return the pilot valve to the desirable "on-speed" condition. Conversely, lowering the rack increases the spring force and consequently the required rotational speed of the flyweights required to return the pilot valve to the on-speed condition.

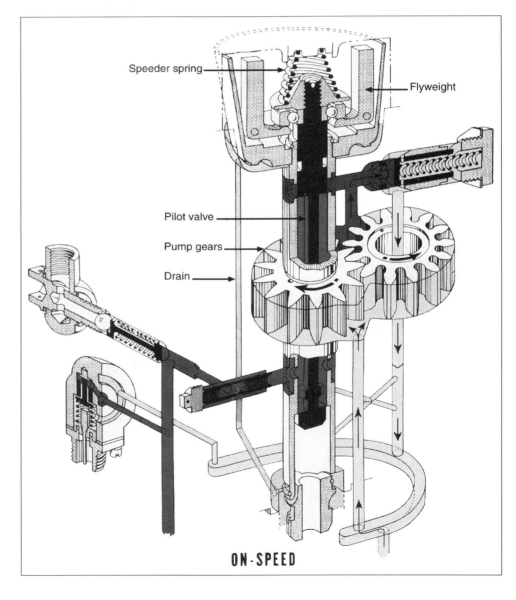

In effect, the flyweights are continuously opposing the force of the speeder spring and seeking a point of balance related to the tension on the spring.

John DeJoris, president and co-owner of Aircraft Propeller Service Inc. in Wheeling, IL, says that governors, in general, are pretty forgiving and for the most part are maintenance-free. He explains that technicians in the field often have questions related to one of three categories: a prematurely feathering prop (on feathering-type aircraft) the surging engine, and various oil leaks that may occur on the governor.

Premature feathering

Most small engines supply oil to the propeller through the front main bearing journal. DeJoris says if that journal clearance gets too great, it dumps more oil back into the case than what's going to the propeller, and servo pressure to the governor is lost.

"The end result on a twin," he explains, "is that the prop will move into feather on or before shutdown." And on single-engine aircraft with noncounterweighted props, the prop goes into low blade angle. On twin-engine aircraft, it can progress to the point where you just throttle back and the prop goes into feather prematurely. "This gets critical in flight, particularly in a go-around situation—you don't want the engine going into feather; you could be in real trouble."

DeJoris says that the governor must get a nominal 50 psi of input pressure from the engine, and the governor typically boosts that pressure to 290 to 300 psi. "If you don't have the nominal 50 psi of input pressure," he says, "it won't be able to boost it to 290.

"Under normal conditions with counterweighted props, when you pull the throttles back, the blades should go to the low stop with servo pressure. The counterweights are trying to pull the prop into the high pitch range and ultimately feather, and you've got oil servo pressure driving the piston to the low stop. But with oil bypassing back to the engine, there won't be enough pressure to hold it against the low stop. If the propeller piston isn't pushed lower than the latch angle, feather will occur on shutdown."

DeJoris continues, "This also can cause a split in the engine rpm on a twin-engine aircraft under certain conditions."

Surging problems

Ken Lazar, accessory shop supervisor for Aircraft Propeller Service, says that "surge problems related to the governor are commonly a result of wear in the flyweight area of the governor. More specifically, surging is usually a result of wear of the flyweight hinge pins. Wear of these pins," he continues, "can result in the weights moving further to the outside of the housing and resulting in an overspeed condition, or, can prevent the weights from moving all the way out and result in an

The relief valve seat has pounded into the housing which wears a groove in the drive gear.

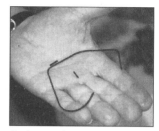

O-ring seals can become brittle over time and result in leakage.

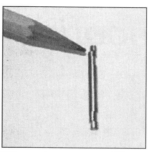

Hinge pin wear, as with this pin, can cause the flyweights to hang and result in erratic governor operation.

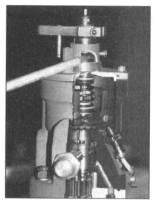

It's not uncommon for O-ring wear in the control shaft area to result in leakage.

underspeed condition. Wear can also cause fluctuations as the flyweights will have a tendency to hang up.

"I've also seen situations where a hinge pin has broken and has torn up the housing. On twin-engine aircraft, this will cause a prop to feather," he says.

"If the governor is found to be OK, the surging problem is probably related to the propeller pitch change mechanism."

Leaks

Lazar says that common areas of leakage are with the O-ring in the control shaft area, or the seal on the pump body.

"In most situations," he says, "there's really nothing you can do to replace seals in the field unless you have a test bench. Replacing the seals means having to disassemble major components that will require testing upon reassembly.

"The best bet with any leak," Lazar says, "is to return it to an overhaul facility capable of verifying the proper operation of the pump.

"Also," he says, "with the pump body seal, splitting the pump body (on some models) requires a bit of force, and you risk sliding the pilot valve and speeder spring

Accessory Technology

Typical Propeller Governor Troubleshooting Chart

TROUBLE	PROBABLE CAUSE	REMEDY
LEAKAGE		
Between control shaft packing nut and head.	Damaged packing nut lock gasket.	Replace packing nut lock gasket.
	Loose control shaft packing nut.	Tighten control shaft packing nut on all models.
Between control shaft packing nut and control shaft.	Damaged packing nut bushing.	Replace packing nut bushing or packing nut assembly on all models.
	Damaged packing washer.	Replace packing washer on all models.
Between head and body.	Damaged head-body gasket.	Replace head-body gasket on four-position head models, replace head-body seal on 360-degree head models.
	Loose head-body attaching nuts.	Tighten attaching nuts on all models.
At high pressure swivel fitting.	Damaged swivel fitting gaskets.	Replace swivel fitting gaskets on all feathering models.
	Loose swivel clamp bolt.	Tighten swivel clamp bolt on all feathering models.
Between transfer valve plug and governor body.	Damaged transfer valve gasket.	Replace transfer valve gasket on all double capacity models.
	Loose transfer valve plug.	Tighten and resafety transfer valve plug on all double capacity models.
Between accumulator line fitting and governor body.	Damaged gasket.	Replace accumulator line fitting gasket on all accumulator models.
	Loose accumulator line fitting.	Tighten accumulator line fitting.
Between relief valve housing and governor body.	Damaged relief valve tab lock gasket.	Replace relief valve tab lock gasket on all models which incorporate a cartridge-type relief valve.
	Loose relief valve housing.	Retighten valve housing on all models which incorporate a cartridge-type relief valve.
Between relief valve housing and relief valve plug.	Damaged relief valve plug lock gasket.	Replace relief valve plug lock gasket on all models which incorporate a cartridge type relief valve.
	Loose relief valve plug.	Retighten plug on all governors which incorporate a cartridge-type relief valve.
Between relief valve plug and body.	Damaged relief valve plug lock gasket.	Replace gasket on all plug-type relief valve single and double capacity governors.
	Loose relief valve plug.	Retighten plug on all plug-type relief valve single and double capacity models.
Between propeller line screened plug and body.	Loose plug.	Tighten plug on all single capacity models.
Between body and base.	Damaged body-base seal.	Replace body-base seal on all models.
	Loose body-base stud nuts.	Retighten body-base attaching stud nuts on all models.
Between governor base and engine mounting pad.	Damaged governor mounting gasket.	Replace governor mounting gasket on all models.
	Loose governor attaching nuts.	Retighten nuts and palnuts.
	Warped governor.	Lap governor base on all models.
	Warped engine mounting pad.	Consult engine manual.
Between governor base, and pressure cut-out switch	Damaged pressure cut-out switch gasket.	Replace cut-out switch mounting gasket on all feathering models.
	Loose pressure cut-out switch attaching screws.	Tighten switch attaching screws on all feathering models.
	Blow holes or cracks in any part of the governor housing.	Replace damaged part.
INABILITY TO ATTAIN TAKEOFF RPM ON THE BLOCKS. NOTE: With takeoff manifold pressure, it is impossible, in some installations, to obtain takeoff rpm on the blocks.	Wrong high rpm setting on governor.	Reset governor external high rpm adjusting screw. On electric heads, adjust high rpm limit switch setting. Reset on test rig if available.
	Incorrect rigging of control system.	Adjust control system.
	Low engine power.	Consult engine manual.
	Erroneous reading tachometers or manifold pressure gauges.	Calibrate or replace instruments.
	Sticky pilot valve.	Remove head, clean pilot valve with crocus cloth. Check for straightness of pilot valve, and if bent, replace.
	Faulty aircraft electrical system on electric head installation.	Check electric head circuits, check potential of battery, and check control wiring to electric head.

adjustment rack shaft to the position where it disengages the control shaft. You'll then need to start from scratch and completely disassemble the unit and reassemble it in a sequential manner.

"One of the big ironies with the regulations," Lazar continues, "is that we can't keep an O-ring or seal on our shelf in dry/protected storage for more than four years, but yet, you can have one on your Part 91 airplane that's in your propeller or governor indefinitely. There's no requirement to ever look at it or change it under Part 91."

Contamination factors

DeJoris says that another problem that's not thought about is that the oil supply is from the engine. "Engine oil maintenance is critical to the longevity of the pump. In fact, Woodward has a flat policy that any metal contamination found in the engine oil necessitates overhaul of the governor.

"McCauley and Hartzell, however, don't specify any actions, and so we recommend taking them apart and performing a flushing and visual inspection.

Typical Propeller Governor Troubleshooting Chart
Continued

POOR SYNCHRONIZATION	Sludge in governor pilot valve or relief valve.	Disassemble and clean.
	Burrs on pilot valve lands.	Disassemble and clean with crocus cloth.
	Backlash in governor control system.	Rerig or adjust control system.
	Short control lever making fine adjustments of speed impossible.	Replace control lever.
	Excessive deflection of structure supporting controls when subjected to control or flight loads.	If possible, relocate control linkage to avoid high deflection points.
	Erroneous reading tachometers.	Calibrate or replace instruments.
	Sticky pilot valve bearing.	Remove head and pilot valve. Clean or replace bearing. (Check speeder spring for burning.)
	Bent pilot valve.	Remove head and replace pilot valve.
	Excessive internal leakage in governor.	Check on rig and make necessary part replacement.
	Galled or corroded speeder rack and bore.	Remove head, clean up and lubricate rack and bore.
EXCESSIVE OVERSPEEDING ON TAKEOFF	Wrong setting on governor.	Reset governor. Use test rig if available.
	Too rapid opening of throttle.	Advance throttle evenly and slowly.
	Damaged or incorrect gasket between governor base and engine mounting pad.	Install correct new gasket.
	Sticky governor pilot or relief valve.	Disassemble, clean, and check for burrs. Replace pilot valves if found bent.
	Erroneous reading tachometers or manifold pressure gauges.	Calibrate or replace instruments.
FAILURE TO FEATHER	Aircraft batteries low.	Recharge or replace batteries.
	Faulty aircraft electrical system.	Check wiring system pertaining to feathering pump and control circuit.
	Failure of push button to remain engaged.	Check battery or pressure setting of cut-out switch. Check hold-down coil circuit.
	Failure to remain feathered with push button failing to disengage.	Reset cut-out switch and check electrical control circuit.
	Sheared coupling in feathering pump.	Replace coupling.
	Restricted oil supply to feathering pump.	Check feathering pump inlet lines for foreign material, and bleed line. Periodically drain condensation from line between oil supply tank and feathering pump on systems which do not incorporate the feathering line bleed feature.
	Defective feathering pump.	Replace pump.
	Damaged high-pressure oil supply line to governor.	Replace high-pressure oil supply line.
	High-pressure transfer valve stuck in closed position.	Disassemble governor, remove and clean pilot valve, and then reassemble and test governor.
	Batteries low.	Recharge or replace batteries.
	Faulty electrical system.	Check control and power circuits of feathering system.
	Damaged or incorrect gasket between governor base and engine mounting pad.	Install correct new governor mounting gasket.
	Sheared coupling in feathering pump.	Replace coupling.
	Restricted oil supply to feathering pump.	Check feathering pump inlet lines for foreign material, and bleed line. Periodically drain condensation from line between oil supply tank and feathering pump on systems which do not incorporate the feathering line bleed failure.
	Defective feathering pump.	Replace pump.
	Incorrect operating procedure.	Manually hold in push-button control switch until propeller begins to windmill.

"Many technicians will send in the propeller to have it flushed out and inspected, but they don't remember to send in the governor. They forget that the oil's got to go through the governor to get to the propeller.

"The only time that you won't have any contamination is if you have an engine failure where the oil stops circulating, or an oil leak where all of the oil drains out of the engine and it is shut down before any damage occurs and there's no chance to circulate debris. But that's not very common."

He explains, "Typically, we don't like to put governors that aren't running correctly on our test benches because if they are making metal or are contaminated with something, it will get into our test bench fluid supply.

"Instead, we prefer to tear down the prop governor and inspect the individual components for evidence of damage."

Other areas of concern

Lazar explains there are other areas of concern related to removal, installation, and inspection of governor systems: "A problem we see out in the field occasionally is the prop governors adhering themselves to the case. A technician uses a hammer or heavy item on the governor and damages it. In fact, we see hammer marks on the housings all the time, and there's really no excuse for this.

"First of all, when installing a governor, you really need to spray a little release agent on the gasket for easy removal because the governor is probably going to be on there for a while. And then with a governor that's actually stuck, you should be sure to use a soft enough medium (such as a plastic mallet) to break it loose so that you don't damage the housing.

Accessory Technology

"With aircraft that have external oil lines to the prop, you've got to pay particularly close attention to proper installation and condition. If the output line of the governor breaks, you're going to dump oil just like that. The governor is pumping approximately 8 quarts a minute, and it will empty the entire contents of the engine almost instantaneously."

DeJoris says that one of the big problems in the industry is overhaulers that don't comply with all of the service bulletins recommended by the manufacturer.

"Our position on overhauling the governor is that if you do an overhaul, you need to bring it up to the latest manufacturer's specs, and that includes all Service Bulletins and ADs. Some shops only do the ADs, but to us that's not right.

"We feel that if you haven't complied with all bulletins at overhaul, you haven't overhauled the product.

"Anything less is a repair and not an overhaul."

November/December 1994

Accessory Technology

Tire care and maintenance
Everything you need to know

By Greg Napert

The aircraft tire is capable of withstanding high speeds, intense loads, sudden impacts, extremely abrasive surfaces, and is expected to perform until worn to its limits without sacrificing any of its properties.

Although it may seem that little can be done to assure long life and proper operation of tires, there is much that can be done.

The ingredients to proper tire maintenance begin with understanding its construction. Add a pinch of understanding about proper tire inflation, a handful of knowledge about tire balance, and a few miscellaneous pieces of information related to operating temperatures and tire replacement, and you have a well-baked scheme for keeping tires in top shape.

The following ingredients for understanding and maintaining tires is edited information taken from Goodyear's Tire Care and Maintenance Manual. Although the information in this text was written for Goodyear tires, it is applicable to all aircraft tires.

Preventive maintenance

Keeping aircraft tires at their correct inflation pressures is the most important factor in any preventive maintenance program. The problems caused by underinflation can be particularly severe.

Underinflation produces uneven tread wear and shortens tire life because of excessive flex heating. Overinflation can cause uneven tread wear, reduce traction, make the tread more susceptible to cutting and increase stress on aircraft wheels. It is recommended that only dry nitrogen be used for tire inflation as nitrogen will not sustain combustion and will reduce degradation of the inner-liner material due to oxidation.

Inflation pressure

Ideally, tire pressures should be checked with an accurate gauge on a daily basis. Pressures on high performance aircraft should be checked before each flight. Check only cool tires — at least two to three hours after a flight.

Use an accurate gauge, preferably the more precise dial type. Inaccurate gauges are a major source of improper inflation pressures. It's important, although rarely accomplished in practice, to check tire pressure gauges periodically and to calibrate them on a regular basis.

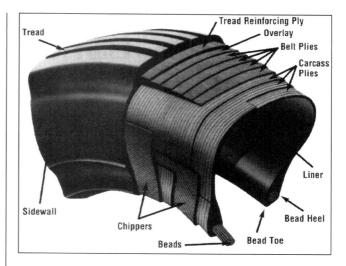

The inflation pressure recommended by the air-frame manufacturer should be used for each tire. Be particularly careful to determine if "loaded" or "unloaded" inflation pressures are specified.

When a tire is under load, the gas chamber volume is reduced due to tire deflection. Generally, if unloaded pressure has been specified, that number should be increased by 4 percent to obtain the equivalent loaded inflation pressure.

Adjusting for temperature

When tires will be subjected to ground temperature changes in excess of 50°F (27°C) because of flight to a different climate, inflation pressures should be adjusted for the worst case prior to takeoff. The minimum required inflation must be maintained for the cooler climate; pressure can be adjusted in the warmer climate. Before returning to the cooler climate, adjust inflation pressure for the lower temperature. An ambient temperature change of 5°F (3°C) produces approximately 1 percent pressure change.

Keep in mind that excess inflation pressure should never be bled off from hot tires. All adjustments to inflation pressure should be performed on tires cooled to ambient temperature.

Accessory Technology

Cold pressure setting

The following recommendations apply to cold inflation pressure setting:

1. "Minimum pressure" for safe aircraft operation is the cold inflation pressure necessary to support the operational loads as determined by the formula under "Unloaded Inflation" or as specified by the airframe manufacturer.
2. The loaded inflation must be specified 4 percent higher than the unloaded inflation.
3. A tolerance of minus zero to plus 5 percent of the minimum pressure is the recommended operating range.
4. Consult the table if in-service pressure is checked and found to be less than the minimum pressure.

Mounted tube-type tires

A tube-type tire that has been newly mounted and installed should be closely monitored during the first week of operation, ideally before every takeoff. Air trapped between the tire and the tube at the time of mounting could seep out under the beads, through sidewall vents or around the valve stem, and result in an underinflated assembly.

Mounted tubeless tires

A slight amount of gas diffusion through the casing of tubeless tires is normal. The sidewalls are purposely vented in the lower sidewall area to bleed off trapped gases, preventing separation or blisters.

A tire can lose as much as 5 percent of the initial inflation pressure in a 24-hour period and still be considered normal.

Tire stretch

The initial stretch or growth of a tire results in a pressure drop after mounting. Consequently, tires should not be placed in service until they have been inflated a minimum of 12 hours, pressures rechecked, and tires re-inflated if necessary.

Nylon flat-spotting

Nylon tires on aircraft left stationary for any length of time will develop temporary flat spots. The degree of this flat-spotting depends upon the load, tire deflection, and temperature. Flat-spotting is more severe and more difficult to work out during cold weather.

Occasionally aircraft can lessen this condition. If possible, an aircraft parked for long periods (30 days or more) should be jacked up to remove weight from the tires. Under normal conditions, a flat spot will disappear by the end of the taxi run.

Cold weather precautionary hints

Aircraft parked and exposed to cold soak for a period of time (one hour or more) should have tire pressure checked and adjusted accordingly. Tires will have taken a nylon "set" and experienced a pressure drop.

Additionally, high-speed taxis and sharp turns should be avoided to prevent excessive sideloading.

An important fact to remember is that every 50°F (30 C) change in temperature will result in a corresponding 1 percent change in tire pressure.

Special procedures after a rejected takeoff

Goodyear recommends that tires subjected to above normal braking energies during a rejected takeoff should be removed and scrapped.

Even though visual inspection may show no apparent damage, tires may have sustained internal structural damage that could result in premature failure. Also, all wheels must be checked in accordance with the applicable wheel overhaul or maintenance manual after a rejected takeoff.

Matching dual tires

When new and/or retreaded tires are installed on the same landing gear axle, the diameters should be matched within the Tire and Rim Association inflated dimensional tolerances for new and grown tires to ensure that both tires will carry an equal share of the load.

Protecting tires from chemicals and exposure

Tires should be kept clean and free of contaminants such as oil, brake fluid, grease, tar, and degreasing agents which have a deteriorating effect on rubber. In the event of exposure to any of these contaminants, wipe with denatured alcohol; then wash tire with soap and water immediately.

Aircraft tires, like other rubber products, are affected to some degree by sunlight and extremes of weather. While weather-checking doesn't impair performance, it can be reduced by protective covers. These covers (ideally with light color or aluminized surface to reflect sunlight) should be placed over tires when the aircraft is tied down outside.

Chevron cutting

Many major airports throughout the world have modified their runway surfaces by cutting cross grooves in the touchdown and rollout areas to improve water runoff. Cross grooves vary in size and shape.

This type of runway surface can cause a pattern of chevron-shaped cuts in the center of the tread. As long as this condition doesn't cause chunking or cuts into the fabric, the tire is suitable for continued service.

Accessory Technology

Safety precautions with split wheels

An inflated tire is a potentially explosive device. Mounting and demounting of aircraft tires is a specialized job that is best done with the correct equipment. The following precautions are advisable in handling both tube-type and tubeless tires, especially those with high inflation pressures.

- Inspect fusible plugs. Fusible plugs are used on the tubeless wheels of high-performance aircraft to relieve excessive pressure created by excessive brake heat These plugs are generally not removed during a routine tire change unless defective or if the wheel is subjected to degreasing and cleaning. The plugs, however, are always removed and inspected during wheel assembly overhaul. The wheel manufacturer's maintenance/overhaul manual typically offers directions for inspection, removal, and installation of fusible plugs.

- Prior to removing the wheel/tire assembly from the aircraft, completely deflate the tire with a deflation cap. It is good practice to deflate the tire before removing the axle nut. When all pressure has been relieved, remove the valve core. Valve cores under pressure can be ejected like a bullet. If wheel or tire damage is suspected, approach the tire from the front or rear, not from the side (facing the wheel). A tire/wheel assembly that has been damaged in service should be deflated by a remote means. If this isn't possible, the tire/wheel assembly should be allowed to cool for a minimum of three hours before the tire is deflated.

- Take special care when encountering difficulty in freeing tire beads from wheel flanges. Even with tire tools, care must be taken to prevent damage to beads or wheel flanges. On small tires, successive pressing with a 2-foot length of wood close to the bead or tapping with a rubber mallet is generally sufficient. On large tires, hydraulic or mechanical bead-breaking press may be required. If using a "bead breaking" press, take care to prevent further movement of the tire bead after it is broken away from the bead seat area.

Inflation pressure action table

Tire Pressure	Recommended Action
100 to 95 percent of service pressure	Reinflate to specified service pressure
95 to 85 percent of service pressure	Reinflate and record in logbook Remove tire if pressure loss is greater than 5% and reoccurs within 24 hours
85 to 70 percent of service pressure	Remove tire from aircraft (See note)
70 percent or less	Remove tire and axle mate from aircraft (See note)
Blown fuse plug	Scrap tire. If blown while in service (rolling), scrap axle mate also

Note: Any tire removed because of low inflation pressure should be inspected by an authorized retreader to verify that the casing has not sustained internal degradation. If it has, the tire should be scrapped.

Bead lubrication in mounting both tubeless and tube-type tires

It is often desirable to lubricate the toes or inner edges of the beads of a tire to facilitate mounting and seating of the beads against the wheel flanges. A light coat of talcum powder or approved liquid bead lubricant can be used.

Wheel and tire assembly balancing/landing gear vibration

It is important that aircraft wheels and tires be as well balanced as possible. Vibration, shimmy, or out of balance is a major complaint. However, in most cases, tire balance is not the cause.

Other items affecting balance and vibration are: installation of wheel assembly before full tire growth, improperly torqued axle nut, improperly installed tube, improperly assembled tubeless tire, out of balance wheel halves, poor gear alignment, bent wheel, worn or loose gear components, or flat spotted tire.

In addition, pressure differences in dual-mounted tires and incorrectly matched diameters of tires mounted on the same axle may cause vibrations or shimmy. The following points should be paid attention to with regard to wheel balance.

Balance marks are placed on many tubes to indicate the heavy spot of the tube. These marks are often paint stripes about 1/2 inch (1 cm) wide by 2 inches (5 cm) long. When a tube is installed, this balance mark must be aligned with the "light spot" balance mark of the tire (red dot). If the tube has no balance mark, place tube valve adjacent to the tire balance mark (red dot). When mounting tubeless tires, the balance mark on the tire should be aligned with the wheel valve, unless otherwise specified by the manufacturer. If a tire has no balance mark, place tire serial number at the wheel valve.

With some split wheels, the light spot of the wheel halves is indicated with an "L" stamped on the flange. In assembling these wheels, position the "L's" 180 degrees apart. If additional dynamic or static balancing is required after tire mounting, many wheels have provisions for attaching accessory balance weights around the circumference of the flange.

Nylon flat-spotting

Nylon tires on aircraft left stationary for any length of time will develop temporary flat spots. The degree of this flat-spotting depends upon the load, tire deflection, and temperature. Flat-spotting is more severe and more difficult to work out during cold weather.

Typically, however, under normal conditions, a flat spot will disappear by the end of the taxi.

Accessory Technology

Tread wear

Inspect treads visually and check remaining tread. Tires should be removed when tread has worn to the base of any groove at any spot, or to a minimum depth as specified in aircraft manuals. Tires worn to fabric in the tread area should be removed regardless of the amount of tread remaining.

Uneven wear

If tread wear is excessive on one side, the tire can be demounted and turned around, providing there is no exposed fabric. Landing gear misalignment causing this condition should be corrected.

Inflation pressure loss in tubeless assemblies

Since there are many causes for inflation pressure loss with a tubeless assembly, a systematic troubleshooting approach is advisable for minimum maintenance costs.

Moreover, when chronic but not excessive inflation pressure loss exists, other factors such as inaccurate gauges, air temperature fluctuations, changes in maintenance personnel, etc., may be the source. If a definite physical fault is indicated, a troubleshooting procedure similar to the one outlined below is recommended:

Valve

Before deflating and removing tire, check the valve. Put a drop of water on the end of the valve and watch for bubbles indicating escaping air. Tighten the valve core if loose. Replace valve core if defective and repeat leak test to check. Check the valve stem and its mounting for leaks with a soap solution. If a leak is detected, wheel must be disassembled and a new valve stem installed.

If valve stem threads are damaged, stems can usually be rethreaded, inside or outside, by use of a valve repair tool without demounting tire. Make certain that every valve has a cap to prevent dirt, oil, and moisture from damaging the core.

Fusible plug

The fusible plug may also be defective or improperly installed. Use a soap solution to check fusible plugs for leaks before removing tire. Leaks can usually be pinpointed to the plug itself (a poor bond between the fusible material and the plug body)... or to the sealing gasket used. Be sure the gasket is one specified by the manufacturer... and that it is clean and free of cuts and distortion.

If excessive heat has caused a fusible plug to blow, the tire may be damaged and should be replaced. After a fuse plug in a wheel blows, the wheel should be checked for soundness and hardness in accordance with the applicable wheel maintenance/overhaul manual.

Release plug

The inboard wheel half may contain a pressure release plug, a safety device that prevents accidental overinflation of the tire. If the tire is overinflated, the pressure release plug will rupture and release the tire pressure. A soap solution can be used to check a release plug to determine whether or not it is defective.

Wheel base

Gas escaping through a cracked or porous wheel base is usually visible in an immersion test. Consult the wheel manufacturer's manual for rim maintenance and repair.

O-ring seal

A defective seal between the wheel halves can usually be detected in an immersion test by bubbles emerging through the center of the wheel. Check to see that wheel bolts are properly torqued.

Bead and flanges

Check the bead and flange areas of a tire for leaks before demounting. This can be done either by immersion or by using a soap solution. Any of the following factors can cause gas loss:

1. Cracks or scratches in wheel bead ledge or flange area.
2. Exceptionally dirty or corroded surface on wheel bead seating surfaces.
3. Damaged or improperly seated tire bead.

Tire and inner liner

Before demounting, use an immersion test to determine if the tire itself has a puncture. If a puncture is found in the tread or sidewall, mark it before demounting tire.

Carcass vents

All tubeless tires, 8-ply rating and above, have been vented in the lower sidewall area. These vents prevent separation by relieving pressure buildup in the carcass plies and under the sidewall rubber. These vent holes (marked by green colored dots) will not cause undue air loss. Covering them with water or a soap solution may show an intermittent bubbling, which is normal.

Air retention test

When no leaks can be found on the prior checks, an air retention test must be performed. The tire should be inflated to operating pressure for at least 12 hours before starting the test. This allows sufficient time for the casing to stretch, but can result in apparent air loss. The tire must be reinflated after the stretch period to operating pressure. Allow the tire to stand at constant temperature for a 24-hour period and recheck pressure. A small amount of diffusion is considered normal. However, an inflation pressure drop of more than 5 percent of operating pressure indicates excessive vent leaking.

Accessory Technology

Tube inspection and repair

Since there are only two reasons for air loss in a tube-type—a hole in the tube or a defective valve or valve core—finding an air leak is usually simple.

As with a tubeless tire, the first step is to check the valve and replace the core if it is defective. If the valve is airtight, demount the tire, remove the tube, locate the leak (by immersion if necessary), and repair or replace the tube.

When inspecting a tube to decide whether or not it is the cause of the leak, use only enough pressure to round out the tube. Excessive inflation strains splices and may cause fabric separation on reinforced tubes.

Reuse of tubes

A new tube should be used when installing in a new tire. Tubes, like tires, grow in service, taking a permanent set of about 25 percent larger. This makes a used tube too large to use in a new tire which would cause a wrinkle and lead to a leak.

Typical tread wear

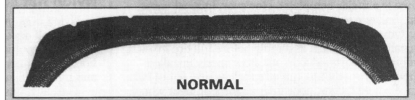

NORMAL

Even tread wear on this tire indicates that it has been properly maintained and run at correct inflation pressure.

EXCESSIVE

Worn to the breaker/carcass plies, the tire should not be left in service or retreaded.

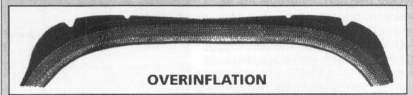

OVERINFLATION

Continuous overinflation accelerates center tread wear. It reduces traction while making tread more susceptible to cutting.

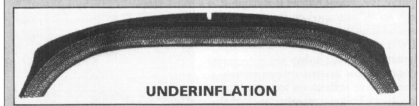

UNDERINFLATION

Excessive tread shoulder wear results from chronic underinflation and increases the chance of bruising sidewalls and shoulders and shortens tire life because of excessive flex heating.

Accessory Technology

A few tire definitions to keep you straight

Tread
Made of rubber, compounded for toughness and durability. The tread pattern is designed in accordance with aircraft operational requirements. The circumferential ribbed tread is widely used today to provide good traction under varying runway conditions.

Sidewall
A protective layer of flexible, weather-resistant rubber covering the outer carcass ply, extending from tread edge to bead area.

Tread reinforcement
One or more layers of nylon fabric that strengthen and stabilize the tread area for high-speed operation. Also serves as a reference for the buffing process in retreadable tires.

Breakers
Reinforcing plies of nylon or aramid fabric placed under the tread rubber to protect carcass plies and strengthen and stabilize tread area. They are considered an integral part of the carcass construction.

Plies
Alternate layers or rubber-coated nylon fabric (running at opposite angles to one another) provide the strength of tire. Completely encompassing the tire body, the carcass plies are wrapped around the wire beads and back against the tire sidewalls (ply turnups).

Beads
High-tensile strength steel wires embedded in rubber, the beads anchor the carcass plies and provide firm mounting surfaces on the wheel.

Apex strip
A wedge of rubber affixed to the top of the bead bundle, serving as a filler.

Flippers
These layers of rubberized fabric help anchor the bead wires to the carcass and improve the durability of the tire.

Chafers
Protective layer of rubber and/or fabric located between the carcass plies and wheel to prevent chafing.

Bead toe
The inner bead edge closest to the tire center line.

Bead heel
The outer bead edge that fits against the wheel flange.

Innerliner
In tubeless tires, this inner layer of low permeability rubber acts as a built-in tube and prevents gas from seeping through casing plies. For tube-type tires a thinner rubber liner is used to prevent tube chafing against the inside ply.

Tire inspection criteria

Tread conditions

Cuts
Penetration by a foreign object. Remove the tire for further inspection for any cuts into the carcass ply or for cuts extending more than half of the width of a rib and being deeper than 50 percent of the remaining groove depth.

Tread chunking
A pock mark condition in the wearing portion of tread... usually due to rough or unimproved runways. Remove if fabric is visible.

Spiral wrap
Some retreads have reinforcing cords wound into the tread which become visible as the tire wears. This is an acceptable condition and not cause for removal. The wrap reduces chevron cutting and tread chunking.

Tread separation
A rather large area of separation or void between components in the tread area due to loss of adhesion. Usually caused by excessive loads or flex heating from underinflation. Remove immediately.

Groove cracking
A circumferential cracking at the base of a tread groove; remove if fabric is visible. Can result from underinflated or overloaded operation.

Peeled rib
Usually begins with a cut in tread, resulting in a circumferential delamination of a tread rib, partially or totally to tread fabric ply. Remove and replace.

Rib undercutting
An extension of groove cracking progressing under a tread rib; remove from aircraft. Can lead to tread chunking, peeled rib or thrown tread.

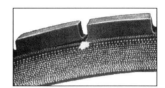

Thrown tread
Partial or complete loss of tread down to tread fabric ply, undertread layer or carcass plies. Remove and replace.

Tread rubber reversion
An oval-shaped area in the tread similar to a skid, but where rubber shows burning due to hydroplaning during landing. Usually caused by wet or ice-covered runways. Remove if balance is affected.

Blister
A void within the tread or sidewall rubber. Remove and inspect.

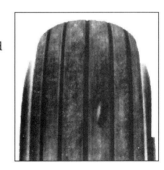

Accessory Technology

Chevron cutting
Tread damage caused by running and/or braking on cross-grooved runways. Remove if chunking to fabric occurs, or tread cut removal criteria are exceeded.

Sidewall conditions

Cut or snag
Penetration by a foreign object on runways and ramps; in shops, or storage areas. Remove and replace if injury extends into fabric.

Ozone or weather checking/cracking
Random pattern of shallow sidewall cracks. Usually caused by age deterioration, prolonged exposure to weather or improper storage. Remove if fabric is visible.

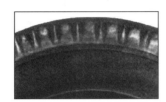

Radial or circumferential cracks
Cracking condition found in the sidewall/shoulder area; remove and replace if down to fabric. Can result from underinflated or overloaded operation.

Sidewall separation
Sidewall rubber separated from the carcass fabric. Remove immediately.

Bead conditions

Brake heat damage
A deterioration of the bead face from toe to wheel flange area; minor to severe blistering of rubber in this area; melted or solidified nylon fabric if temperatures were excessive; very hard, brittle surface rubber. Tire is to be scrapped.

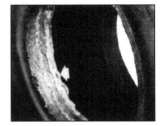

Kinked bead
An obvious deformation of the bead wire in the bead toe, face or heel area. Can result from improper mounting or demounting and/or excessive spreading for inspection purposes. Tire is to be scrapped.

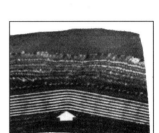

Carcass conditions

Inner tire breakdown
Deterioration (distorted/wrinkled rubber of tubeless tire inner liner or fabric fraying/broken cords in tube-type tires) in the shoulder area—usually caused by underinflated or overloaded operation. Tire is to be scrapped.

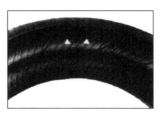

Impact break
Rupture of tire carcass in tread or sidewall area, usually from extreme hard handing or penetration by foreign object. Tire is to be scrapped.

September/October 1994

Accessory Technology

Aviation ignition exciters
Construction, operation, and maintenance

By Ted Wilmot

The exciter is the heart of the ignition system. It provides the necessary energy in sufficient quantity to allow reliable light-off of the fuel-air mixture.

Exciters are extremely rugged and specialized power supplies with a number of unique design features designed for the hostile engine environment.

An exciter is one of four primary components in most modern ignition systems. The input power supply, either DC or AC, depending upon the design of the individual aircraft, supplies current to the exciter, which fulfills the same role as the coil in an automobile engine.

Most modern exciters are DC powered due to airframe/Electronic Engine Control (EEC) integration. The EEC monitors many engine parameters, including the ignition process, and regulates such items as air/fuel mixture and spark discharge. Low voltage DC is easier to control with EEC circuits than the relatively high voltage alternating current supplied on PMA (Permanent Magnetic Alternator) AC applications.

Exciters are also classified into Low Tension (LT) and High Tension (HT) categories. Those with output voltages below 10 kilovolts (10,000 volts) are generally considered *Low Tension*. Typical output voltages for these exciters are in the 3 to 5kV range. Voltages above 10kV are considered *High Tension*, with typical output ionization voltages in the 15 to 30kV range for an igniter plug with a 0.050-inch gap.

To best understand how an exciter works, think of it as a transformer. The exciter receives input voltage from a DC source (battery) in the 10- to 30-volt range or from an AC source (Permanent Magnetic Alternator) in the 30- to 700-volt range. By the time an electrical pulse is discharged from the exciter, through the leads and to the igniter, voltage may range from 3,000 to 30,000 volts, depending upon conditions such as igniter plug wear and spark gap width.

One important note: Peak output voltage from ignition exciters never exceeds the value necessary to jump the igniter gap and create spark. In other words, an exciter capable of delivering 30kV may only have to deliver 15kV when using a new (small-gap) igniter plug, or if combustor pressure is low, such as when the engine is idling.

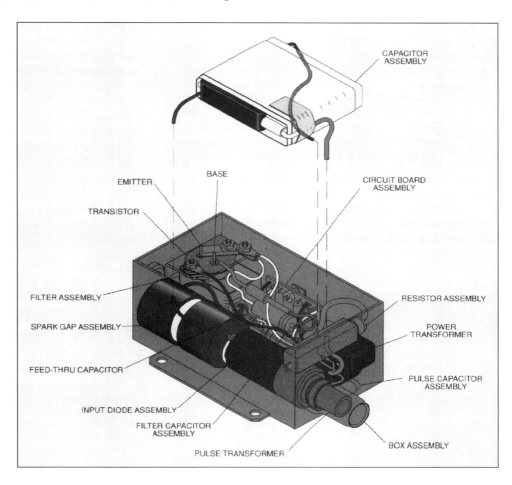

All exciters are designed with four basic circuit function blocks:
- Electromagnetic Interference (EMI) Filter
- High Voltage Power Supply
- Energy Storage and Switching
- Output Pulse Forming Network (PFN).

Each functional block of exciter design and its purpose is as follows:

EMI filter— Exciters, and more specifically aviation ignition systems, are notorious generators of radio waves, a result of the high magnitude and high-frequency currents and voltages produced. The emissions can interfere with the normal operation of avionics, EEC, and other sensitive electronic instrumentation outside the ignition system. Therefore, such emissions must be minimized and contained within the exciter and other ignition components.

Also, since ignition systems employ sensitive electronic circuits, EMI generated from outside sources must be prohibited from entering and interfering with exciter operation. Therefore, the first block in exciter design is an EMI Filter.

EMI Filters are generally built into separate metallic compartments or enclosures within the exciter case. This provides increased protection between the internal exciter components that may be susceptible to outside noise.

Likewise, the filter protects systems external to the exciter from radiated and conducted emissions generated by the exciter circuitry. Radiated noise is contained within the exciter by effectively creating a short circuit to ground within the exciter case. Since the exciter enclosure is a completely sealed metallic structure, radio emissions are shorted out by the exciter case. In essence, the exciter case acts as a single conductor that is short circuited to itself.

High voltage power supply stage— The high voltage power supply is the heart of every exciter. The circuit converts input energy into high-voltage direct current, which is stored in the main storage capacitor.

High voltage energy storage offers several distinct advantages: First, if the tank voltage is at least 3kV, minimal output wave shaping components are required for low tension operation. This means the electrical pulse generated by a low tension exciter will directly fulfill low tension igniter plug [surface coating (engobe) or homogeneous semiconductor pellets (such as silicon carbide)] spark generating requirements without need for additional pulse forming circuitry. Second, volumetric efficiency of the energy storage capacitor is improved at high voltages.

Because stored energy is proportional to the square of the voltage, a twofold increase in stored voltage will yield a fourfold increase in stored energy for a given capacitance.

All exciters use magnetic components (transformers) to accomplish voltage transformation. Circuit design and construction are dependent upon input voltage—AC or DC. AC exciters rely on transformers and voltage doublers to transform voltage, while DC exciters use DC-DC converter units, or choppers, to achieve high voltage generation. DC-DC converters are required since transformers, by definition, operate on AC voltage only. The converter "chops" the input power to simulate an AC electrical wave.

Energy storage and switching— This stage is the business end of every exciter. The energy storage and switching section stores the incremental charge provided by the transformer or DC-DC converter. This allows a large amount of energy to be accumulated for rapid discharge at the igniter plug.

A tank capacitor is used to store energy. The energy storage and switching components include the tank (storage) capacitor and a switching device.

Most exciters use mica capacitors for energy storage. Mica capacitors are capable of working voltages in the order of 20kV. However, most exciter circuits use capacitors capable of approximately 3kV storage potential.

Three kilovolt storage, or tank voltage, is considered an optimum number for several reasons. Among these: spark gap and semiconductor switching components generally cannot tolerate higher voltages in compact aviation-oriented packages. Also, the energy storage (volumetric) efficiency of compact aviation-type mica capacitors is at a maximum (with a reasonable life expectancy) at 3 to 3.5kV. Also, the operational temperature range of mica is compatible with aviation-grade exciter environmental conditions — -55 to 121°C (-67 to 250°F.)

Conventional exciters use spark gaps to accomplish the switching function. Spark gaps are very resilient devices, capable of withstanding extreme environmental fluctuations, but have relatively short operational lives of about 30 million sparks. Spark gaps are relatively insensitive to temperature, making the exciter more tolerant of high temperature environments. However, because of their limited life, relatively bulky size, and inefficiency, gaps are being replaced with solid state semiconductor switching devices called thyristors. Unlike spark gaps, semiconductor switching devices are not life-limited when used within their design limits.

Output wave shaping— The output stage performs a vital function: shaping the discharge current waveform. By shaping the discharge waveform, peak power and delivered energy can be tailored to meet specific engine requirements and protect the energy storage/switching components.

By controlling the rate of energy delivery, plasma (spark) achieved at the igniter plug gap may be short in duration but have extremely high peak power (heat). Or the spark may have a long (burn) duration. With a long burn, peak power is diminished—but delivered energy with respect to time is actually increased slightly because of lower losses in the ignition system. As a rule of thumb, output durations on the order of 50 to 150 microseconds are desirable and are considered average. The duration will vary depending upon ignition lead length, igniter plug design and engine requirements such as fuel atomization, mass airflow, combustor geometry and operating altitude.

Of course, regardless of the design of the exciter, its sole mission is to supply the ignition system with proper power to begin the combustion process. The ideal exciter would deliver infinite energy at an infinitely high spark rate, but in the real world issues such as power handling, thermal management and igniter wear rates must be considered.

Spark rate is the number of sparks over a unit of time. The number of sparks per second required for ignition varies greatly depending on engine design. Small turbine engines that accelerate quickly require higher spark rates—2 to 6 sparks per second; larger turbine engines only require 1 to 2 sparks per second to ensure light-off within the ignition window.

General maintenance tips for turbine ignition systems

Exciters are usually the most expensive component of the turbine ignition system, so it's important that positive steps be taken at regularly scheduled maintenance intervals to ensure maximum exciter life.

The exciter is only one of the three major components of the ignition system; however, all three components must be in good condition to prevent adverse or rapid wear of the other two.

For example, an igniter that has exceeded its recommended wear limits may still spark, but its increased voltage requirements will unnecessarily stress the lead and exciter, shortening the useful lives of both components. A lead that can no longer maintain solid contact with the igniter or exciter connector will eventually cause voltage to arc internally due to excess contact wear—thus welding the connectors and leading to rejection of otherwise good parts. In order to obtain maximum ignition component life, tips as outlined below should be followed.

Exciters
1. Exciters can store a lethal dose of electrical energy. Always wait several minutes after ignition system shutdown before handling any turbine ignition components.
2. Make sure connectors are clean and free to grease, oil or thread lubricant. All of these compounds are highly conductive and can lead to flashover.
3. Never attempt to open or repair an exciter without reference to the specific component maintenance manual for the unit. These manuals are available from the exciter or engine manufacturer.

Igniter connectors and leads should be free of dirt, oil, grease, and lubricants. All of these compounds are highly conductive and can lead to flashover.

Leads
1. Inspect the braided lead conduit for broken or cut wire strands, especially in areas where routed around sharp surfaces. The conduit provides protection against radio frequency interference (RFI) as well as a path to ground for expended electrical energy. The lead should be repaired or replaced if the conduit is cut, or if there are more than about four broken strands.
2. Lead connectors should be free of dirt, oil, grease, and lubricants. Additionally, operators should ensure proper spring tension is maintained by lead sockets.

Accessory Technology

3. Weld and braze joints on connector elbows should be inspected for cracks, which sometimes are the result of vibration caused by cowling impingement. Should this condition be noted, check appropriate application data (manufacturer's catalog or engine IPC) to make sure the proper lead is installed.

Igniters

1. Check to make sure the igniter is within recommended wear limits by using an erosion gauge. If not, the igniter should be replaced to protect against flame-out, to maintain structural integrity and to lengthen exciter life.
2. Igniter connectors should be free of dirt, oil and grease, and anti-seize lubricants. Connector pins or terminals that show obvious pitting probably have experienced arcing because of insufficient lead contact. Severe pitting requires igniter replacement and a close inspection of the lead for possible replacement. Continued operation in this condition will result in premature exciter failure.
3. Do not attempt to reinstall a dropped igniter. The ceramic insulator will have cracked internally, leading to flashover and severe ignition system stress.

July/August 1994

Ted Wilmot is manager of Electrical Product Development at Champion Aviation Products, Cooper Industries, in Liberty, SC. He has been actively involved in the turbine engine component and accessory business for over seven years.

Accessory Technology

Lubricants as tools

By Eric Kornaw

Most aircraft technicians recognize that having the right tool for a particular job is of paramount importance. As a result, many technicians have toolboxes the size of mini-vans packed with every conceivable type of hand tool.

An A&P wouldn't even think about installing an MS or AN fastener with a set of vice-grips (vice-grips do have their place). But when it comes to selecting a lubricant for use on an aircraft, these same technicians will select an aerosol of "general purpose" lube and expect it to be suitable for every purpose. It just isn't so. Just as wrenches come in various shapes and sizes, the lubes and compounds we use have specific properties and uses.

Specifications

Specifications are used by aircraft manufacturers as means of denoting what types of products are suitable for use on aircraft. If specifications did not exist, maintenance manuals would have to list the trade name and vendor for each product authorized for use on an aircraft. Anyone who has tried to track down a product whose maker has changed its name or gone out of business knows what a problem this can be.

A specification is nothing more than a document listing minimum conditions which a product must meet. Rarely do specifications for lubricants specify formulations. Usually consumable material specs list required properties such as viscosity, corrosion resistance, oxidation resistance, rubber swell, etc. The specification may require that products which meet these minimums be compatible with each other... then again, they may not. That's why mixing lubes from different manufacturers is not recommended.

Types of specifications

Consumable Materials (oils, greases, hydraulic fluids, sealers, etc.) are generally described by military, federal, industry, or aircraft manufacturer's standard.

Military & federal specs

The basic identification for these documents is the spec number, i.e., MIL-G-81322, VV-P-236. The symbol "DOD" replaces the symbol "MIL" in new and revised specifications and standards covering "hard metric" or "hard converted" items (for all you metric system aficionados).

Revisions are indicated by a suffix letter to the basic number, except for AN standards and Qualified Products lists (QPL) which indicate revisions by a suffix number (e.g., MIL-G-81322D, VV-P236A). Amendments are indicated by a suffix number in parenthesis to the basic number (e.g., MIL-G-81322D(2) indicates Amendment #2 to revision D of MIL-G-81322).

In general, revisions are issued when circumstances dictate changes or adjustments to the properties, minimum attributes, or test procedures. Amendments are issued to correct or clarity language in the existing publication. For government use, the latest revision or amendment always supersedes prior issues.

With few exceptions, most producers of consumable materials manufacture products only to the current revision. There are, however, some civilian applications where products are manufactured to superseded standards acceptable to or preferred by the aircraft builder.

Industry specs

Industry specifications are those which are issued by groups such as the Society for Automotive Engineering (SAE) and carry numbers with prefixes such as AMS or NAS. Aircraft and engine manufacturers are also involved in issuing specifications for materials used on their aircraft. Boeing, for instance, prefixes its specs with the letters "BMS" (Boeing Material Specification).

Qualified vs. conforming specs

Some military and federal specifications require that products purchased to spec by the government be included on that specification's Qualified Product List. The identifier for a Qualified Product List (QPL) consists of the letters "QPL," the basic number of the specification to which it is related, and a dash number to indicate the revision status (e.g., QPL-81322-15 is the 15th revision of the QPL for MIL-G-81322).

To quality for inclusion on a QPL, a product must be tested and passed by the agency designated in the specification. These products are referred to as "qualified."

Qualification ensures that the government has tested representative material and that the product meets the specification requirements. Many, but certainly not all, aircraft category specifications have Qualified Products Listings.

Those specifications which do not require QPLs generally leave the process of testing and certification to the manufacturer of the product. These products are sold to conform to the specification and are usually referred to as "conforming products." When using conforming products, you are relying on the integrity of the product's manufacturer when it says that its lube or sealer conforms to the specification.

Accessory Technology

There are also instances where a material manufacturer will certify that its product meets the performance requirements of a specification where a QPL exists, even though its product is not included on the QPL. These too are sometimes referred to as conforming products.

There may be good reasons why a manufacturer does not submit a product for qualification. In some cases the cost of qualification is prohibitive or the manufacturer may be concerned over the loss of trade secrets during testing. Some products may meet the composition and performance criteria, but may not meet the specific packaging criteria of a specification. For example, there are specifications for hydraulic fluids (MIL-H-5606) which make no provisions for packaging in 1-gallon screw-top containers. A fluid packaged this way would not technically meet the spec, while the same fluid packaged in hermetically sealed quart containers would qualify.

Remember, though, when you accept a conforming product instead of qualified product, you are accepting the responsibility for its suitability for use on the aircraft.

Equipping the consumable material toolbox

Below, are some of the chemical "tools" which are commonly specified for use on aircraft. This list isn't meant to be all inclusive, but it's a starting point for someone starting to equip his box with the basics.

Aircraft greases

There are four or five greases commonly specified for use on aircraft. In addition, there are a number of specialty greases used in special situations such as oxygen lines and trim actuators. Greases are generally not required to be compatible by specification, so it's not a good idea to mix brands from different manufacturers or products made to different specifications.

Greases are mostly oil with a thickener added to hold it in place. The oil does the lubricating; the thickener keeps it in place. Greases are used when the weight or complexity of a pressure lubrication system is not practical. Greases can also seal, cushion, and absorb shock loads.

Compatibility problems can arise between base oils, or the thickening systems used. Most aviation greases are composed of synthetic oils thickened with inorganic thickeners such as bentonite clay. They are generally *not* compatible with the soap-thickened greases in general automotive use. If you inadvertently mix greases, it is good practice to disassemble the components and clean out the mixture. When this is not practical, purge the fitting with grease until it appears clean, and decrease the interval between lubrication.

MIL-G-81322—grease, aircraft, general purpose, wide temp range—The most widely used grease on aircraft today. Good heat resistance. Mobilgrease 28 (red in color), Aeroshell 22 (caramel to brown), and Royco 22 (dark brown) are commonly available qualified products. Supersedes MIL-G-3545, MIL-G-7711, and MIL-G-25760.

MIL-G-81827—grease, aircraft, high load capacity, wide temp range—Usually manufactured by adding molybdenum disulfide to MIL-G-81322. For high load situations. Black in color. Mobilgrease 29 and Royco 22MS are qualified products.

MIL-G-23827—grease, aircraft and instrument, gear and actuator screw—These greases are required to have better low temp performance than MIL-G-81322. Aeroshell 7 and Royco 27 are commonly available. Supersedes MIL-GT-3278.

MIL-G-21164D—grease, molybdenum disulfide. For low and high temperatures—Usually manufactured by adding molybdenum disulfide to MIL-G-23827 greases. Aeroshell 17 and Royco 64 are commonly available.

MIL-G-3545 (obs)—grease, aircraft, high temperature—Here's a case where a specification that has technically been superseded by MIL-G-81322 is preferred by some aircraft manufacturers for use as a wheel bearing grease. Aeroshell Grease 5 is the easiest to find.

Other greases

Other greases that you might want to include on your list are as follows:

MIL-G-25537—for helicopter oscillating bearings
MIL-G-6032—for fuel system valves and gaskets (Plug Valve Grease)
MIL-G-4343—pneumatic system grease
MIL-G-83261—extreme pressure, anti-wear (This one's expensive so unless you need it, forget it)
MIL-G-27617—fuel and oxidizer resistant grease (oxygen system use)

Oils

Remember that the most important property of an oil is its viscosity. There are some lubricants which are "slippery" but do not have the viscosity to adequately cushion and protect moving surfaces. Many of the penetrating oils fall in this category.

These products can free up components, but unless they leave behind an adequate film of the proper viscosity, they actually lead to increased wear. A number of the products commonly found in the hangar are not actually qualified as lubricants, but as corrosion preventive compounds.

Accessory Technology

MIL-L-7870—general purpose lubricating oil with good low temperature performance. Has a very low evaporation rate and good rust protective properties. The most commonly specified general purpose oil. Has a viscosity of about 10 cs. (a little thicker than turbine oil).

VV-P-800—general purpose oil similar to MIL-L-7870, specified by some aircraft manufacturers instead of MIL-L-7870. This oil is also water displacing. Its viscosity is slightly higher than 7870.

Other oils

The other products commonly specified by airframe manufacturers include mineral oils in various viscosities such as SAE 10, 20, and 30 weight. In addition some manufacturers include corrosion preventive compounds as recommended lubricants.

Summing it up

Most aircraft manufacturers approve lubricants and consumable materials by specification. Many of the chemical products touted for use on aircraft do not meet any particular specification. Some products which do meet the specs are tested and qualified to the specification. Other products are not qualified but conform to the specification. It comes down to reading the maintenance manual, determining what the aircraft manufacturer specifies, and then choosing the correct material for the application. Check the can the next time you pick up a lubricant. You may be surprised that the product you've been using doesn't meet the specification called out in the maintenance manual. *May/June 1994*

Eric Kornaw is an aviation maintenance instructor at Cincinnati Technical College and is president of ACI (Aviation Consumables, Inc.) in Cincinnati, OH.

Accessory Technology

Slip sliding away
Escape slides are misunderstood

By Greg Napert

Escape slides are a critical item that have a large potential for saving lives in the event of an aircraft accident. Too often, however, they are open to abuse that should be cause or concern. Escape slides have got to work at that critical moment. And it's the technician's responsibility to assure that all systems are go when the slides are needed most.

Technicians at BFGoodrich Aerospace, Aircraft Evacuation Systems in Ontario, CA, a repair station that repairs and overhauls escape slides and rafts, say there are many incidences where rafts are mistreated in handling, servicing, and shipping.

Often, they say, these items are subjected to abuse by people servicing the aircraft. Service trucks or catering people pounding items against the door and damaging the valve and/or gauge or pulling and tugging at the girt bar (the bar that locks the slide in place for deployment prior to takeoff) are all too common. So it's important to perform regular inspections of these areas to assure that no damage has taken place between overhaul cycles.

The girt and the girt bar are subject to considerable amount of wear and tear. The bar is constantly being installed and removed, so there's wear and tear on the bar and the fabric. It's a good idea to keep an eye on the fabric to make sure that it's not excessively worn or that fasteners aren't excessively corroded. Also, make sure the release latch cable that secures the slide to the aircraft is in good condition. If you see any problems developing in these areas, the slide should be pulled and inspected by an approved facility or the airlines slide shop.

Accidental deployment

It's not uncommon for escape slides to be accidentally deployed. When this happens, the slides must be disconnected from the aircraft so that another unit can be installed. The deployed slide must then be overhauled and inspected. An avoidable mistake that happens when many of these slides are removed is that the slides are often dropped onto the ground without any concern for the CO_2 cylinders, valves and gauges, which often results in damage.

Further, the inflated slide typically needs to be moved to an area where it can be packed for shipping. In a hurry, ground crews drag the slide to the nearest hangar for repair and deflation. This dragging often results in severe damage to the slide.

Many of these escape slides cost upwards of $60,000. If one is moved, it should be lifted off the ground and protected in a fashion that will preserve the integrity of the slide. The preference would be to lift the slide off the ground or to deflate the slide prior to moving it.

Damage is also inflicted to the escape slides during the deflation process.

The best method for releasing the air is by opening the flapper valve on the aspirator when possible. Some aspirators, such as the 747 BFGoodrich slide, however, will not allow this. In that case, the best method would be by opening the inflate/deflate valve. A shop vac or other vacuum source will speed up the deflation process. Remove the aspirator or pressure relief valves only as a last resort due to the increased chance of parts loss and/or damage.

The aspirator exists to allow ambient air to mix with the CO_2 and nitrogen from the cylinder to more quickly inflate the slide/raft. The CO_2 and nitrogen by themselves don't produce enough volume to inflate the slide. Some of the slides inflate in as little as three to four seconds.

BFGoodrich technicians say that they have seen instances where airline maintenance personnel had difficulty deflating the slides and used a knife to slit the slide open.

Although most any type of damage to these slides can be repaired through replacement of complete panels, the additional cost of repair for this type of damage is unnecessary.

The overhaul/inspection process

An escape slide requires a complete overhaul every three years despite its condition.

At the time of overhaul, the slide is unraveled and inflated to inspect for any damage, wear, or seam separation. The CO_2/nitrogen cylinders are life limited so dates are checked to determine whether or not they need to be taken out of service.

Usable CO_2/nitrogen cylinders are hydrostatically tested, the valves and regulators inspected, overhauled and tested, and mechanical restraints (frangible links) are inspected. Additionally, any emergency supplies are inspected and replaced.

Folding and other tips

A critical step in the overhaul process is the proper folding of the slide. Folding has a direct impact on proper deployment of the slide. Improper deployment can mean total ineffectiveness of the slide, particularly if the slide fails to inflate straight out from the aircraft. If it deflates beneath the aircraft, or against the ground, it's as useless as not inflating at all.

Folding methods are very dependent on the overhaul manual and the design of the slide.

It's important that all of the wrinkles are taken out and that the air is completely removed so that the folded slide fits into the compartment in the door. This is accomplished by maintaining a vacuum on the slide to continue to remove any existing air.

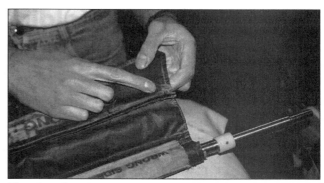

The girt takes the most abuse and should be inspected often.

Although the best method for releasing air on most slides is by opening the flapper valve on the aspirator, there are some models, such as this 747, that won't allow it.

One final thing to be cognizant of at all times is the danger of activating one of the gas cylinders. The rafts can inflate with explosive force and, if activated at the wrong time, could mean damage to equipment or bodily injury.

The cylinders should have a safety pin installed when possible during handling and shipping of the escape slides (some systems are not equipped with safety pins). The company says it often finds safety pins aren't installed incorrectly on slides that are received, and that these are the equivalent of a loaded gun ready to fire.

A good habit to develop is to check and double check the safety pin prior to packing and/or shipping the slides. *March/April 1994*

Accessory Technology

Typical escape slide folding sequence...

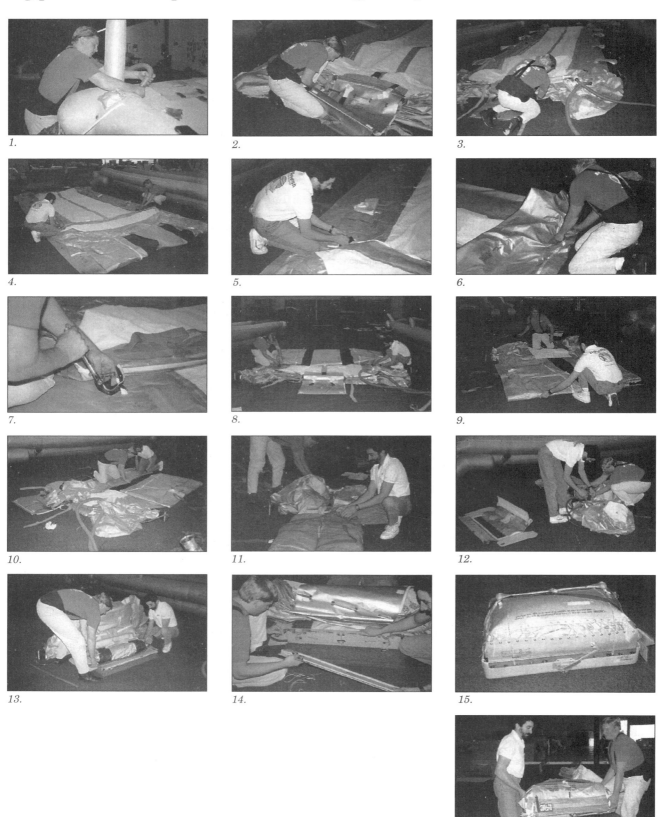

1.
2.
3.
4.
5.
6.
7.
8.
9.
10.
11.
12.
13.
14.
15.
16.

Accessory Technology

The composite propeller
Accepted but misunderstood

By Greg Napert

Composite propellers were initially introduced in the '40s. The blade was a phenolic, ground adjustable propeller that offered a weight savings to aircraft owners.

But the widespread utilization of composite technology didn't take off until the current decade with the development of new materials such as Kevlar® and carbon fibers. Today, composite propeller blades are quite common, particularly for applications on larger propeller-driven aircraft where weight savings can result in reduced fuel consumption, lower horsepower requirements and better overall performance.

These propellers are not only a weight-savings proposition; composite blades are a horsepower-saving proposition as well. Instead of swinging around metal propellers with higher horsepower engines, lower horsepower engines can be utilized to produce the same thrust, and, therefore, an aircraft can be designed with smaller engines.

However, composite propeller blades are understood by few technicians in the aircraft maintenance field.

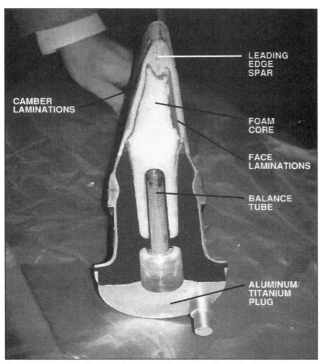
Cross section of typical composite propeller blade.

Jeffery Slattery, superintendent of the composites division of Hartzell Propeller Inc., in Piqua, OH, says that "we have seen things as crazy as someone applying a coating of Bondo® to their blades. We caught it because the blades weighed 2 1/2 pounds more when they came in than when they went out. When we peeled the paint off, we found layers of primer and Bondo on the blades."

The skills and equipment needed for performing maintenance and repairs to composites are fairly simple, and don't require a great amount of experience, but some basic knowledge and understanding is needed before repairs are attempted.

First, the technician must understand that the technology is completely different from aluminum blades. You can't apply any of your maintenance and repair knowledge of aluminum propeller blades to composite propeller blades.

Next, you've got to learn the makeup and structure of the particular blade that you're working on. Composite blades vary from model to model, and different repair and inspection criteria may apply from blade to blade.

If you don't know what you're doing, it's best to get some expert advice. Call the factory or an approved service center for information, or ship the blades out for repair. Composite blades are very expensive, so it's not worth risking a guess at a repair scheme.

How the blade is manufactured

Understanding how the blade is manufactured is important to understanding its makeup and how to repair it. Hartzell, for example, begins the manufacturing process with an aluminum/titanium plug at the root end. They then pour a high temperature urethane foam core to support the materials (Kevlar or carbon) which are then layed up over the foam core. Each blade, says Slattery, consists of laminates on both the face and camber and composite or metal spars (depending on the manufacturer) on the leading and trailing edges of the blade. After layup, the blade is cured in a close tolerance metal mold. After molding and inspection, an electroplated nickel erosion shield is bonded to the blade.

One of the reasons that composite blades are so expensive is because of how labor intensive the manufacturing process is. The materials are typically layed up by hand and can require up to 80 layers of fabric that are each individually cut and trimmed. Vacuum bagging or compression techniques are required at various stages

The Best of Aircraft Maintenance Technology Magazine   **175**

for debulking (getting the air out between layers); then four to six hours of curing time are required as it's heated and pressed in the finish mold.

And even after the blade is cured, it requires trimming, sanding, inspection, erosion edges, painting and final machining of the root—all hand labor that consumes hours of time. The Kevlar, carbon and electroformed nickel lead edges are also very expensive, which also adds to the final cost of the blade.

Hartzell recommends 3,000-hour TBOs on its propellers, but different operators will get extensions from the FAA based on their operating experience. If you do everything you need to get an extension and you maintain it properly, the manufacturer will typically support the operator. In general how the operator treats the blades and its history with them will determine how often the blades need to be overhauled.

"So much of the damage that you find brings a great deal of panic to the operator because they don't really understand what they're seeing, and these damages are usually very easily fixed," says Slattery.

Typical damage

He offers an example of a typical damage. "Here is an example of tip damage due to a bird strike. This blade has had the tips wiped out. On a metal blade, you'd have to grind it down and you might come out below the minimum dimensions and have to scrap it. With this composite blade, you simply rebuild the damaged area by removing some of the damage and 'laminate' in a patch to take up the area that's missing."

The one thing that you've got to keep in mind when making composite repairs is that every layer of material has its own unique fiber orientation. The fibers are oriented to provide whatever structural support is needed for that direction. So when you develop a repair scheme that involves anything that is structural, engineering has to get involved to incorporate fiber orientation. Those laminates have a reason for being aligned the way they are.

Repair schemes vary depending on where it is on the blade. If you have a deep gouge in the root area of the blade, where it's a high stress area, the repairability goes down quite a bit. At the tip, the repairability goes up because it's not as critical.

"Layered repairs are typically in a category that only the factory or an approved repair center can do. It means that the prop is being rebuilt structurally and the strand orientation is important," he says.

Slattery says that it's impossible to design and develop a manual that covers every possible type of damage, so if something is outside of normal limits, the damage must be evaluated on a blade-by-blade basis.

From an economical standpoint, explains Slattery, it always makes sense to repair a composite blade vs. scrapping it. If you add up the high materials cost and the high amount of labor involved, it's an expensive

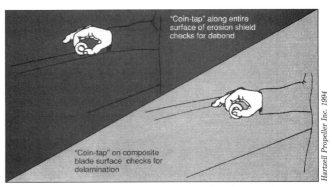

A large washer or similar object is a valuable tool for detecting debonds in the propeller. Variations in the sound, produced by tapping firmly with the washer, provide clues as to the soundness of the blade.

blade, and if there's anyway that you can put a few hours of labor and save the blade, it's worth it.

In fact, he says, engineering says that many repairs are actually stronger than the original blade after the repair is complete. So it's not like you have to live with a less structurally sound blade afterward. After a repair, the blade should be as good as new or better.

Due to the nature of the erosion shield on the leading edge, it's quite common to have to replace it. It's designed with a 40- to 60-thousandths thickness at its midpoint, and when it gets to the point where it has worn out or is damaged beyond limits, you simply pull it off, throw it away, and install a brand-new one. This restores the prop to a like-new condition, and you're ready for another two to three overhauls. Minor repairs are allowed in the field, but replacement must be done by an authorized service center.

The edge is typically removed at the service center by applying heat very sparingly and moving the torch rapidly along the edge until the adhesive softens and the edge comes loose. "It's very easy to scorch the material and damage the propeller doing this, so you've got to have some experience or proper training before you attempt it. I've seen leading edges where technicians stop-drilled the cracks. This is not an appropriate repair. In fact, you may not have to do anything at all with cracks in the erosion shield, depending on the crack," he says.

Field maintenance

"General maintenance is very important to the blades," says Slattery. "If you see a little damage, address it right away; don't let a small delamination grow into a larger one and get out of hand so that the repair costs get out of hand."

Joe T. Hahn, senior technical support representative for Hartzell, says that "the most important thing with a composite propeller is to do an accurate inspection.

Accessory Technology

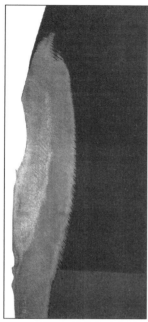

Preparation of trailing edge damage involves removing the damage and sanding the surface to accept layers of fabric.

Really examine what type of damage there is, and at the same time make sure you understand: what kind of damage you have, how large it is, how long it is, the extent of it, the nature of it. Is it a compressed area or sheared area where material has been cut away? That's the first step, inspecting the damage you have."

Slattery says "We've seen several things out in the field that kind of make us nervous, but one of the big ones is when someone soaks a deice boot in solvent to remove it. You can do that with metal blades, but on composite blades, you basically have to peel them off using a knife to break the bond.

"The problem with soaking it," he says, "in anything is that you're allowing the chemical to break down the paint, or more importantly, flow up into the foam core through the balance tube. When the solvent enters the foam core, it begins to eat away at the laminations and delaminate the blade. The balance tube is supposed to be sealed, but it isn't always perfectly sealed. The solvent is thrown out toward the tip and usually begins to delaminate the tip first."

Most solvents that are applied to the outside painted surface won't harm the blade. Slattery recommends using things like acetone, or solvents that evaporate quickly so that they have little or no effect on the surface.

Hahn says that another thing to "keep an eye out for are any dark areas that appear as a dark brown or black spot that appear to be heated. These are often the result of lightning strikes. A way to verify that it's a lightning strike is to check the hub for magnetization by placing a compass or gaussmeter nearby. If the hub is magnetized, you can be pretty sure that it is.

"We've also seen cases where grease leaks from the hub and into the balance tube, then into the core of the propeller, and this results in the prop delaminating from the inside of the blade and working its way out. Again, the inspection for this is a tap test for bad spots."

Trailing edge damage is quite common as well, he explains. "If you have damage on the trailing edge, you need to inspect it closely, remove the damage, and repair it by applying successive layers of material as specified in the repair manual, until it is rebuilt to above its original dimension. You then place a piece of plastic over it and apply pressure by either using a vacuum bag or by placing two flat surfaces over the repair and clamping it together."

Pressure is applied to remove any of the entrapped air as the adhesives cure. "There's some room for creativity with these repairs; applying pressure is the objective, and any way that you can do that will suffice. You then sand it to restore its original shape," he says. "You need to use the applicable maintenance manual to determine if damaged areas are airworthy or unairworthy."

Hahn says that one of the most important tools that the technician should use to inspect the propeller is a big coin or a round-edged steel washer. This is used to perform tap tests to search for delamination. The tap test may not pick up every problem, but it'll show you the ones that need immediate attention.

Hahn says that there are fancy tap hammers that you can buy for this purpose, but they're not necessary. A good size washer or similar object will do. "This is about the best test that you can do. When you run into a structurally deficient area, you hear it. Sometimes debonds are visible and sometimes not, but they're almost always audible."

For minor repairs, the tools that you'll need are quite basic. Kahn says that they should include: sanding blocks, files, sandpaper, scrapers, and a dremmel tool with various attachments, and a drum sander is handy (hobby tools lend themselves well to these repairs); plastic bowls, mixing sticks, tongue depressors, a scale for weighing the materials, chopped fibers, tools for measuring damage areas, etc. Also various clamps are handy for applying pressure.

An important point to remember is to make sure that you're using proper materials to make repairs, he says. One of the ways to assure that is to use the correct manufacturer's maintenance manual and order supplies from the manufacturer only. Materials from other sources probably haven't been tested properly and are probably not approved for use. Kahn says that by ordering from the factory, you also ensure these materials are fresh as well.

Many of the materials have shelf lives, and you need to make sure that this material hasn't exceeded its shelf life. Depending on how often you make repairs, it's a good idea to order only what you need from the factory as it's needed. That way you don't waste any by letting it age beyond its service life.

"Remember that for mixing purposes, repair materials are measured by weight," he says, "not volume. This is very critical for attaining the correct mixtures. If you don't mix it correctly, it won't cure properly and you've wasted your time. You need to have a good accurate (preferably digital) scale to weigh these adhesives."

In the manual, there's an abbreviated procedure for spot touch-up. It involves a primer-filler to fill in small types of damage, pinholes and irregularities. You then let it dry and sand it off so that it's level with the rest of the area. You then apply your finish coat as specified in the manual which may be polyurethane, vinyl, etc., and if it has a p-static (anti-static) coat, it will need that applied as well.

"Try to avoid the temptation to use unapproved finishing products that won't last and keep in mind that you may affect the balance by performing a repair. If you paint one blade, paint all the blades on the propeller so it remains balanced," he says.

Erosion shield repairs are a common requirement, he says. There's actually an allowable amount of debond of the leading edge erosion shield as long as there are no cracks.

He explains that erosion shield debonds are sometimes repairable. One type of repair for this is to carefully drill a small hole in the shield over the debonded area so it just penetrates the shield. The leading edge is a very beefy area so if you do drill slightly into the composites, it won't affect it structurally. You then apply a vacuum source to the hole and draw adhesive in from the debonded edge of the shield. This ensures that the adhesive is entirely under the debond. You then apply pressure to the shield with a clamp until it dries.

"What you don't want to do is continue to operate the airplane with a debond in the area of the crack. There is a certain amount of debond that's allowable, but if it's tied in with a crack, that's an unairworthy condition," he says.

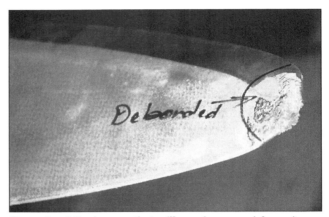

Example of tip damage that will require a special repair designed by the factory. Essentially, multiple layers of material will need to be removed and the tip will be completely rebuilt.

Reducing the cost of maintenance

Slattery says that there are some things that you can do to reduce the cost of maintaining composite props.

For instance, if an operator has a group of pilots that love reverse, they're going to add a lot of overhaul costs to the propeller because the leading edge is not going to last and need frequent repair.

If you see this type of abuse, you should feed this information back to the pilots or the person in charge of the operation and tell them that they can bring down their overhaul costs by proper operation. Just for example, the approximate cost of replacing the leading edge on a Beech 1900 blade is $625. Multiply this times four blades per engine and two engines and that's an additional $5,000 added to the overhaul costs.

There are many customers that have learned that they can save significant dollars on overhaul costs and have changed their operating procedures to accomplish this. *January/February 1994*

Accessory Technology

Troubleshooting the Bendix DP fuel control system

By Dennis Dryden

While all problems associated with the Bendix DP fuel system, used on Allison 250 Series engines, can't be resolved in the field, a better understanding of the systems and some simple troubleshooting tips can help save the customer valuable downtime and expenses.

The following is a brief description of the operation of the DP fuel control system and a few items of troubleshooting in areas where technicians seem to have the most trouble.

Operation

Regardless of the airframe/engine installation, all Bendix pneumatic control systems used on helicopters operate on essentially the same principles.

The fuel control is primarily a gas producer control which controls the engine during starting, idling, acceleration, deceleration and shutdown.

During starting, the fuel control schedules a rate of fuel flow increase in direct proportion to the compressor discharge pressure increase. Loss of, or restriction to, the compressor discharge pressure signal to the control will cause a starting problem.

At idle, the fuel control runs the engine at a requested N1 (gas producer) speed. The power turbine governor is normally not allowed to override (exception is the C18).

On all Allison 250 engines except the C18, the governor reset function is deactivated by the Pr-Pg valve at N1 throttle angles up to approximately 10 degrees above idle. This is a mechanical valve actuated by a plunger, which rides an external cam on the fuel control throttle shaft.

During acceleration, the fuel control again schedules a rate of fuel flow increase according to compressor discharge pressure increase, until the power turbine and

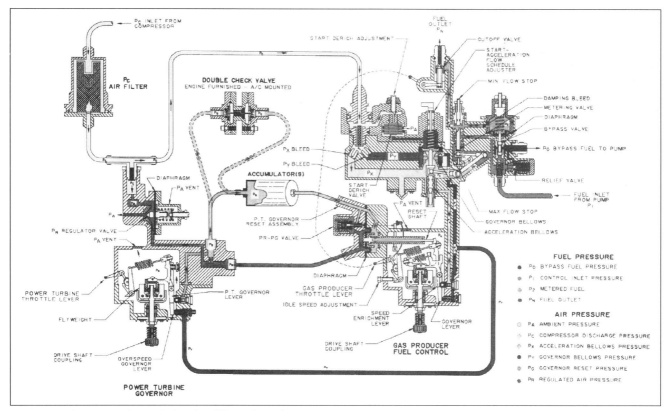

Fuel control system schematic for the Allison A250-C20.

Accessory Technology

rotor reach a speed which the power turbine governor recognizes as its area of responsibility. The governor then effectively begins to control the N1 speed by modifying the N1 speed requested input from the operator (twist grip or N1 throttle).

The pilot's input to the fuel control will still be requesting something in excess of 100 percent N1 speed. This request is established by the max N1 speed stop on the fuel control and is the N1 topping limit. The governor will continue to modify that request so that fuel flow is regulated to meet the requirements to hold the power turbine and rotor at the required speed for the operating condition, whether on the ground, in hover or in flight.

If collective pitch angle is increased, the governor coordinating linkage in the aircraft moves the governor throttle lever to a higher angle. This changes the reset signal to the fuel control and schedules more fuel flow to support the increased load and prevent the reduction of N2 and rotor speed.

During deceleration, the fuel control reduces the fuel flow to a set schedule which will allow for the most rapid reduction of N1 rpm without allowing the engine to flame out.

At shutdown, the cutoff valve is closed by the twist grip movement to the cutoff position. On Bendix controls, it's not necessary for the fuel pointer to be indicating zero degrees. Cutoff angle will normally be between 10 degrees and zero. On an aircraft the only important rigging points to check are idle and the minimum and maximum throttle stops.

The governor is primarily a power turbine speed control. It compares power turbine speed (N2) to a load requirement (collective pitch) and produces a signal which resets the fuel control accordingly. It controls the engine in flight by constantly varying the reset signal to the fuel control so that it can deliver the correct fuel flow to maintain the power turbine and rotor at the required speed for flight regardless of load change (collective pitch for example).

The fuel control and governor respond to only four inputs: throttle position, engine rpm (gas producer and power turbine), engine compressor discharge pressure and engine torque oil pressure on some twin-engine installations. They don't recognize or respond to engine combustion temperature.

It's important, especially when troubleshooting, to keep in mind that the fuel control and governor are two separate units that are pneumatically interconnected. Leaks or restrictions in these pneumatic lines, or circuits, may cause a reduction in performance to either or both units.

Problem: Engine will not start

Probable cause	Remedy
Air in the gas producer fuel control and lines.	Purge air from system at fuel nozzle and try a second start.
Faulty circuit to ignition unit.	Listen for ignition operation. Observe for fuel vapor coming out of the exhaust. Check input power to ignition unit. Isolate and replace defective part.
Faulty ignition exciter.	Listen for igniter operation. Observe for fuel vapor coming out of exhaust. Replace with known satisfactory unit.
Fuel nozzle valve stuck.	Replace fuel nozzle.
Fuel pump inoperative. (Fuel vapor will not be observed leaving the exhaust.)	Check pump for sheared drives or internal damage. Check for air leaks at inlet or fluid leaks at outlet.
Fuel nozzle orifice clogged.	Replace nozzle.
Water or other contaminant in fuel.	Check a sample of fuel from the bottom of the tank.

Start problems

Hot starts

With "hot starting" situations, if the peak turbine operating temperature (TOT) occurred suddenly after light-off at low N1 rpm (20 percent range), the cause may be due to fuel pressure disturbances generated in the fuel pump.

A rapid rise in TOT following a delayed light-off could indicate a faulty ignition component allowing fuel to collect in the combustion chamber between throttle movement and ignition spark. High TOT just after light-off with slow acceleration up to 30 to 35 percent N1 can be caused by a low battery or starter problem. The difference in TOT between a good battery/starter combination and a weak one can be as much as 100 C. High TOT at a steady state rpm indicates internal engine or air leak problems.

The fuel control schedule can be checked by motoring the engine and catching the fuel control output in a beaker (baby bottle) for 15 seconds. Compare the amount captured to a known standard (ref. customer service letter CSL1049). Such a standard can be established for any engine model by accomplishing this check on a number of known "good starting" engines, and recording the results.

Slow starts

In slow start situations, TOT and N1 relations, again, go a long way toward defining the problem. Slow starts and low TOT in the light-off (to 20 percent) can be caused by leaking Pc air line, blocked Pc filter or low fuel control schedule.

Accessory Technology

Slow starts between 35 percent and idle can be caused by contamination in the fuel control air circuits or low fuel control schedule. If normal starting adjustments, i.e., start derich and start/acceleration adjustment on the 250 engine have no effect, a great probability of air leaks or contamination exists. In some cases, the adjustments should be returned to their nominal settings and a fuel flow schedule check (baby bottle) should be made to find a new starting point.

Another possibility is a faulty power turbine governor. To determine if this is the problem, cap off Py line at the fuel control unit to isolate the problem.

No starts

Failure to start can be caused by a number of problems, the above chart covers most situations related to no starts.

Drooping and topping problems

Problems in this area are generally described as an inability to reach, or to hold, the maximum allowable N1 rpm with a resulting decrease in N2/Nr at lower than maximum allowable TOT or torque.

There are three primary limiting factors which independently relate to this regime of engine operation.

(1) N1 (or Ng) rpm. The maximum allowable limit. It is adjustable in the field at the fuel control maximum throttle stop screw.

(2) TOT, EGT, TIT, MGT. Different engine manufacturers use different designations. All refer to the temperature at the station in the turbine section which each manufacturer uses as his limiting reference. It's not adjustable and is not the fuel control responsibility except during acceleration.

(3) Output torque or horsepower. On most helicopters, this limit is imposed by the aircraft manufacturer and relates to the maximum safe power level which the transmission gear train/rotor system can tolerate. It's also not adjustable. The maximum fuel flow stop setting on the fuel control is intended to prevent inadvertently exceeding this acceleration.

There are several reasons for an engine not able to reach the maximum allowable limit. There are three parameters which apply to the engine's operating mode. Some of these are:

- Pneumatic leaks in the Pc/Px/Py/Pg plumbing connecting the engine fuel control and governor systems.
- Contamination partially blocking one of the screens, channels, bleeds or orifices in the fuel control, governor pneumatic circuits.
- Improper adjustment of the fuel control maximum throttle stop, or the throttle lever not reaching the stop because of rigging error, linkage wear, etc.
- Internal wear or malfunction in the fuel control or governor, resulting in reduced maximum fuel output capability.
- Low fuel pump output.
- And finally, partially clogged fuel manifold (discharge nozzles).

Fluctuations

There are two areas where fluctuations occur: rpm and torque.

On any problems related to fluctuations or oscillations, the first suspect is the indicators. Remove the indicators and substitute a known gauge to verify proper operation.

With rpm fluctuations, the governor should be checked first. Deactivate the governor by disconnecting the Pr line. Note, however, that because of the physical location of the N1 throttle, deactivation of the power turbine governor on the AS350D is not recommended. However, if it's done, the pilot must be aware that you are doing a test run and the engine must be controlled with the N1 throttle. If the problem still is present, it's probably fuel control related. If the problem is isolated to the fuel control or the governor, certain areas on these units can be field cleaned.

If field cleaning or adjustment doesn't correct the problem, replacement will be necessary.

Although these are the most common problems associated with the system, other situations may occur with which you may need assistance. If so, feel free to call AlliedSignal Controls and Accessories, West Coast Support at (213) 849-3961. *November/December 1993*

Dennis Dryden is the program manager for fuel metering at AlliedSignal Controls and Accessories, West Coast Support Operations in Burbank, CA.

Accessory Technology

Carbon brake repair
Plenty of attention and wise maintenance programs are required for performance and economy

By Greg Napert

During the '70s, aircraft manufacturers introduced carbon brakes as an option over conventional steel brakes. Introduction into the marketplace moved quite slowly at first; however, in the late '80s carbon began appearing on a variety of corporate commuter and corporate aircraft.

Today, carbon is being introduced on new aircraft brakes because, according to the manufacturers, it offers a weight savings over steel, the wear rate is slower, and performance at high temperatures is better than steel.

However, progress in the application, performance and resulting acceptance of carbon has been slow primarily due to the relative high cost of carbon. There were also some initial problems from a technical standpoint that have affected the acceptance of carbon by many operators.

Nonetheless, manufacturers still maintain that with improving manufacturing processes, solutions that have been designed to iron out the problems, and maintenance programs that have been implemented to extend the life of carbon discs, carbon brakes are a smart, economical and safe option.

Two-for-one refurbishing of brake discs (left) results in a disc that is the same thickness as a new disc (right).

Chattering brakes can result in battering of the channel clips.

Paul Bureau, general manager for Ryder Aviall's Wheel and Brake Shop in Atlanta, which provide repair services for carbon brakes, says that you've got to measure the value of the increased performance against the increased cost. He explains that "a benefit of carbon over steel is that it does not fade. The hotter the carbon gets, the less it wears. It actually becomes harder at higher temperatures. The exact opposite of steel."

Bob Hansen, production manager for Aero Tire & Tank, Inc. in Dallas, TX, similarly says that "for commercial use, carbon is probably the best thing that ever came out. The landings and the extended life of the carbon brakes are a benefit to commercial airlines, mainly because they can take advantage of refurbishment and two-for-one refurbishment programs."

Money-saving programs

In spite of the fact that carbon brakes are more expensive upfront than steel, manufacturers have introduced programs over the past few years to get more life out of existing carbon brakes, thereby, making them more affordable.

Among the programs are: *two-for-one* where two carbon discs that are worn beyond service limits are sanded to one-half the thickness of a new disc and clipped together to create a disc that is the thickness of a new one; *carbon redeposition*, in which carbon is redeposited onto worn discs to rebuild them to new dimensions (this program is only offered by brake manufacturers at this time); and *reclassification* programs, where discs that are only partially worn are combined with other discs that have a significant amount of life left to produce a serviceable brake stack that will yield more life.

Bureau suggests that the most economical way of operating is to incorporate as many of the options (new, refurbished, reclassification, etc.) on an aircraft as possible as components wear. There are many aircraft, for instance, particularly in airline operations, where you have one aircraft with one or two brakes that are refurbished, one brake that's reclassified and another that's new.

He says that there are a few things to keep in mind when looking at these programs. For one, don't expect two-for-one refurbished brakes to last as long as new. The nature of carbon discs is such that the material on the outside of the disc is more dense. The dense carbon which exists on new discs is removed when the discs are sanded to 50 percent of the original. Bureau says that after grinding off half of the disc, the material that you're left with is more porous than the original material.

Because of this, even though the two-for-one disc is dimensionally the same as a new one, it's not going to yield you as many landings. It's going to give you somewhere in the area of 50 to 60 percent as many landings as a new disc, even though it's dimensionally the same.

Reclassification, however, incorporates discs that have 70 to 80 percent of life remaining, and the only material removed may be what's required to true them up. This leaves you with a disc stack that will give you as many or more landings than a two-for-one refurbishment.

Hansen explains that it's his opinion that "refurbishment really only works well on the commercial side of the business due to the fact that they have many times more aircraft from which they are salvaging carbon, and with which they can build overhauled brake stacks or use in reclassification programs. If you're a corporate operator, though, and you have one airplane with carbon brakes, say a G-IV or G-III, you only have two brakes on that aircraft, and a two-for-one's really not going to do you much good.

"You've got to understand that there is a rejection rate; some of that carbon will not be able to be used due to the rejection rate. And even without rejecting any of the discs, the best you could hope for is to end up with one refurbished brake stack, and you still have to buy a new brake stack for the other side."

Working out the bugs

One of the main squawks when carbon was first introduced was severe chattering. The chattering problem has been reduced, however, as operators have been educated on the care and maintenance of carbon and as new technology is being introduced into the design of the brakes.

Hansen says that the least expensive solution to chattering is to simply retrue the surface of the discs. "Uneven wear results chattering, and uneven wear can be caused by a piston or pistons dragging. Don't assume that the carbon is the problem, you've got to look a little farther than just the carbon," he says.

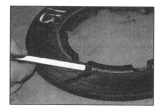

Circumferential grooves are machined into the outside diameter of the rotors and the inside diameter of the stators to prevent chattering.

He says that another common cause for chattering is the design of the rotors and stators. "The rotors are allowed to float, and this floating," he says, "results in chatter because the discs are not tracking properly.

"To fix this problem, a circumferential groove repair has been introduced. The grooves are machined into the outside diameter of the rotors and the inside diameter of the stators. And as the discs start to wear, a groove and slot is worn into the inside diameter and outside diameter of the rotors and this keeps the rotors in line with the stators essentially creating a pocket that the stators and rotors run in which keeps them perfectly aligned."

Maintenance wise

Bureau says that there are a number of items that you need to keep an eye on in order to get the maximum use out of a set of carbon discs. Among them, make a regular habit of checking the clips and rivets on the discs for looseness or wear. Chattering can result in these loosening up and if loose, they will beat the slots until the corners break or something falls off. Any metal that comes loose and lodges between the rotors and stators will really tear up the carbon.

If aluminum clips on the discs are worn through, the channels in the discs can become destroyed quite quickly, he explains. Some of the discs have a relocation option for the channel and clips. Gulfstream, he says, has a repair that allows for remachining of new channels between the existing channels. The old channels are then covered with aluminum covers as a safety precaution so that the wheel is installed correctly on the aircraft.

Accessory Technology

Cracks in the carbon are another thing that you need to look for, he says. "You're not likely to see them on a new disc, but it's not uncommon to see them on refurbished discs that aren't machined correctly. The two surfaces of the two-for-one discs which are mated against each other have to be perfectly flat. And if they aren't, you end up with cracks. Unevenness on the mating surface of the two discs will more than likely cause the discs to crack."

Hansen says he sees quite a bit of cracking on the discs that he's involved with. "On the 767, right now, you're looking at approximately a 25 percent rejection rate out of a set of discs. In the manuals there's a requirement to check each disc for what they call 'linear faults,' " he says, "which is nothing but a fancy word for a crack. No doubt that we do see cracks."

Hansen says that "linear faults" can exist in any direction and at any place on the disc, and they must be inspected visually to determine that there are no cracks. Cracks are especially common in the drive slot face, but you have to take off the clips that line the channels and look in behind them in the drive face area. There again, we have to go in and do a dimensional inspection.

Another thing to be concerned with is corrosion. "Corrosion appears as a white powder and if allowed to get bad enough, the disks will actually start delaminating.

"If you notice just a small amount of white powder, keep an eye on it to make sure that it doesn't progress. If it looks like it's progressing in any way, get the heat stack off of there and inspect it, because something is fixing to break loose."

Cleaning can cost you

Hansen says that they have a hard time getting through to customers the idea that carbon brakes need to be covered up when washing the aircraft. Soap can be very detrimental to the performance of the brakes.

"What happens is the soap itself gets into the carbon and makes the carbon quite sticky after the heat is applied and it causes violent chattering. At that point you've got to remove the brake stack and clean them out with alcohol or whatever's recommended in the maintenance manual.

"Deicing fluid and Skydrol® can also cause problems," he says. The deicing liquid or Skydrol gets to the brakes and causes the very same effect, and you'll have to pull the discs and clean them to stop them from chattering.

He says, "Bendix requires us to do a cleaning with isopropyl alcohol or ethyl alcohol and scrub each individual disc surface; then we take the discs and bake the alcohol out of the carbon at 500°F for seven hours.

"If you want to clean the brakes and brake housings, I recommend using a cloth dampened with alcohol and wipe them down as best you can, or if you have to you can use a toothbrush with alcohol to scrub out the cracks and such. Just try not to get anything on the carbon—don't let the carbon absorb anything."

September/October 1993

Accessory Technology

A few D.C. generator basics

By Greg Napert

Proper care and maintenance of D.C. generators is even more critical today than it has ever been. The reason: Parts for these older power supplies are becoming scarce, and along with that the cost of replacement parts for generators is on the rise.

In fact, even though it's possible to purchase many of the internal replacement components such as brushes, brush holders and related components such as voltage regulators, it's actually impossible to purchase a new housing or armature assembly, and it follows that completely new generator units aren't available either.

Aside from the option of switching the aircraft's generator to a modern-day alternator power source, the technician can only keep up with a good maintenance program designed to extend the life of the generator as long as possible.

Mike Strickland, chief inspector for Electrosystems Inc. in Fort Deposit, AL, says there are a few items that are often overlooked by the technician. One of them, he says, is the buildup of carbon between the commutator segments due to brush wear.

"What happens is over a period of time, the carbon from the brushes cakes up between the commutator segments, and this causes a failure of the armature because of the commutator segments shorting together.

"Most people check the bearings and brushes, but they're not checking to see how clean the commutator segments are. We've received many cores and all that was wrong with the generator is that the commutator wasn't properly cleaned.

"The armature can't always be saved during the rebuild," he says. "Some of the older generators, have had the commutator overhauled so many times that it's below limits and needs to be replaced."

Strickland says that another item that should be inspected regularly is the brush holders. "The insulation," he says, "on the positive side can sometimes crack and the brush holder can short against the housing. If this insulation is cracked in any way, you'll have a direct short, and the unit won't work at all."

Opening it up

Strickland says that repairs done out in the field often require pulling the end housings from the alternator to replace, clean or repair components. Any time the end housings are removed, he says, the bearings should be replaced. The reason, he explains is that when you pull the housing, you put a side load on the bearings with the puller, and you can distort the side of the bearing race, which can cause premature failure.

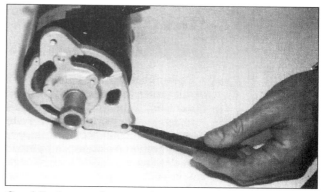

Carefully inspect the mounting flange areas for cracks.

Many technicians comment that following that logic, the bearings would be damaged during installation as well. But contrary to this thinking, Strickland explains that when you press the bearing on properly, you're pressing on the inner race and the outer race simply moves with it. However, when you remove the housing with a puller, you're pulling on the outer race, and the inner race resists movement because of the press fit on the armature shaft.

With the unit open, Strickland says, the insulation on the field coils should also be looked at closely, especially on older units where it has become dry and brittle.

If there's insulation missing, you need to inspect the windings for signs of shorts. Shorted coils will typically be evident by a dark burned looking area. Also, check for any kind of nicks, or if insulation has been worn off any of the wires. If this happens, the field coils must be replaced. Coils showing no sign of damage can be rewrapped with insulation, tested and reused.

Burned insulation and premature failure of a generator can sometimes be because of improper cooling. On units that are equipped with cooling shrouds, Strickland says to be careful when installing the cooling shroud on the end of the generator. "It is possible," he says, "to line up the cooling shroud onto the unit so that the cooling holes in the frame don't line up. This restricts proper airflow. Always verify that the shroud is lined up correctly."

The Best of Aircraft Maintenance Technology Magazine

Accessory Technology

Proper belt tension

Extending the useful life of the generator can be as simple as assuring that the drive belt is tensioned properly. If the belt tension isn't sufficient, it can slip and wear and not produce the proper output, and if it's too tight, it can place an excessive side load on the bearings and cause premature failure.

"The bearings," says Strickland, "just aren't made for some of the side loads that are placed on the unit as a result of overtensioning."

"As an overhaul facility, we often see generators come in with the bearing on the drive end of the unit completely destroyed, and the other bearing's perfect."

Units that are overtensioned, explains Steve Inabinet, director of technical services for Electrosystems, are also more susceptible to cracking in the mounting flange areas. If you find that a bearing has failed on a generator unit, take a close look at the mounting flanges to make sure that they are not damaged.

Voltage regulators

The voltage regulator, in most cases, is a separate unit from the generator; without it, the generator could not properly do its job. Besides controlling voltage output of the generator, the regulator prevents reverse current flow, provides overvoltage protection and, depending on the sophistication of the regulator, can perform a number of other functions as well.

Many of the regulators that are currently on the market are units that are built with a number of relays and points that open and close as necessary. Points are susceptible to wear and tear, so it's not uncommon to see voltage regulator failures as a result of burnt or worn points.

Inabinet says that "typically, the voltage regulator is replaced more often than the generator itself." He doesn't recommend any field repairs to these types of voltage regulators, because of the nature of the construction.

Tip Poorman, general manager for Electrosystems, says another problem that is sometimes created as a result of the voltage regulator failing or with units that aren't properly reverse current protected, is that reverse current can reverse the polarity of the field. If this happens, the generator will not produce voltage and you may need to "flash" the field. Flashing is a process by which you run current through the generator to correctly polarize the field.

Poorman explains that some technicians flash the field as a matter of practice every time they install a generator to assure that it's correctly polarized and that the unit has residual magnetism.

Finally, Poorman says that it's also important for the battery to be in good shape for proper operation of the entire electrical system. "The voltage regulator doesn't always do its job. The battery serves as a large capacitor and can absorb spikes in the electrical system and protect the avionics and other components. If it isn't in top condition, it may fail to protect the electrical system," he says.

Accessory Technology

Upgrading to an alternator

The price and availability of replacement parts for generators can sometimes warrant investigating upgrading the aircraft's electrical system from a generator to an alternator.

Not only does upgrading make sense from a cost standpoint, but there's also an advantage in using alternator power in terms of weight savings and increased power output. Additionally, due to fewer moving parts, and its solid-state nature, the alternator is simply more reliable than the generator.

Tip Poorman, general manager of Electrosystems, explains that currently, "it's particularly common in the aircraft refurbishing business to convert to alternators. Especially if they are upgrading the lighting and avionics systems. This results in a requirement for more electrical power, which in many cases can only be obtained from an alternator."

He says that "with an alternator, you can get a 30 percent increase in power output with a 30 percent decrease in weight."

He also says that there are some difficulties to consider when upgrading to an alternator; among them are: "The FAA requirements to approve a new accessory to a particular engine. They usually require 150-hour test for operational characteristics, cooling studies and so forth. It also will require different bracketry, ring gear, etc. It's not just a matter of swapping one for the other," he says.

"The other part of the picture can be the difference in the shape of the alternator. Sometimes you have to make modifications to the cowling because of the shape. You'll need to fit it into the space available, and provide the proper cooling to the alternator.

Poorman explains "that there are STCs available, but they're typically held by companies that specialize in overhauling certain types of aircraft that they see as being economically feasible.

"The engine manufacturers have made new brackets and component kits to adapt alternators.

"If there is an existing STC or the engine manufacturer has a bracket or service kit available, the cost of doing the conversion can be very attractive. In fact, I think it makes a lot of sense.

"Eventually, operators are going to be forced to convert to an alternator because the availability of repairable generators is dwindling." *July/August 1993*

Accessory Technology

Raw data
A basic understanding of avionic inputs

By Jim Sparks

Avionics. The word alone has been known to cause great anxiety especially when accompanied by discrepancy. Even the simplest definition, electronics as applied to aviation, doesn't inspire a high degree of confidence.

Basic flight instruments supply raw data to the flight crew as well as autoflight systems. To successfully trouble-shoot complex flight systems, it's important to understand the processing of raw data.

In the early days of navigating aircraft, the major instrument used for determining a course was the "IRON Compass," more commonly known as railroad tracks. In today's world, where visual reference cannot always be utilized, other inputs must be used to calculate aircraft position. Nature supplies many devices to allow those calculations to take place.

Air data

Air data is a term that can be broken down into three areas; static pressure, pitot pressure and temperature. Static pressure by definition is exerting force by reason of weight alone—without motion. The weight of air at sea level during standard day conditions is 14.7 psi.

As altitude increases, the weight of air decreases because there's less air pushing down. By incorporating a sealed flexible aneroid capsule with a specific internal pressure (14.7 psi), surrounding air pressure is reduced and the capsule expands. As pressure on the outside increases, the aneroid gets progressively smaller, so with a calibrated mechanical linkage this expansion and contraction can be accurately displayed as distance above sea level. This type of device is known as an altimeter.

Of course altimeters of today may not incorporate much mechanical linkage, but, instead, use an electronic correction for pressure change. The rate of change in static pressure can be monitored by an aneroid device incorporating a restrictor which allows pressure to equalize inside the aneroid. Compared to outside, this rate of change is displayed as vertical velocity or vertical speed.

A physical property of air is that it can be compressed. By strategically locating an air-sensing probe on the external fuselage airframe, manufacturers can attain an accurate ram-air pressure by using a calibrated aneroid; this pressure can be displayed as airspeed. Since air pressure requires an altitude compensation, the ram-air bellows is surrounded by atmospheric pressure, and the housing of the airspeed indicator is connected to the aircraft static system.

Air temperature is another component that requires monitoring. George Simon Ohm stated that voltage used by a circuit is proportional to the circuit's resistance multiplied by the amperage or current flow. By applying Ohm's law and sending a constant current through a temperature-sensitive resistor (thermistor), the voltage drop across this resistor will be directly related to the temperature.

There are several things that can alter the accuracy or temperature. As the aircraft moves, in many cases at extremely high speed, the air molecules contacting the thermistor become compressed. As air is compressed, temperature increases. Moisture content will also have a noticeable effect. Special probes have been designed to compensate for compression and moisture so that an accurate Static Air Temperature can be referenced. Many of today's aircraft are equipped with Air Data Computers (ADCs). These units are plumbed into the pitot and static systems and also supply the output current for the air temperature sensors.

By observing these three inputs, the ADC can supply very accurate electrical signals to drive altimeters, airspeed indicators and vertical speed gauges. It can also compute true airspeed and calculate Mach numbers.

In addition ADCs can communicate with other aircraft computers like flight guidance or autopilot. It can be a primary input for auto throttles and electronic pressurization systems and can even supply altitude signals for transponders.

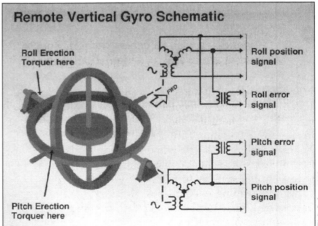

Directional information

Directional information is another requirement for flight. Since the earth is a large permanent magnet with a continuously flowing magnetic field, a permanent magnet that has freedom of movement will try to align with the earth's field. By orienting this free magnet with a compass card, direction can be determined relative to the earth's magnetic poles.

This means a sensing heading can be quite accurate at the magnetic equator, but as latitude increases, heading accuracy decreases. At the magnetic North and South Poles a compass system is of little help.

A flux sensor can be installed in a remote area of an aircraft such as a wingtip or stabilizer. This device consists of a three-legged inductive core with a coil of wire wrapped around each leg. This core is also divided into an upper and lower segment. The segments are joined by a center post wrapped with a coil.

Depending on the orientation of this flux sensor to the earth's magnetic field, the flux flow through the three legs will be different. By applying a low voltage alternating current to the coil around the center post, the three secondary coils will have an induced voltage proportional to the flux flow through that segment. This three-phase low-voltage output signal is then supplied for reference to adjust compass display instruments.

The Horizontal Situation Indicator (HSI) and the Radio Magnetic Indicator (RMI) are the two most common displays for compass information. However, in many cases the flux valve does not directly supply these instruments.

Both HSI and RMI can be considered gyro instruments. A gyro is a spinning mass. Once rotation has begun and speed reaches operating levels, it becomes quite difficult to move the gyro from its original plane of rotation. This spinning mass is housed in a framework which has full freedom of movement.

As the aircraft changes position, the gyro tries to remain stationary. The aircraft actually moves about the gyro. The position of the framework is electrically sensed by a synchro circuit. The output from the gyro unit is then supplied to the HSI and RMI. The spinning mass has no way of determining where it is; all it can do is spin relative to the direction from which it was started.

It's the job of the flux valve to supply information regarding heading to align the gyro initially and keep it slewed to the earth's magnetic field. Most systems incorporating a magnetically slaved gyro also have a means of breaking this communication line. The FreeSlaved or DGMAG switch, as it is commonly identified, can be used in areas of magnetic disturbance or around the earth's magnetic poles to allow the directional gyro alone to run the heading instruments. Once out of these polar realms the switch can be returned to the mag or slaved position, allowing the flux sensor to automatically reposition the directional gyro.

Gyro slaving is required periodically during flight due to gyro precess; that is, gyros tend to travel in the direction of rotation, which makes their uncorrected information accurate for, at best, one hour, but in most cases only 20 minutes.

Well-controlled air pressure and filtration is critical to maintain proper gyro operating speed. Particle contamination can be very detrimental to operating mechanisms in an air gyro, so proper maintenance on air filtration and moisture removal systems are required. Manufactured maintenance procedures and operational tests are essential to maximum gyro life.

A remote gyro (that is, one not contained within the instrument) can be used to drive an electrical ADI. This type device is more accurate and typically requires less maintenance than an air-driven unit. Electrical gyros most commonly use AC electric power to drive the spinning mass, and these Vertical Gyros (VGs) utilize a synchro output (three-wire AC) to drive a follow-up synchro housed within the ADI. A VG can be used to sense movement around roll, pitch and the yaw axis. A gyro flag found in the ADI, when in view, signifies the gyro is not up to operational speed or the displayed information does not agree with gyro information.

A test switch that may be found on some indicators will usually introduce a pitch and roll command to the ADI display. When the display does not agree with the gyro, the gyro flag comes into view. Operational speeds of an electrically driven spinning mass can be in excess of 20,000 rpm and generally require several minutes to become oriented. Also, a coast-down time that may be in excess of 10 minutes might be required before the gyro will cage (become mechanically supported after aircraft shutdown).

This coast-down time is important to observe prior to repositioning the aircraft, as gyro tumble could lead to mechanical damage. If immediate aircraft movement is required, it might be better to restore gyro power. In addition to supplying the ADI, a VG can also communicate with autoflight systems (flight director and autopilot) as well as provide stabilization for radar systems.

Attitude Heading Reference Systems (AHRS) is a means of locating a VG, DG and associated components in a stand-alone container. Installation of an AHRS unit requires a leveled platform. Location and power requirements are varied by manufacturer.

Laser gyros are becoming widespread in airline and corporate aviation. These devices eliminate the spinning mass. The ring laser compares the relationship of two light beams. As aircraft position changes, the phasing of the light beams will shift. With no moving parts the basic laser gyro is very reliable.

Accessory Technology

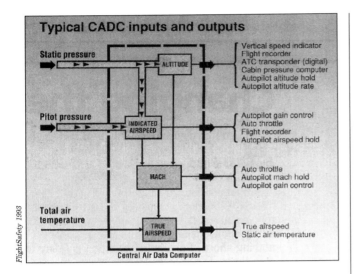

The Federal Aviation Regulations (FARs) incorporate procedures to ensure accurate display of raw data. Altimeter systems require certification every 24 months. This is documented in FAR 91.411. The procedures to accomplish this check is in FAR 43, appendix E. In addition to altimeter certification, it's also required that the static port and area surrounding must be free of defects that could alter airflow. If a static port is heated, the heating circuit also requires an operational test. Compass systems as well as altitude indication usually require functional tests at an annual interval.

Troubleshooting raw data systems can be facilitated by understanding display. When flags are in view, it's imperative to realize these advisories can indicate lack of power or erroneous information. Communication with the flight crew and a thorough description of the fault can lead to a quick, accurate and cost-effective fault evaluation.

Black box avionics systems, even though highly sophisticated, have no capability of observing what goes on in or around the aircraft. These black boxes depend on external sensors to supply all required data to perform required tasks. When a discrepancy occurs and a black box appears to be at fault, all inputs or sensors to that box need to be checked. Without valid inputs, displayed information or loads won't be accurate.

May/June 1993

Jim Sparks is an instructor for FlightSafety International in Little Rock, AR. He has over 12 years of maintenance instruction and holds an A&P and an FCC certificate.

Accessory Technology

Ultrasonic testing basics

By Greg Napert

Although ultrasonic testing has been around for a long time, its use in general aviation has gained great popularity over the last 10 years. Manufacturers are more commonly recommending ultrasonics as a method of monitoring cracks, inspecting for corrosion and thickness testing as a more economical means of maintaining aircraft.

Dan Nichols, NDT manager for KC Aviation in Appleton, WI, says that "there is not one aircraft that I work on that does not require some type of ultrasonic testing for some part or component on the aircraft. Nichols explains that in theory, "ultrasonics is simply measuring the time that it takes for sound to travel through a material. And that time can be expressed in anything, inches, millimeters, feet, or can be shown as a graphic representation."

The transducer, used to transmit ultrasonic vibrations, is no more than a crystal that's been cut to a specific thickness. The thickness of the crystal determines its frequency, and the frequency determines the depth of the penetration and the sensitivity of the test.

According to FAA AC 43-7, crystals, when subjected to an alternating electric charge, expand and contract under the influence of these charges.

Conversely, these materials, when subjected to alternating compression and tension, develop alternating electric charges on their faces. This is referred to as the piezoelectric effect.

Ultrasonics, explains Nichols, is used for very specific applications on an aircraft. If there are large areas that need to be inspected, ultrasonics is typically not a method of choice.

Wave forms

There are basically two wave forms for ultrasonic testing that are commonly employed in aviation: longitudinal and shear. Longitudinal or compressional waves are used for thickness gauging, or for defects that are known to be parallel to the surface. These waves are sent directly into the material and are reflected directly back to the transducer.

"A second method called shear waves," says Nichols, "sends sound waves into the material at an angle, typically 45- to 70-degree angles. Shear waves are used for inspecting weld joints, or areas where the transducer cannot be positioned directly over the area to be inspected because there's a rib or stringer or something in the way.

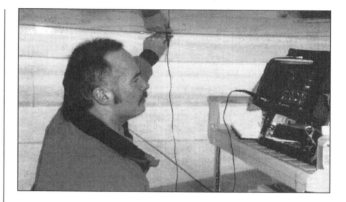

KC Aviation NDT technician Mark Peters calibrates equipment and performs inspection on a Gulfstream G-IV.

"By using shear waves, we can bounce the waves off of the defect so that they return to the instrument and are displayed.

"Typically, we inspect with more than one angle to cover various defects. The waves are typically sent out in a broad enough pattern so that they will catch a return signal even if the defect isn't exactly perpendicular to the wave. A wide band of waves is sent out so it's forgiving enough that you don't have to be totally accurate in sending out the waves," he says.

"In some cases, shear waves are inappropriate. In the case of corrosion, we want to use compression waves. Especially if the corrosion is intergranular or exfoliation. Exfoliation or intergranular causes a layering effect, and if shear waves are used, the sound will bounce between the layers of corrosion and won't be returned to the transmitter."

Reference blocks

Reference or calibrating blocks are a required item for ultrasonic testing. There are several different types of basic reference blocks, says Nichols.

"You have to make sure that you're actually calibrating for the type of material you're working on. You're actually calibrating for a characteristic of the material, which is the velocity that the sound travels through the material. Plastics, particularly, can vary widely from type to type so it's important to have the same exact type of plastic for reference," he says.

Aluminum is not so critical. The velocity of aluminum is the same for all intents and purposes, and the different types of aluminum won't alter the reading enough to matter.

"One of the features that modern ultrasonic units have is the ability to place a numeric value on the velocity. So if you know what the velocity of the material is that you're measuring, you can set that into the equipment to calibrate it. It's still good to calibrate this equipment using standards also, because you are verifying that the equipment is accurate," he says.

Practical use

Besides testing for defects, cracks or inclusions, ultrasonics is also used for thickness gauging of aluminum, plastics, windows, etc. One application that's quite common is where you clean up corrosion or scratches on critical surfaces and use ultrasonics to determine if the material is still within limits, Nichols explains.

"For thickness testing," he says, "we use a special ultrasonic thickness gauge. It's designed specifically for testing thickness and that's all we use it for. The same gauge is used for all materials that require dimensional testing.

Typical application of shear waves to detect flaw or damage.

"There are some materials that you can't use ultrasonics on. Very porous materials, such as Kevlar for example, are not able to be tested.

"Honeycomb structures, however, are very easy to test for disbonds," says Nichols. "We use a very low frequency, typically a 1 MHz transducer, and as long as it's not a Kevlar skin, we couple the transducer to the skin. We're not only able to get through the composite, or the aluminum skin, we're also able to penetrate through the honeycomb part.

"You can tell if the honeycomb is either broken down or disbonded from the surface material because the signal will change drastically. You either get a signal, or no signal at all. With a 1 MHz transducer, if there's a disbond, the skin and the disbond are too thin for any measurement to take place so there will be no signal (clean screen presentation). Conversely, if it's bonded, the screen will be filled with echos, repeat echos from the honeycomb core. So a clean screen is no go, and a screen full of echos is go—a very basic go/no-go inspection.

"Air leak detection is another application of ultrasonics. A leak of air creates a pressure differential. And every time you have a pressure differential, you have sound. And this differential creates ultrasonic waves. The instruments for detecting this sound simply convert these ultrasonic waves into audible sounds."

Equipment considerations

Nichols says that with ultrasonic equipment, "the latest technology is digital technology. Additionally, the equipment really needs to have a damping control, especially if you're planning on testing composites. We find that a lot of equipment doesn't have it. A damping control dampens the signal from the transducer going into the amplifier. It acts like a filter that filters the signal."

AC 43-7 says that there are several methods of observing and recording ultrasonic response patterns such as: CRT, indicating lights, alarm lights, alarm devices (bells, buzzers, etc.), go/no-go monitors and others. These methods may be used in combination to suit a particular need. *March/April 1993*

A look inside the Bell 206L transmission
1,500-hour inspection

By Greg Napert

One of the most popular, yet simple transmissions that exists on today's light helicopters is on the Bell 206L.

Emile Mouton, overhaul shop supervisor for Petroleum Helicopters Inc. (PHI), in Lafayette, LA, says that the number of Bell 206L transmissions that PHI inspects and overhauls on a day-to-day basis is why the company knows what to keep an eye on during a typical inspection.

Pat Clay, technician for PHI, says that there are basically two 1,500-hour inspections performed during the service life of the 206L transmission, and an overhaul at 4,500 hours. At each 1,500-hour inspection, the main points of concern during the inspection are the splines on the inside of the top case, and wear/or damage to sun gear splines.

Disassembly

Before splitting the transmission, you've got to be sure to remove the main oil jet and oil filter. Once the top case is removed, you have complete access to the ring gear and planetary assembly.

The teeth on the upper case, Clay says, have a tendency to crack to the point where you can just grab a piece of a tooth and break it off. You've got to look closely at these teeth. There are a certain number of teeth that are allowed to break, he says, depending on how many are broken within a given area and the extent of the damage on each individual tooth (not more than six damaged teeth in any 180-degree segment of the splined area, no more than three damaged teeth within any group of 15 consecutive teeth).

Clay says that Bell provides information specifically on the extent of damage allowed on individual teeth and other criteria. "What's interesting," he says, "is that the teeth in this area seem to break within the first 1,500 hours if they're going to break; then there are no further breaks during the life of the transmission.

"The teeth on the ring gear and planetary/sun gear occasionally are damaged, but it's not that common," Clay says. "These Bell gears are made pretty well, and there are not many problems with them. We typically give the gears a good once-over inspection to be sure that there's nothing that's obviously wrong and then set them aside for reinstallation."

The gear teeth in the top case half must be inspected closely for damage.

Most of the gears in this transmission are made with a black-oxide coating. The coating makes it somewhat difficult to inspect because of the minimal reflection of light. A tip that Clay says is important to follow when inspecting the surface of these gears is to use a white piece of paper and place it between each gear tooth to reflect light onto the adjacent tooth. The mirrorlike surface of the gears reflects the white of the paper and allows you to more easily view wear patterns, scratches and other damage that may exist.

Another area that has to be examined closely is where the sun gear fits into the bevel gear shaft. "Bell has had a few problems with the sun gear and the bevel gear," Clay says. "Basically the tolerance was too great, and the gears would bang together when the engine started up."

To resolve this, Bell went from an oil-lubricated sun gear, to a grease-lubricated gear with two O-rings and silver coating on the splines of the gears. Because the silver is soft, it's common for it to flake off and get into the oil. This isn't a problem, Clay explains, because according to Bell, the silver is soft enough that it won't cause damage to the gears. It will simply work its way through and around the gears without any consequence.

Loose plating should be removed with Scotchbrite®, however, and no type of chemical removal or cleaning of the silver is allowed. Clay says that Bell recommends if you notice any silver flakes or slivers in the oil, you need to change the oil, change filter, clean oil screens, flush the transmission and refill the transmission with oil.

The newest revision to the sun gear is the use of improved materials and tighter tolerances. Bell also eliminated the O-rings and grease and went with oil lubrication again. The older sun gears that still have the grease lubrication may be difficult to remove due to the holding power of the grease and O-rings. "If you grab the sun gear with your hand and can't pull it out, you're going to have to remove the lower mast bearing support and press the sun gear out from the bottom," Clay says.

The area of the sun gear that has the silver plating must be checked carefully for minimum dimensions and the corresponding gears on the main or input gearshaft also need to be carefully checked.

All of the gear spline wear measurements are taken by placing gauge pins 180 degrees apart and measuring from one pin to the other. You've got to be careful, Clay says, when measuring with gauge pins to be sure of two things. First, you want to make sure that you're measuring exactly 180 degrees apart. To ensure this, always count the number of teeth in either direction. Also, you want to be sure that you're using the correct size gauge pin as recommended by Bell. If you're using a gauge pin that's three thousandths off, you may get a reading that's below minimums and reject a part that's good.

"Since we remove the sun gear during this inspection, we also take the time to check the external dimension of the splines on the sun gear and the inside diameter of the internal splined gear shaft where the sun gear contacts," Clay says. "It's not required as part of the inspection, but we like to make sure that the parts are still sufficiently within limits to make it another 1,500 hours, particularly since we've seen problems with wear in this area."

Another item that is taken care of during the 1,500-hour inspection, Clay says, is the oil pump drive. The drive has a rubber seal that frequently leaks a small amount of oil. If there are any signs of leakage, it's a good time to replace the seal. To do this, the oil pump has to be removed.

"Some people change this seal regardless of whether it's leaking or not," Clay says, "because there's a good chance that it'll start leaking within the next 1,500

A piece of paper will reflect the light to more effectively inspect the gear teeth.

hours. Also, the oil pump drive coupling, which is removed with the oil pump, is known to wear. So the splines should be inspected closely."

Assembly

"You've got to be careful when you're pressing any of the assemblies back together," Clay says. For example, Clay explains, when you're pressing the input gear shaft back into the case, you need to make sure that the gears are properly mating with the bevel gear. If you don't, it's easy to snap one of the gear teeth off.

The best way to ensure that the teeth are mating properly is to continuously rock the bevel gear back and forth. "When you're pressing the input gear shaft in, the ring gear will contact the bevel gear shaft first," he says, "so you've got to rock that gear back and forth; then as you continue to press it down, it will contact the oil pump drive gear."

Installing the grease-packed sun gear into the input shaft can be quite tricky. A means of releasing the excess grease is needed; otherwise, you'll never be able to press it into the proper position. This is done by using a couple of pieces of safety wire and placing them between the O-rings and the input gear shaft to release the excess grease.

"Any O-rings that are disrupted during the inspection must be replaced with new," Clay says. "To eliminate leaks, we like to apply a small amount of sealant

(MIL-S-8802) on the O-rings to help seal the case. Bell recommends installing the O-ring dry, but we've found that the sealant helps ensure that the case doesn't leak.

"For most shops, it's fairly common for these transmissions to leak after assembly," he says, "but we ensure that they don't by pressurizing the case. If any leaks are found, they're corrected before the transmission is put back into service."

Measuring backlash is necessary on all drives. A couple of special tools are required to accomplish this. Make sure to take the measurement at the exact point specified on the tool. "In the event that you have a larger backlash than allowed, you should check to be sure that the bevel gear is seated properly and that the shaft is seated in the bearings well, then you've got to make sure that you have the correct shim behind the bevel gear. You may be able to adjust this shim .0005 according to a formula in the manual.

"The other shim that may be adjusted, within certain limits, is the shim under the duplex bearings. If you look on the case, there are some readings; then you are given a formula in the book and if you fall within a specific range of shims you may be able to adjust it," he says.

Clay explains that some of the backlash readings may come out close to the maximum allowable, but the important thing is that the wear isn't progressing. If you have a reading that's marginal, and the wear shows signs of progressing, you may want to consider making further repairs.

Saving time

One thing that some shops don't realize is that Bell allows you to run slightly over the inspection interval (100 hours over, so long as you don't exceed 1,600 hours) so that you get the maximum time out of the transmission.

If you perform the inspection before the 1,500-hour time limit, you could potentially lose 20 to 30 hours on the transmission over its life.

Since the overrun is allowed, and there are very few problems with this transmission, it's wise to use the additional time that Bell allows.

Gulf Coast corrosion protection is part of PHI's maintenance program

Emile Mouton, component overhaul supervisor for PHI, says that the company is particularly challenged with its helicopter maintenance because of the severely corrosive operating environment that most of its helicopters operate in.

The corrosive environment offers many challenges that aren't normally encountered, so many steps have to be taken to avoid corrosion like putting a sealant (MIL-S-8802) on everything.

But the company has an advantage as well; that is, the number of helicopters that it operates provides the opportunity to work with OEMs on corrosion prevention methods.

"We need to take a few extra steps to protect these transmissions," Mouton says. "For example, we put a transmission varnish available from Sikorsky on the through bolts and dip them to the bottom of the threads to protect them from corrosion.

"We also use a two-part sealant to seal all mating surfaces on the outside, all boltheads and anywhere water can collect and remain. The sealant does a great job of protecting the transmission against the environment but it creates a little additional work during inspection and overhaul."

Pat Clay, technician for the company, says, "You've got to scrape the sealant off of the boltheads and away from the parting surfaces in order to split the case. Be careful, however, when removing it. Use a plastic or plexiglass scraper to remove it; use of a steel knife, razor or other metal instrument will scratch the case and set up stress areas that may cause cracking."

Clay explains that a common area to see corrosion of the Bell 206L transmission is in the flange area between the top case half and the gear shaft case. The reason, he says, is largely because of dissimilar metals—the top case half is aluminum and the gear shaft case is magnesium.

Clay says that Bell provides a coating to spray on the magnesium surface of the gear shaft case to prevent corrosion. "We've found that it's not very effective and through experimentation have found that a product provided by Aerospatiale called Graphoil-D (P/N 148) works much better," he says. "As a matter of fact, we haven't noticed any corrosion in this area when Graphoil-D is used; it's a kind of slippery Teflon coating that remains on the part and protects it quite well."

"We're very corrosion minded," Mouton says, "we're in the worst environment you could ask for. If we can't paint it, we seal it—everything has got to be sealed when operating in the Gulf. We've got to take additional steps to protect our equipment beyond what the manufacturer recommends as adequate for most environments."

January/February 1993

Accessory Technology

Safety wiring basics

By Greg Napert

Everyone who makes their living as a technician has experienced "breaking-in" their hands on safety wire. You either learn quickly to respect its razor-sharp edges, or lose enough blood to supply a small blood bank for a year.

Once you get the hang of it, however, it's quite common to gain a sense of pride over your safety wiring skills. Besides contributing to aesthetics, a good safety wiring job can add to safety and reduce hazards for those who tread in the same territory during future maintenance and/or inspections.

Unfortunately, human nature tends to make people complacent when people become too familiar with something—this can result in skipping over basics and forming bad habits.

For instance, how many of you use the automatic twisting feature of safety wire pliers when twisting the end tail on a safety?

If you go back to Basic Safety Wiring 101, you'll remember that the tail of the safety should be twisted counterclockwise in order to assure that the loop around the bolthead remains in place—while the portion between the fasteners is twisted clockwise. Safety wire pliers will always twist the wires in the clockwise direction. So when you get to the tail of the safety, you've got to reverse the direction of the twist—which means getting the ol' wrists into the act and reversing the direction of the twist.

Another item commonly overlooked is the recommended number of twists per inch. Too many twists per inch results in the safety wire being too tight, which means the wire is overstressed and therefore weakened; not enough twists, and the wire is too loose, which may allow the fastener to release. The recommendation is four to five complete revolutions of the pliers per inch, or eight peaks per inch. Take a look at your typical twists, and see if they conform to this recommendation; you may want to ease up a bit on the number of twists, or tighten up if not enough.

In any case, an occasional brush-up of some of the basics is a good way to identify problems and prevent you from developing bad habits.

Here's a quick overview of some basic rules for safety wiring:

- Double-wrap method is preferred over the single-wrap method for wiring turnbuckles.
- A pigtail of 1/4 to 1/2 inch (three to six twists) should be made at the end of the wiring. This pigtail

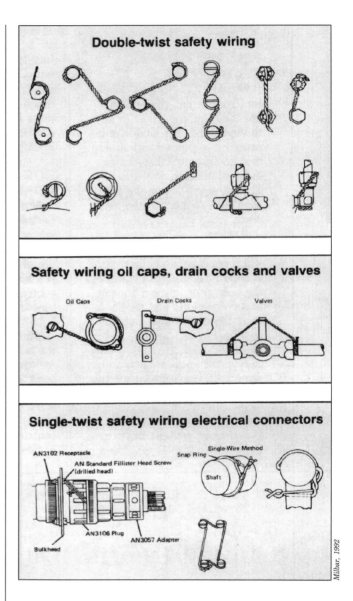

 must be bent back or under to prevent it from becoming a snag.

- When castellated nuts are to be secured with safety wire, tighten the nut to the low side of the selected torque range, unless otherwise specified, and if necessary, continue tightening until a slot aligns with the hole.

- All safety wires must be tight after installation, but not under such tension that normal handling or vibration will break the wire.
- The wire must be applied so that all pull exerted by the wire tends to tighten the nut.
- Twists should be tight and even, and the wire between the nuts as taut as possible without overtwisting.
- The safety wire should always be installed and twisted so that the loop around the head stays down and does not tend to come up over the bolthead, causing a slack loop.
- Inside diameter snap rings may be safety wired; inside diameter snap rings must never be safe-tied.
- Before any screw, nut, plug, etc. is wired, check for proper torque.
- Never reuse safety wire.
- Use the size and type safety wire called for in applicable specifications.
- Whenever possible, use the double-twist safety wiring method. Limit single twist to small screws located in closely spaced, closed geometric patterns, or parts in electrical systems or parts that are difficult to reach.
- Avoid kinks, nicks or stretching in the wire.
- Always recheck the fastener to assure that the pull exerted by the wire will tend to tighten the nut. (The wire should leave the fastener in a clockwise direction for right hand threads.)

New technology in safety wiring

Recently, a company called Bergen Cable Technologies introduced a system for safetying nuts and bolts.

The principle, developed in cooperation with G.E. Aircraft Engines (GEAA), essentially utilizes the same principle of applying a wire to one or more fasteners in such a way, that if the fastener begins to loosen, the wire tightens and prevents the fastener from backing out.

The difference with this new system, though, is that it eliminates the need for twisting wire, which can be both time-consuming and produce inconsistent results.

The patented safety cable system consists of three basic components:

Safety cable—which is available in two diameters, .021 and .032 inch, and various lengths from 9 to 24 inches, in stainless steel and Inconel. The assemblies come with a fitting swaged at one end and the opposite end fused to prevent the cable from fraying prior to installation.

Safety cable ferrules—which are designed to lock the cable in position. They are prepackaged in a spring-loaded disposable magazine.

Bergen Terminator™ crimping tool—available in various diameters and lengths. The crimping pressure is factory preset for consistency. The tool tensions the cable, crimps the ferrule and cuts the cable flush with the ferrule.

To safety a bolt pattern, the operator simply slides the cable through the bolts in the correct fashion, inserts the cable into the crimping tool, applies tension to the cable, and squeezes the tool to crimp the ferrule and cut the excess cable.

Bergen Cable, 1992

The company says that the Terminator is designed to minimize operator training and is equipped with a "cycle lockout" to ensure that the ferrule is fully crimped before the tool will release. Should the user stop midcycle, the cable will not release from the tool until the cycle is completed (or cuts the cable away). The company also says that the tool is ideal for use in tight places where accessibility is minimal.

Bergen says that this system recently received FAA approval as a standard part based on SAE specification (AS4536). Approval for use on an installation that currently calls for standard safety wire, however, must be obtained on a case-by-case basis from the FAA. Comparative data will be available in the near future, explains Paul Messina, product manager for Bergen, that will aid in expediting these approvals.

November/December 1992

Accessory Technology

Do you know Eddy? Current, that is
Basic principles behind an underutilized NDT method

By Greg Napert

One of the simplest, yet misunderstood methods of NDT (non-destructive testing) in aviation is eddy current testing. The process, which involves generating a magnetic field at the tip of a probe to detect flaws in metals is one that has become highly developed in the last decade.

Although it's a simple process, the procedure does require a good basic understanding of the principles behind the test. Also, due to its level of sophistication, eddy current equipment is now applied in more ways than ever; therefore, more skill is required to interpret results and to screen out erroneous readings.

If you think that eddy current is used only to test for cracks—think again. The testing processes have been developed so that the equipment can also be used to test for material thickness, hardness, corrosion and more. And the areas that it can be applied to are limited only by the ability to access the area.

Basic principles

Troy Woller, instructor at Zetec, an eddy current equipment manufacturer and training school in Issaquah, WA, explains that conventional eddy current testing uses alternating current sent through a tiny coil at the end of a probe to produce an alternately expanding and collapsing magnetic field.

This magnetic field, referred to as the primary field, when placed near a conductive object, generates small currents in the object called eddy currents—the same principle behind the generation of electricity.

The eddy currents in this object, in turn, (using the right-hand rule, a basic law of electricity) produce a small magnetic field of their own, called the secondary field, which opposes the primary field.

It's this opposition to the primary field that produces an impedance, or opposition to flow in the test coil. This opposition is then monitored and the results interpreted.

Variations in test materials including: type of material, thickness, hardness, permeability, cracks and corrosion affect the conductivity of the material and subsequently, the amount of eddy current flow. For example, says Woller, "if we have a break in a material, the eddy currents experience more resistance to flow. This, in turn, reduces the amount of secondary magnetic field, which in turn reduces the impedance of the coil. It's this change in impedance which tells us something about the materials that we're testing."

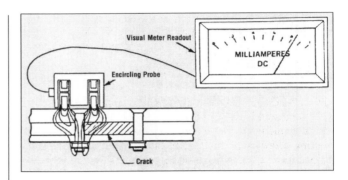

In order to determine that there's actually something wrong with a particular indication or reading, you need a standard to test the object against. This can be done in various ways. The most basic equipment employs a single coil at the tip of a probe. This arrangement is referred to as "absolute."

In order to determine deficiencies in materials, a reading from a calibration standard must be taken first. The reading obtained from standard is noted and then is used as a comparison for the test piece.

A more complex arrangement of coils, known as "differential," consists of two or more coils electrically connected to oppose each other. According to Woller, the result of placing opposing coils on identical test objects is a net output of zero, or no indication.

The arrangement of the coils in a differential setup depends on the application. It's desirable, in some cases, to rearrange the coils in a setup known as "self-comparison." With this setup, the coils are physically located next to each other at the end of a probe.

As the probe is passed along a surface, each coil continuously produces a signal that opposes the other. Inconsistencies in the material are indicated when one of the coil's signals is altered due to a flaw in the material. This arrangement is insensitive to test object variables that occur gradually, such as slowly changing wall thickness, diameter or conductivity.

A more common variation of the differential arrangement, known as "external reference differential," also uses two or more coils electrically connected to oppose each other. However, one of the coils in this setup is used simultaneously on a reference object, while the other is used on the test object.

Accessory Technology

Because this arrangement is used to detect differences between a standard object and test object, it's best for use in comparing conductivity, permeability and dimensional measurements.

Woller says that most of the equipment on the market today uses the external reference differential coil arrangement to achieve accurate readings.

Test equipment

The most simple eddy current test equipment, commonly referred to as metered instruments, depends on only the impedance (one value) to evaluate materials. Even though this equipment is valuable, its use is limited. Woller explains that "metered equipment can only tell you that there's something wrong with an object. It doesn't tell you specifically what's wrong."

More sophisticated eddy current equipment, referred to as "impedance plane" devices, takes eddy current testing a few steps further. In addition to monitoring the impedance of the coil, these instruments also monitor another variable—the time that it takes for the magnetic field to penetrate the test object. This additional information provides a critical variable to the eddy current equipment known as the "phase angle."

In order to display this information you need to have a display that's capable of displaying two values, in other words, an instrument that will graph or plot the information on a two-dimensional surface.

Impedance plane instruments display and define variables that provide information such as depth of penetration, and allow the user to separate and define variables present in the test object that could not be seen using metered equipment.

In addition to allowing the user to determine things such as skin thickness and depth of penetration, the impedance plane device also minimizes the potential for erroneous errors in readings.

For example, Gary Davis, instructor for Trans World Airlines says that "impedance plane devices give you one reading when you unintentionally lift a probe off the surface. This is referred to as lift-off and is indicated on the instrument panel by a straight line to the left. If you run across a crack, however, the reading is portrayed as a line in a different direction. With a metered instrument, he explains, you wouldn't know the difference and may assume that there is a crack in a part when there really isn't. It allows you to discriminate between actual cracks and human error."

Applications

Generally speaking, the application of eddy current testing to large surface areas is impractical. The cost of equipment needed to scan large surface areas makes it ideal for testing small surface areas only.

Additionally, most of today's equipment is designed to be used on non-ferrous materials. Woller explains that although it's possible in some cases to apply eddy current to ferrous materials, the power requirements needed to effectively use it for ferrous test objects preclude most equipment from being used.

However, an example of one piece of eddy current equipment, that is designed specifically for detecting cracks in ferrous welds, is available from KrautKramer Branson. The company's WeldScan is designed for detection of surface-breaking cracks in ferrous welds.

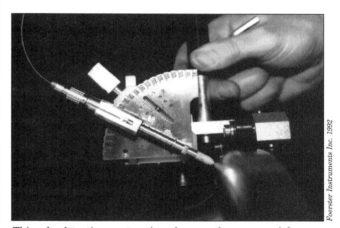

This wheel testing system is only one of many special application probes designed for use in aviation.

Test probe development has allowed application of eddy current to areas that previously couldn't be inspected. Special probes have been developed for scanning the inside diameter of holes, including bolt holes.

Zetec, for instance, makes a special probe coil that spins as it is withdrawn from the hole at a uniform rate. The probe is capable of being indexed so that any flaws that are found can be located and identified.

Another eddy current equipment manufacturer, Foerster Instruments Inc., has developed a universal wheel testing system for testing wheel rims (bead seat area) of aircraft. A gear mechanism design makes it possible to accurately inspect the radius and adjacent areas of wheel hubs.

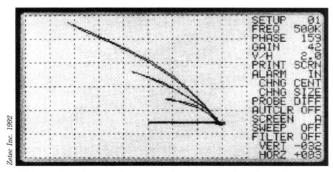

This is the plot as shown on the screen of an impedance plane display. Cracks are shown at 1.0, 0.4 and 0.2mm deep.

Accessory Technology

For testing bolt holes, probes have been developed that are essentially a miniature coil mounted on a single thread of a plastic rod that's machined to the specifications of the threads being inspected.

Some of the basic probe designs available include:

- **Probe coils** — also referred to as surface coils, probe coils, flat coils or pancake coils. As the name implies, these coils are used to scan the surface of test objects. They are available in various shapes to aid in inspecting irregular areas. Pencil-shaped probes are common to overcome a variety of obstacles. Probe coils are also used where high resolution is required because of the ability to shield the coil. A desirable trait for probe coils is to be spring-mounted so that a preload is placed on the tip of the probe as the scanning is performed. This makes it less likely that the user will unintentionally lift the probe off the surface of the test object.

- **Encircling coils** — also referred to as OD coil and feed-through coil. This type of coil is used to inspect tubular and bar-shaped products that are fed through the coil at a high rate of speed. It's important, says Woller, to keep objects centered in this type of coil. Because a larger area is examined (the entire circumference at any given point) at one time, the sensitivity is lowered, and uniform readings are difficult to obtain.

- **Bobbin coils** — also called ID coils and inside probes. These coils are used to inspect the inside diameter or bore of tubular test objects. These coils are inserted with semiflexible cables or blown in with air and retrieved with an attached cable.

Selecting equipment

Of the many different manufacturers of eddy current equipment, most provide two basic types of equipment, metered and impedance plane. However, Bill Chevalier, marketing director for Zetec, says that there are many differences to consider when selecting equipment, some subtle and some not so subtle.

Some of the important items to look for when purchasing or selecting equipment to use, he says, include:

- **Suitable frequency range** — The frequency range affects the depth of penetration of the test equipment. A wide range to select from will allow you to be selective in the penetration of various test objects.

- **Resolution of the equipment** — This affects the ability to discriminate between signals of interest and unwanted signals, such as noise, and affects the accuracy of the tests. The resolution is affected by many factors such as instrumentation design, ability to filter the data and signal mixing of data.

- **Availability and variety of probes** — For the widest variety of applications it's desirable to have a wide range of probes to select from. Typically, one manufacturer's probes are capable of being used on a different manufacturer's equipment with adapters.

- **Type of monitor** — Manufacturers incorporate various types of monitors such as CRT (Cathode Ray Tube) which have good resolution but can be easily damaged, and LCD (Liquid Crystal Display) which are harder to damage and also provide good resolution. Chevalier says to make sure that the LCD displays are equipped with a heater so that they perform reliably in cold temperatures.

- **Field adaptable** — Check to see if equipment is portable, rugged, battery-operated, comes with a detachable battery pack (so that depletion of the battery doesn't take the tester out of service) and is UL listed (which assures the unit is safe for use in hazardous environments such as inside a fuel cell).

- **Memory** — Many manufacturers are adding the capability to their equipment to store programs for easy recall.

- **Customer service** — Look for a manufacturer who'll provide you with support and loaner equipment in the event that your equipment needs to be sent out for servicing.

It's important to note that there is a big difference in price between the metered and impedance plane equipment. This price disparity makes understanding the application of the equipment very important when determining the type of equipment you need.

Finally, purchasing the equipment is useless unless you learn how to use it properly. Therefore plan on an investment in training, as well as an investment in the equipment. Not only do most of the manufacturers offer courses on their equipment, but there are many schools throughout the country that offer the courses as part of their NDT curriculum. *September/October 1992*

Accessory Technology

More than meets the eye
Today's borescope can be an important timesaving and money-saving tool, but know what's out there before you buy

By Greg Napert

Today's borescope can be a very versatile tool when it comes to engine inspection. In some cases, the cost savings that can be realized make the instrument a necessity.

Recent advancements in borescope technology have given the instrument additional value. Besides having the ability to aid in visual inspection of internal components, the borescope can also be used as a tool for retrieval of lost parts, eliminating the need for engine disassembly in the event of a mishap.

Additionally, advancements in video borescope (videoscope) technology have given the technician the ability to document inspections and share this information long distance through the use of computers, telephones and other communication equipment. There are still some limitations, however, and it's important that a good amount of consideration be given to what equipment is purchased before you're stuck with something that can't be used.

Among some of the questions that you should ask when making purchases are: Does the manufacturer supply specialized inspection equipment for the model(s) of engines that you work on? Is the resolution of the equipment good enough to detect the type of damage you're looking for? Is the equipment versatile enough to be used for other types of inspections? Does the manufacturer provide support for its equipment?

Basically, borescopes can be broken down into three categories: rigid borescopes, flexible fiberscopes and videoscopes.

Rigid borescopes

Because of their relative simplicity (straight tubular design incorporating relay lenses), rigid borescopes are capable of giving you the highest quality resolution. They are also the least expensive, but their use is somewhat limited. They can only be used where there's no need to maneuver the instrument around obstacles. Typical application includes inspection of cylinder walls, inspection of the internal diameters of straight shafts or assuring the proper seating of turbine assemblies during assembly.

Rigid borescopes are available as a single length borescope, or with telescoping or extender sections to accommodate varying depths. Also, various viewing heads are available to allow you to see objects at multiple angles or various magnifications.

Flexible borescopes

Flexible fiber-optic borescopes, also known as fiberscopes, utilize a flexible fiber-optic image bundle to transmit an image to the eye.

The image bundle is composed of thousands of glass-fiber strands capable of snaking into hard-to-reach areas. The resolution of this image bundle is dependent on its construction. Alignment, tightness of packing and amount of fiber breakage can all affect the transmission of the image and typically, the higher the quality, the higher the price.

Because of the fiberscope's flexibility, the ability to articulate (bend) the tip of the instrument is necessary to manipulate the probe for viewing and for guiding it into position. Most fiberscopes have the ability to be articulated in at least one plane (up and down) and some can be articulated in two planes which makes it possible to rotate the tip in a complete 360-degree arc.

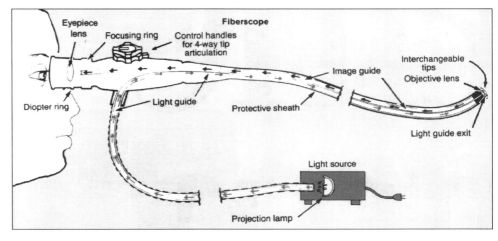

Accessory Technology

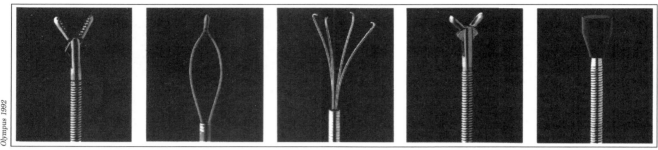

Examples of the power of working channel borescopes and samples of the different tools available. The tips provide a set of tools to be used at the tip of the borescope.

To assist in working the borescope into critical areas of jet engines, some manufacturers supply sheaths, which are prebent tubes that guide the borescope into a predetermined area. The sheaths are available for specific engine models and for various inspection areas on these engine models. Additionally, some fiberscopes have focusing capability, and some don't.

Accessories and attachments are available for the fiberscope that include camera adapters, corner sensors, various light sources and more.

Working-channel scopes, which are flexible borescopes with a hollow channel into which an instrument can be inserted, can be extremely helpful when working with engines. Various working tips, retrieval tools, cutters and magnets can help in the retrieval of nuts and bolts that would otherwise require the disassembly of an engine.

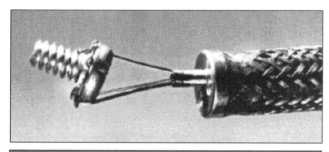

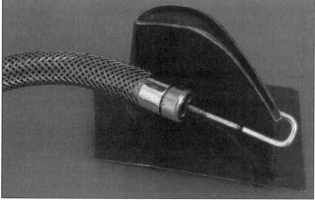

Tools such as the four-prong retriever (top) and the retrieval hook (bottom) can be invaluable timesaving devices.

One example of efficient use of the working-channel-type borescope, according to Olympus Corporation, is in inspecting the Pratt & Whitney F100 engine. "A hook is placed onto the tip of the borescope," says the company, "and instead of entering the engine from the rear and going through the third and fourth stages—a long, tedious process—the inspector enters through the first stage and hooks the scope on the trailing edge of a first-stage blade. Then, the turbine is manually rotated and the scope tip is pulled into the engine. This eliminates 29 separate insertions and many hours of work."

In addition to working-channel borescopes, measuring fiberscopes are available which allow you to accurately measure damage and make determinations on continued operation.

Another type of fiberscope is the ultraviolet (UV) fiberscope. The UV scope is used in conjunction with dye penetrant to test components for cracks.

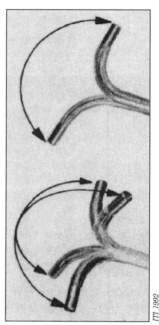

Single plane articulation (top) and dual plane articulation (bottom) allow the user to maneuver the borescope into otherwise hard-to-reach areas.

Videoscopes

The videoscope, which essentially consists of a miniature TV camera at the end of the insertion tube, sends an image to the eyepiece, or monitor, through a wire—instead of through fiber optics or a series of lenses.

The advantage with the videoscope is that length of the insertion tube isn't limited by the quality of the image or the cost of the optics. At a certain length, fiber optics become economically impractical to manufacture.

Accessory Technology

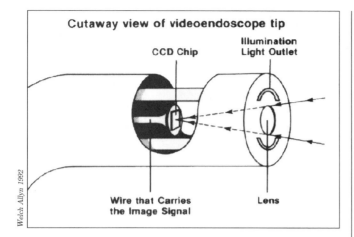

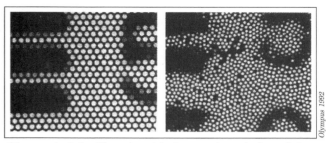

Alignment of the fibers is critical to good resolution of the image. Photo on right is example of poorly aligned fibers; note the indistinct edges of the image.

Although videoscopes haven't been considered appropriate for engine inspection in the past, the quality of the image and reduction in the size of the electronics have made them more compatible for this task.

In addition, advances in positioning methods have added to the flexibility of the videoscope. For example, Welch-Allyn recently released the "Flying-Probe," that incorporates a jet microthruster to aid in the positioning of the probe tip. The jet force at the tip of the probe quickly pulls the instrument through cavities to locations previously unreachable by traditional borescope techniques.

The potential length of the videoprobe (up to 70 feet) makes it ideal for use in areas such as inside fuel cells and underneath floor panels.

Illumination

Despite the quality of the borescope you're using, any borescope is virtually useless without a good light source.

According to Olympus Corporation, "the true performance of a light source is the result of all of the components in the illumination system: the type of lamp used in the light source, its effective reflection and concentration of light on the light guide plug, the efficiency of the scope's illumination system, including the light guide bundle and the windows that bring the light to bear on the subject and finally, the ability of the scope to transmit the reflected light from the subject into a bright image at the eyepiece... In other words, the scope and the light must work together as an efficient system."

There are a number of light sources, however, and there are some that are considered brighter than others. Keep in mind that videoscopes and video systems used in conjunction with fiberscopes or rigid scopes will require more light than the human eye.

Among the brightest sources of light, though not the most common, are the xenon and the metal halide lamps. Both of these are discharge-type lamps. Incandescent lamps are more popular, however, typically generating an average of 150 watts through a fiber-optic light guide to the tip of the borescope.

Guide tubes, such as the one pictured above for the PT6, are available for various engine models.

Around aircraft, it's also a good idea to make sure that the light source you're using is rated as an explosion-proof light source, particularly if you're using it near fuel tanks. Another consideration is the portability of the light source. If you require a battery-powered light source, the intensity of the light will not be as great as a plug-in model.

Other light sources available include UV, for performing dye penetrant inspections, and strobe lighting sources, for freezing motion.

Regardless of type of borescope that you're considering, it's important to talk to as many sources as possible prior to making any selection. It's also wise to contact the manufacturers and get as much detailed information as possible on the type of equipment that they offer.

July/August 1992

Accessory Technology

Precision bearing inspection

By Don Ross

From the turn of the century, anti-friction bearings have become an important part of almost every mechanical device developed by man. Their importance is particularly evident in modern aircraft where thousands of different sizes and types of bearings are utilized.

In an effort to extend bearing life and achieve the highest reliability, state-of-the-art bearings require the best materials and lubricants, exacting tolerances and rigid inspection procedures.

Too often, precision high-speed bearings don't make it to their full design life. They're prematurely removed from service for reasons that are preventable. Part of the problem lies in not recognizing the importance of bearing workmanship, proper inspection techniques and sound decision-making.

An engine's expected service life depends on the thoroughness and accuracy of a quality bearing inspection program.

The purpose of this article is to help standardize the interpretation of discrepancies found during bearing inspections and reduce the rejection rate of precision bearings based on maintenance manual and standard practice procedures.

Removal/installation

The greatest single source of unnecessary bearing damage occurs during bearing removal and installation. Using adequate care during these operations can't be overstressed.

Whenever possible, well-defined bearing removal areas should be established. These areas must be kept clean and protected against exposure to moisture, abrasive materials and corrosive fumes.

Proper tools that are clean and in good condition must be available. Worn or broken tooling or improper removal techniques can lead to bearing rejection or failure.

Along with keeping a clean shop it's important to compile complete and accurate inspection records. Such records expedite inspecting or reworking and lend assurances to a thorough and carefully monitored bearing inspection program.

In some cases, access during unrelated repair, for instance, a visual integrity inspection, is all that's required. A visual inspection should be accomplished without magnification, except where indicated to further define a suspected defect.

A scheduled inspection most often requires bearing removal and replacement with new bearings or complete disassembly, cleaning and inspection in order to certify them for continued service.

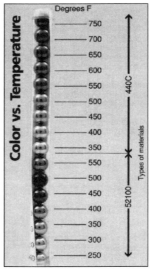
Color change due to specific temperature application on two types of bearing material.

Bearing integrity is very important. Specific serial numbered bearing components must be maintained as a unit. During disassembly, identify and tag matched items for processing. Don't commingle matched items with other like items.

Although it's not necessary to return individual rollers to the exact cage pocket they were removed from, it is critical to maintain proper thrust positions usually marked on raceways.

Demagnetization

Before bearings can be cleaned or inspected they need to be demagnetized. Bearings become magnetized over a period of time due to general rotation, welding without proper grounding, aircraft radar operation and even lightning strikes.

Most bearings, manufactured of steel alloys, accumulate a certain amount of residual magnetism during use and tend to retain steel chips and particles during the cleaning process.

Cleaning

The success of any bearing processing program depends on the effectiveness of the cleaning procedures. Bearings must be cleaned before inspection and relubrication. Individual bearings vary widely in the type and amount of contamination present. The cleaning results obtained by a given cleaning procedure will also vary, and close control of the process and evaluation of cleaning results must be maintained.

A solvent bath and rinse can usually clean cold section bearings. However, chlorinated solvents tend to absorb water and form harmful acids that may corrode bearings. Hot soapy water with a soft bristle brush works well on silver cages. A sonic cleaner is best used for non-separable bearings.

Bearings that have been cleaned of all protective films are particularly susceptible to corrosive attack. To protect these bearings from finger acids, it's important to wear clean latex gloves. It's also essential to handle and store unprotected bearings and assemblies in low humidity environments.

During engine operation, bearings can take on different color changes due to staining, overtemperature, chemicals or oxidation. In most cases, the colors are due to oil staining. This is seen frequently on bearings that operate in the engine's hotter zones. An oil-stained bearing is acceptable and doesn't need to be cleaned to its original condition. The colors that are of greatest concern are those that are due to overtemp.

Discoloration or bluing of the bearing's exterior usually indicates overheating. Likewise, bubbling or flaking of the silver plating on the cage also indicates excessive operating temperatures.

If an overtemp condition is suspected, clean the bearing in a caustic solution to determine if a true overtemp condition exists.

Heat discoloration of bearing surfaces ranges from light straw to blue or purple. Purple or blue discoloration which still exists after the bearing is properly cleaned in a caustic solution is cause for rejection.

Use care when evaluating discolored bearings.

Inspection

Bearing inspections should be conducted in a clean room, well-lit and humidity-controlled. Careful inspection techniques will detect most all normal surface defects or material decomposition.

Some bearings don't have to be disassembled. These "non-separable" bearings may vary from engine to engine. For instance, the same exact bearing part number used in several locations has different loading characteristics in different environments. What may be a "throw away" or "time change" item in one engine may be an "on-condition" item in another.

For non-separable bearings, a detailed visual inspection must be performed to ensure the cage is free of cracks and loose or missing rivets. A bearing analyzer can be used to amplify the vibration noises that rolling elements generate when there's a surface defect or obstruction on the contacting active surface.

Properly understanding the bearing analyzer's readings can help you determine if the problem is an outer raceway defect, an inner raceway defect, a ball defect or a combination of the three.

Depending on maintenance manual criteria, the bearing is thoroughly examined with the aid of a 6 to 20 power microscope, for early stages of bearing distress, damage and defects. During the inspection, you must identify the defect, determine the cause and recommend the necessary corrective action to be taken.

A specific radius ball-tipped stylus or scriber is used to help determine the acceptability of a visual defect. When used correctly, a stylus of known radius can be used to accept or reject bearings with various surface defects.

A lightweight, aluminum-handled stylus should be used. The stylus should not be firmly grasped but guided with the finger and thumb over the area to be inspected. The only pressure exerted on the defect to be checked should be the weight of the stylus.

In order to help standardize the interpretation of discrepancies found during a bearing inspection, it's mandatory to follow correct maintenance manual procedures appropriate for your engine.

Preservation

After inspection, bearings must be given immediate protection by applying either a lubricant or preservative coating prior to use or storage. The type of lubricant used is dependent on the application, bearing design and operating conditions that will be encountered.

To avoid potential corrosion damage, don't store bearings in an area cooled by an evaporative cooler. Also, to prevent the possibility of magnetizing bearings, never store bearings near electrical equipment.

Bearings are extremely vulnerable in a clean and dry condition and must be handled and preserved properly. Remember, clean and dry bearings plus humidity and fingerprints equals corrosion, pitting and etching.

Don Ross is a training instructor at Garrett General Aviation Services Division in Phoenix, AZ.

Accessory Technology

Typical bearing defects

The following are inspection criteria for bearing component defects:

Pits are small, irregular-shaped cavities in a surface from which material has been removed by corrosive action such as chemical or electrolytic attack or metal fatigue.

Nicks and dents are slight depressions or hollows on the surface made by mechanical injury to the material such as blows or pressure caused by hard objects contacting the finished surface.

Scoring is deep, multiple scratches caused by sliding at the rolling contact surface in the presence of foreign particles usually found on ball, roller and raceway surfaces.

Roller end wear may appear as circular scratches, burnishing, scuffing, frosting or pitting.

Scratches and scuffs are narrow, shallow linear abrasions caused by moving sharp objects or particles across a surface.

True brinelling is a plastic flow of metal characterized by smooth, shiny bottomed indentations in raceways and rolling elements usually equally spaced corresponding to rolling element spacing.

False brinelling is characterized by surface marks or blemishes on balls, rollers or raceways that normally have a polished or satin finish appearance.

Corrosion on surface areas is characterized by a broken, pitted or discolored appearance. Corroded surface areas will exhibit an orange color.

Galling is a more severe condition of fretting where there's a significant transfer of material between surfaces due to welding and breakaway of particles.

Cage defects which are cause for rejection include: silverplate flaking from base metal, peeling and blistering; grossly uneven silverplate wear on the pilot surface; cage pocket wear that has worn through the silverplate; and cracked, bent or broken cages.

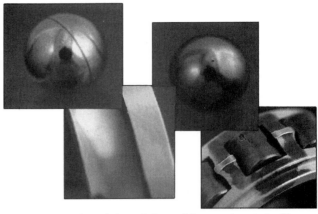

Typical examples of, from left to right: scoring, brinelling, pitting and spalling.

Arcing results in a round, pit-shaped cavity in a surface from which metal has been removed by an electrical arc, causing a temper change in the metal around the pit-shaped cavity. Arcing is caused by lightning strikes, welding without proper grounding, etc.

Cracks and fractures are separations, fissures or ruptures characterized by sharp edges and/or sharp changes in direction.

Skidding is excessive slipping between balls or rollers and raceways. Skidding is evidenced on roller bearings by a burnished or frosted ring around the rollers and/or raceways... on ball bearings by a speckled wear pattern or by a burnished or frosted ring on balls and by burnished or irregular frosted or smeared areas on the inner race.

Spalling is exhibited by irregular-shaped cavities of flaked-out metal from a raceway or rolling element surface with jagged bottoms caused by rolling contact fatigue.

May/June 1992

Accessory Technology

The neglected ELT
Up to 40 percent currently in use may be inoperable

By Greg Napert

Just about everything ever taught to the student A&P about the emergency locator transmitter (ELT) is contained on about half a page in the Airframe and Powerplant Mechanics Airframe Handbook. The information covered is just enough to know that the unit operates on 121.5 MHz, can be tested, and must have the battery replaced at a specific interval.

There's much more to this instrument, though, which when properly functioning can be a lifesaving device. Unfortunately, however, inadequate service and poor training have resulted in many non-functional ELTs and there are aircraft flying around at this moment with these inoperable ELTs.

According to the Radio Technical Committee on Avionics (RTCA) Document DO-182, 97 percent of all ELT activations are false alarms (40 percent of those false alarms are at airports with towers, 37 percent at airports without towers), and the non-activation rate of ELTs in actual crashes is 70 percent. A more common problem, though, is the inadvertent activation of ELTs and subsequent unnecessary search and rescue missions carried out every year. This very problem was probably the most significant factor in prompting the FAA to propose new specifications for the manufacture and installation of ELTs.

According to the FAA's Notice of Proposed Rulemaking (NPRM), issued April 2, 1990, nearly $3.5 million in federal, state and Civil Air Patrol resources are expended every year on ELT false alarm missions. Fortunately for all involved, the laws that regulate ELTs are in the process of changing.

Bob Chambers, chief executive officer of Artex Inc., a company that repairs, sells and manufactures ELTs, says that new legislation will go into effect by April of this year. These new regulations, which will include a new manufacturing specification (TSO-C91a) will affect the manufacturing, installation and maintenance of ELTs in the future.

But even with the final rule going into effect in April, manufacturers don't have to comply until six months later, and owner/operators will have until 1995 before they're forced to replace existing ELTs.

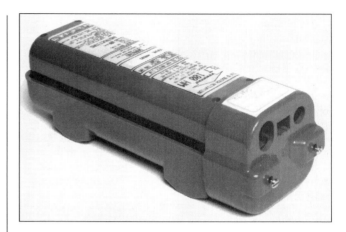

Design changes

Many of the problems associated with false activation and inadequate maintenance are related to the design of the ELT under the old regulations. The most important changes required by TSO-C91a are modifications to the G-switch and recommended improvements for installation.

Problems with current G-switches range from the switch being too easy to actuate, thereby triggering numerous false alarms, to not actuating when necessary. The new design addresses inadvertent activation by changing from an "instant on" switch to one that must be held on continuously for a brief period. A pulse of 100 milliseconds or 1/10th of a second of contact is required before it will come on.

With these new switches, you can't just activate the unit by bumping it with a jolt on a desktop, for instance. You have to more closely simulate the G-forces that take place during a crash. This can be done by pushing it away from your body then pulling it back quickly. The new style reduces the likelihood of activation due to such things as hard jolts, hard landings, touch-and-go maneuvers and quick stops.

Chambers says that many of the G-switches that are activated too easily do so because the springs are actually worn or weakened.

The Best of Aircraft Maintenance Technology Magazine

Accessory Technology

Another reason that G-switches stop working is due to corrosion. ELTs that don't conform to new specifications contain switches that aren't protected from the environment and are exposed to humidity and condensation. Over time, contacts and pivot points corrode and the switch becomes more difficult or impossible to activate. TSO-C91a attempts to alleviate this problem by requiring G-switches that are hermetically sealed which reduces the likelihood of corrosion.

Installation

Another problem with current regulations is that installation standards aren't good enough. Chambers says that most mounting brackets leave a lot to be desired and that many ELTs have killed pilots because of not being securely attached to the airframe.

"They break loose and just shoot right through the cockpit. The six to eight batteries enclosed in the ELT are a concentrated source of weight, and that mass can penetrate just about anything," he says.

Because of the many appropriate installations that exist, TSO-C91a will require mounting brackets to be made to withstand 100 G's of shock and still hold the ELT securely.

"The ideal thing," says Chambers, "is to have the ELT in a good solid mounting frame in the same bulkhead as that of the antenna. Then at least in a crash situation, you have a much better chance of survivability because of the reduced chance of the antenna cable and related components becoming severed from the system." He also says that it's important that the ELTs are positioned correctly. The unit must be mounted so that the G-switch will activate in the event of a crash. Keep in mind also that ELTs installed in helicopters should be mounted according to manufacturer's instructions.

The new rules may also require that a remote cockpit switch be apart of every installation. This would inform pilots of the ELT being activated and give them the ability to switch the unit off. Another option would be a remote audible alarm that would be installed in the cockpit.

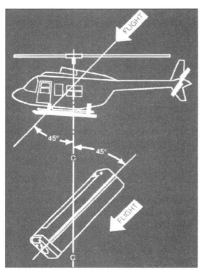

Preferred installation of ELT in helicopters.

Maintenance

Probably the most important changes to the new specifications have to do with maintenance requirements. Chambers says that the new TSO will require aircraft and/or ELT manufacturers to develop new maintenance and inspection criteria. These criteria will address removal of the ELT, inspection of battery, activation of G-switch, possible testing and evaluation of condition of G-switch, and power output of the antenna.

Current inspection requirements simply require the ELT to be checked for battery expiration dates and to switch it on to see if it works. But there's more that can and should be done, says Chambers. "You have to make sure that the unit is functional in terms of its G-switch integrity, its on/off, its power output and that all aspects of the ELT are operational."

The G-switch on TSO-C91 ELTs can be checked by removing it from its mounting and giving it a quick rap against the palm of your hand. When you're testing TSO-C91a units, however, you can only activate it by pushing it away from your body and pulling it back in one swift motion.

Chambers says that, by using this method, you're applying about a 5 to 7 G shock force. But he emphasizes that this is a rather primitive way to check the operation of the switch and that it really should be bench-checked using the proper equipment. Unfortunately, the equipment to check the switches is not readily available to most repair shops.

Besides testing the operation of the switch, Chambers says that the ELT should always be opened up so that the battery can be inspected for leakage. Current production methods for making batteries are pretty good, he says, but it's impossible to guarantee that they won't leak. Also, they may go a year without leaking and then suddenly begin to leak, so it's important to continue to check the batteries, even if they appear OK.

In the event you find a leaking battery, clean up and neutralize the spill as soon as possible and replace the battery. Leaking battery acid will corrode the electronics inside of the ELT and any metal that it comes in contact with. Depending on the type of ELT, it may need to be sent back to the factory for battery replacement.

In addition to checking for leakage, check the battery expiration date on the outside of the housing. FAA regulations require that the expiration date placed on all ELT batteries be one-half of the cell manufacturer's battery life. If it's expired, or if the battery has more than one hour of use on it, the FAA requires that the battery be replaced.

Accessory Technology

Chambers says to be aware of the remaining life on new batteries and new ELTs when purchasing them. ELTs that have been sitting at the distributor for extended periods of time have a reduced lifespan. We try to encourage all of the distributors to keep only a 30-day stock of ELTs so that they aren't sitting around for long, he explains.

Hopefully, expresses Chambers, the new regulations will help improve current problems with ELTs. But the problems probably won't be totally alleviated until tougher maintenance requirements are imposed. As a matter of fact, he says, other countries don't have these same problems. Canada, for instance, requires maintenance on the ELT annually, and has the situation under control.

"Eventually," says Chambers, "I think that the United States should require the technician to send the ELT to an avionics shop at each annual to verify its operation. Canada currently operates this way and has established a good safety record with ELTs." *March/April 1992*

Accessory Technology

Rotor track and balance

By Greg Napert

PROVO, UT—Nineteen years of experience as a helicopter technician stand behind the unwavering voice of someone trying to get a point across. The man—Larry Boyer, maintenance training instructor for Rocky Mountain Helicopters. "There are so many technicians who don't understand the basics of balancing and just go through the motions," comments Boyer.

"If you're going to effectively troubleshoot vibration problems out in the field, you've got to understand the basic theory behind electronic balancing."

What causes vibration

In simple terms, vibration is caused by the uneven distribution of mass around the center of rotation. However, the complexity of the rotor system is such that the distribution of mass can be affected by numerous factors. Aerodynamic forces, for example, can cause an imbalance to occur and lead to unacceptable vibrations. Some of these forces are outside of the control of the technician so it's impossible to completely eliminate all vibration.

In order to systematically troubleshoot the rotor system, you need to understand the items that can be controlled to affect the balance of the rotor system. These include adjusting blade sweep, tracking, adding and subtracting weight, and controlling wear.

It's also important to know what types of vibrations are the most damaging to the helicopter. Boyer explains that high-frequency tail rotor vibrations cause much more damage to the helicopter than the low-frequency vibrations caused by the main rotor. "High-frequency vibes are the Achilles' heel of helicopter maintenance," Boyer says. "They're not always easy to detect or eliminate, and if they're left unattended, can greatly accelerate the failure rate of bearings, cause bushings to quickly wear, and cause components all over the airframe to fail."

Main rotor vibrations, on the other hand, can cause the greatest discomfort to those flying on the aircraft, but the frequency (three to eight cycles per second) is such that they cause little harm to the helicopter.

Chadwick-Helmuth a manufacturer of balancing equipment for helicopters classifies correctable vibrations as: one-per-rev vibrations in the plane of the rotor disc, and one-per-rev vibrations perpendicular to the plane of the rotor disc. "Vibrations in the plane of the rotor disc," says Chadwick, "are induced by mass imbalance. This is correctable by adding (or subtracting) the right amount of weight at the correct locations, or by sweeping blades

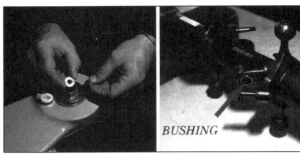

Left. *Balance weights on a Bell 206 JetRanger are added in the blade attach bolts. Other helicopters use different methods for attaching weights.*

Right. *Worn or incorrectly installed bushings on the hub of a BK117 tail rotor can make balancing difficult and result in severe vibrations.*

(on some models). Vibrations perpendicular to the plane of the rotor disc are caused by faulty track, which can be corrected in-flight."

Tracking basics

Because it's very difficult to distinguish which type of vibrations are affecting the helicopter (it may be a combination of both), it makes sense to first be sure that the rotor assembly is tracking correctly before attempting to balance the rotor.

There are a number of methods that can be used to determine if track is correct, but regardless of which one you use, the idea is the same—to get an indication of where the blades are flying in relation to each other. Old methods include the use of a stretched out piece of tape (flag) mounted on a pole. Prussian blue, water-based paints, grease pencils or crayons are used to mark the tips of the blades.

As the helicopter's running, the flag's moved to the blade tips until the blades make contact with the tape. Marks on the flag indicate the track of the blades, and the pitch links are adjusted accordingly. New methods involve using a strobe to freeze the motion of the blades so that you can visualize where the blades are tracking while the helicopter's running. The advantage of the strobe tracking systems is that the rotor can be tracked while the helicopter's in flight and in different flight configurations.

Some helicopters have provisions for mounting mirrors on the tips of the blades. With this system a light is then directed at the mirrors, and its reflections indicate the position of each blade.

Accessory Technology

Trim tabs also affect the track of the blades. They fine-tune the track of the blades in forward flight and are used to correct a climbing or diving blade. If you're making adjustments to the trim tabs, make sure that you don't bend them beyond the limits specified by the manufacturer.

Additionally, keep in mind that not all rotor blades will give the smoothest ride in a tracked situation. Sometimes the blades must be slightly out of track to give the most comfortable ride.

Static balancing

Balancing of the rotor system really starts in the shop with a static balance before it's installed on the helicopter. This is particularly true in cases where it's difficult to balance dynamically, or the rotor head is removed for overhaul.

At Rocky Mountain, a Marvel balancer, manufactured by the Marvel Manufacturing Company, is employed to balance tail and main rotor assemblies. The Marvel is a balancing tool with which the rotor assembly is suspended from a fixture that indicates a balanced condition. To achieve balance, weights are added or subtracted in locations specified by the manufacturer or the blades are swept fore or aft.

Boyer says that static balancing can also be done with a knife-edge balancer, provided that you have the correct cures to support the rotor. "It's still a proven method," he says.

But the static balance doesn't take into account aerodynamic forces that affect the rotor while in flight. You still have to perform a dynamic balance to "fine-tune" the rotor.

Dynamic balancing

The fundamentals of electronic dynamic balancing, says Boyer, start with understanding that vibration is caused by the uneven distribution of mass around the center of rotation. Additionally, you need to understand that the amount of imbalance is a function of the distribution of weight, the relative position of this weight and the speed at which it's traveling.

Simply put, your task is to find the heavy spot on the rotor assembly, and add weight opposite the heavy spot to achieve an equal amount of weight on opposite sides of the center of rotation. Sounds like static balancing, but you're doing it with the rotor head spinning and the effects of aerodynamics influencing the condition of the rotor.

In an unbalanced condition, the rotor causes a vibration in the helicopter. When using an electronic balancer, this vibration is sensed with a device called an accelerometer. The accelerometer sends an electronic signal to the balancing unit as it vibrates. The signal is used to determine the amount of weight in inches-per-second, and the position that the weight needs to be added.

Although many vibrations are picked up from different parts of the helicopter, the accelerometer can be electronically filtered or "tuned" to eliminate all vibrations except the one you're interested in tuning in. Tuning is done by setting the balancing unit to the rpm of the item that you're balancing.

The filtered signal, sent when the heavy spot on the rotor passes the accelerometer, triggers a device which helps the technician determine the position of the rotor. The device for determining the position of the rotor depends on the type of equipment used.

Chadwick-Helmuth, for example, manufactures a balancing unit that uses a strobe to freeze the motion of the blades called a Strobex™. It also can use a magnetic pickup, called a Phazor™ that senses the position of the blade and illuminates the blade angle on a ring of 24 lights. The indicated angle of the blade from these devices, called "clock angle," indicates the position where the weight needs to be added.

Positioning the rotor to the clock angle places the heavy spot next to the accelerometer. Theoretically, weight needs to be added opposite the accelerometer. But there's very rarely a blade exactly opposite the heavy spot. So weights have to be added to two or more blades so that the net effect is placing the weight exactly opposite the indicated "heavy" position. This procedure usually needs to be repeated a few times before an acceptable level of vibration is achieved.

You'll know when you're approaching a near balanced condition, says Boyer, because the clock angles will become erratic—the image of the blade will not be able to be held in a fixed position during the balance procedure.

The latest balancing equipment consists of computerized balancers that are capable of achieving track and balance in as few as two flights. A typical track and balance using Chadwick's Model 8500C, for instance, requires a test flight consisting of hover, forward flight at 130 knots and return to land. During this flight, says the company, the equipment gathers data, determines corrective action and displays the solution. Changes to weight, sweep, pitch link and tab are made as indicated. Finally, a second flight is made to verify the results.

Tips

Difficulty in achieving balance, even when you're properly using your equipment can be caused by a number of factors, says Boyer. Worn bushings and/or bearings are probably the biggest culprit, he says. An indication that an item may be worn is that the clock angles observed during balancing are inconsistent. In other words, every time you stop and start the helicopter, the clock angle will change and the assembly will appear out of balance again.

Excessive wear can make the rotor assembly nearly impossible to balance if not corrected.

Accessory Technology

Boyer points out that vibration is something that typically develops gradually. One of the most valuable things that you can do is develop good communication with the pilot. The pilot can continually monitor the condition of the helicopter and make you aware of developing vibrations.

It helps, he says, when different pilots fly the same helicopter. If the same pilot operates the helicopter all the time, the changes in the condition of the aircraft are so gradual that he can't always tell that there's a problem.

It's also important to develop a good feel for the bushings and bearings during daily inspections. Grabbing the blades by hand and wiggling them around can clue you in to impending failure. This is where experience pays off. Experienced technicians know what the bushings and bearings feel like when they're new and can judge that feel against the deteriorating condition. Manufacturers typically give limits and methods of checking bearings in their maintenance manuals.

Boyer also points out that balancing should be performed on a fairly regular basis to provide continual smooth operation. Erosion of the blade during normal use can contribute to vibration problems. "Remember," he says, "that because of the length of the arm you're dealing with, it doesn't take the loss of much weight to significantly affect the balance of the rotor."

Other items he points to are improper lubrication and improper handling. For example, permitting the blade grips to twist on the Bell 206 Series can cause the blade retention straps to stretch. This results in a longer blade and throws the assembly out of balance.

Also, says Boyer, track and balance will be much easier if you run the helicopter in low wind conditions. Boyer says that "high winds can adversely affect your results. You'll also have problems when performing balancing near a building on a windy day. Air flowing over and around the building will give you extra problems."

Finally, he says, you don't want to be standing inside the arc in which the tail rotor boom swings. Control of the helicopter may be lost during the balance procedure. "I've seen cases where we've lost control momentarily and the helicopter's swung around—balancing equipment and all."

Balancing isn't an easy concept to learn, explains Boyer. You've got to sit down and practice, even if it's only with a desk fan—and eventually, it'll sink in.

Technician's input alters magneto

The 4200/6200 Series Slick magnetos have been around for over 10 years and have remained relatively unchanged. Recently, however, Slick decided to respond to input from technicians and redesigned the magneto. According to the company, engineers compiled a "wish list" of product improvements based on customer input.

As a result of this input, Slick introduced the 4300/6300 Series of magnetos.

The primary feature of the new line is the frame-mounted distributor block and gear assembly. Now, when the distributor is removed, the magneto internals are exposed without altering the timing.

The rotor gear, which slides onto the end of the rotor shaft, features more pronounced alignment marks. A new feature is an alignment mark added to the distributor block. When the block and gear is installed on the frame and properly matched with the rotor gear, the mark on the block and on the rotor gear will align—assuring proper magneto timing.

A unique feature of the rotor is the slots cast into the face of the magneto head. The slots are marked "R" and "L" rotation. With the coil oriented in the 12 o'clock position, Slick's T-150 E-Gap gauge is inserted into the slot that corresponds to magneto rotation. The rotor is turned to hold the tool against the left side of the frame for left-hand rotation, and to the right side for right-hand rotation. The points are then adjusted to where they just open—E-Gap, in other words.

A new housing and frame assembly encase the magneto, says Slick. The mating surface of the two halves is wider for better seals and reduced EMI (electromagnetic interference). The result is better radio noise suppression. Cast into the housing is a groove for an O-ring used for pressurized applications. The new O-ring seals are better and less troublesome to service than earlier gaskets, says the company.

Also, for Lycoming applications, a new solid block clamp is supplied to ensure secure mounting of the magneto to the engine.

Slick recommends inspecting the new 4300/6300 magnetos every 500 hours, and replacing or overhauling them at engine TBO. *January/February 1992*

Accessory Technology

Nickel-cadmium battery maintenance

By Greg Napert

The nickel-cadmium battery came into existence as higher-power, longer-life batteries with short recharge cycles were needed for business and commercial turbine aircraft.

Although similar to the lead-acid battery, in that it consists of cells with positive and negative plates, separators, electrolyte and a similarly shaped container, the nickel-cadmium (referred to as "ni-cad" by most technicians) battery uses different materials in the construction of the cells.

The plates used in the cells are basically two groups of porous nickel plates or mesh screen. These plates are impregnated with nickel-hydroxide for the positive plates and cadmium-hydroxide for the negative plates. The plates are separated with a synthetic material to prevent shorting. The electrolyte, which allows transfer of ions between the positive and negative plates, is a solution of potassium hydroxide and distilled water.

Unlike the lead acid battery, the specific gravity of the electrolyte remains relatively constant. Therefore, the state of charge of the nickel-cadmium battery cannot be determined by using a hydrometer. The only way to measure the state of charge is to observe a measured discharge. The specific gravity of the electrolyte, however, should be checked periodically to ensure that a proper mixture of potassium hydroxide and water exists. The specific gravity typically should remain between 1.24 and 1.32.

The frequency of maintenance intervals for the nickel-cadmium battery varies from installation to installation and climate to climate. Factors that affect this interval, says Paul Scardaville, director of engineering for SAFT America Inc., one of the manufacturers of nickel-cadmium batteries for the aviation industry, include the environmental conditions that the battery is used in, the frequency of use, the aircraft condition, etc. The best way, says Scardaville, to determine how frequently to service the battery is to start with frequent inspections, and then cut them back until you are comfortable with the fact that they are being properly serviced.

Many people make the mistake of waiting until there's a noticeable decrease in the capacity before tending to the battery. At that point, the electrolyte level has probably been diminished and there is already damage done to the cells. It's important, he says, to maintain the battery before the allowable decrease in electrolyte level is exceeded. Running low on electrolyte exposes the separator material in the cells allowing them to break down. This diminishes the overall life of the battery.

Scardaville says that you should use extreme care in making sure not to excessively overcharge the battery on the aircraft. "Probably what has the most impact on the battery is not the method that's used to charge it, but the amount of overcharge that the battery receives. Overcharging causes the electrolyte level to decrease—resulting in shortened battery life."

Servicing

Cleaning of the nickel-cadmium battery should be done only after removing all metal articles from your hands and wrists such as jewelry, watch bands and bracelets to avoid severe burns or electrical shock. Then wipe all surfaces clean using a dry, clean cloth. The hardware should then be inspected for any obvious damage

Accessory Technology

such as loose connectors or flaking nickel coating. Scardaville says that most of the connectors are made of copper with nickel plating. Any hardware that has chipped or missing plating should be replaced with new.

The battery should also be inspected for other damage such as a dented case, damaged vent tubes or damaged latches. Be sure to retorque any hardware per instructions in the manufacturer's maintenance manual.

If no damage is found, the next step is to give the battery a "top charge" which will prepare it for adding electrolyte and get it ready for further testing. To top charge the battery, a current of C ("C" being the rated capacity of battery per hour) divided by 5, for Marathon, and C divided by 10 for SAFT or GE should be applied until the battery voltage reaches an average of 1.50 to 1.60 volts per cell. Don't exceed more than 1.75 volts per cell.

At this point, the electrolyte level should be checked and distilled, deionized or demineralized water added. SAFT recommends checking the level and filling according to its maintenance manual just prior to the end of the charge. Marathon, on the other hand, recommends adjusting the water level immediately after the charge, and GE has more complex instructions on checking the levels. In any case, the electrolyte should be adjusted using the appropriate tools per the instructions provided in the manuals. Overfilling the cells won't allow for expansion of the electrolyte and results in spillage into the battery compartment.

After the electrolyte level has been adjusted, the battery must be checked for leakage from the battery terminals to the case. A reading should be taken from the positive terminal to the case and from the negative terminal to the case; in either instance, the current shouldn't exceed 50MA. Currents of over 50MA can be the result of leaking or cracked cells, but a thorough cleaning may be all that's needed to eliminate the flow of current.

To determine the capacity of the battery, it's necessary to discharge the battery for a specific period of time. The idea is to discharge the battery to see if it in fact can deliver its rated capacity before the individual cells reach 1.00 volt. A typical example being a 36AH battery discharged at 18 amps for two hours. At the end of the two-hour period, the individual cell voltage should be above 1.00 volt. If the cell voltages fall below 1.00, the battery needs to be reconditioned or deep cycled.

To accomplish this, continue discharging the battery. While monitoring the battery, as each battery drops below 0.5 volt, place a shorting clip (Marathon recommends using a 6-inch 8-gauge minimum insulated wire with alligator clips at both ends) across the terminals to stabilize the voltage of the cells. The cells should not be allowed to go to zero or negative voltages.

This procedure should be done to approximately 75 percent of the cells. The last 25 percent of the cells (three or four cells) should be shorted with clips that

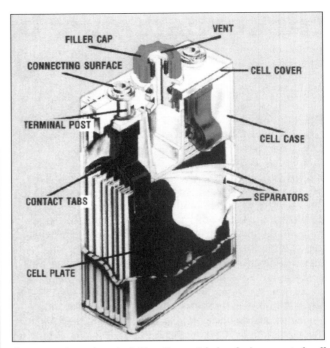

Cutaway view of typical Marathon nickel-cadmium vented cell.

contain a 1.0 ohm, 2-watt resistor. This allows any remaining voltage to discharge without overheating the leads. These shorting clips should remain attached for a minimum of three hours before recharging the battery.

Scardaville says that a common mistake made by many technicians is to deep cycle every battery that comes into the shop. This isn't necessary, he says. In fact, it's actually detrimental to the life of the battery. "The more deep cycles you give it, the shorter the life. During cycling, the positive plate (nickel hydroxide) has a tendency to expand (grow) with the cycling, and as it expands, the active material becomes less available. Even though it's necessary to deep cycle to restore the battery's capacity, you're breaking some of the bonds in the sintered material. The result is loss of conductivity and an increase in internal plate resistance," he says. "Therefore, deep cycle only when necessary."

Charging

After deep cycling of the battery, you're now ready to charge it. There are a number of methods for recharging and normally, manufacturers offer a number of options. Typically, however, they recommend a two-step charge: The first step being a high rate of charge, and the second step being a low rate charge. The battery is charged to full capacity at a low rate to avoid separator damage and reduce the possibility of overheating and excessive gassing.

Aero Quality, an inspection equipment manufacturer and distributor for aircraft batteries, says that during the charge cycle, the cell voltages should be monitored at

15-minute intervals starting 30 minutes prior to the end of the charge. This is important because any cells that are decreasing in voltage near the end of the charge should be removed from service. Those cells that don't rise to the manufacturer's recommended voltage should be top charged for an additional 30 minutes. Cells that fail to reach full potential or rise above the manufacturer's recommended voltages, should be removed from service.

The battery should then be checked for capacity by performing a capacity test. Any battery that fails to pass this test can be deep cycled up to three more times. If the battery cannot be recovered by deep cycling, it's time for a new battery.

Scardaville says that automatic chargers are OK, but he highly recommends constant current-type chargers. He says that the automatic type of charger is incapable of reacting to problems with the individual cells because it measures the overall voltage. If a cell begins to heat up early in a charge, for instance, the charger won't detect it and will continue to apply power. If you do choose to use this type of a charger, he recommends placing the battery on a constant current charger for 15 minutes prior to placing it on the automatic charger. This allows the battery to accept the high rate of charge without early overheating.

Thermal runaway

Under certain conditions during the constant potential charging, some nickel-cadmium batteries have been known to overheat, and sometimes undergo a phenomenon called "thermal runaway." This process, characterized by high currents and battery temperatures, is a result of the failure of the membranes between the plates within the cells, says Scardaville.

"Nineteen-cell batteries draw more current as they near the bus potential and thus, dry out faster. Twenty-cell batteries draw less current as they near the bus potential, and operate at cell voltages below the full gassing point. As a result," says Scardaville, "they are less likely to experience thermal runaway."

Thermal runaway, he explains, is simply the result of oxygen generated on the positive plates recombining on the negative plates. The cell is designed for this not to occur. The membrane (cellophane or polypropylene) is supposed to be a gas barrier, but when this membrane breaks down, thermal runaway occurs. When it begins to occur, the temperature of the cell heats up, and as the temperature rises the cell voltage falls and the current increases. As the current increases, the temperature rises, and the whole process continues in a vicious cycle until the cell is destroyed.

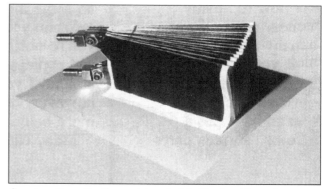

Construction of cell consists of nickel-hydroxide and cadmium-hydroxide plates separated by layers of synthetic materials. This example is a SAFT cell.

Due to the possibility of this happening, it has been made mandatory to have some kind of indicating device in aircraft in the form of a temperature or current indicator. However, the highest possibility of thermal runaway exists during an overcharge condition in the shop. Because of this, it's recommended that the battery be closely monitored while charging. It's also recommended that charging not be attempted when the battery's internal temperature exceeds 145°F, as this is detrimental to the separator membranes.

A few extra pointers

Scardaville says that it is possible to install a cell incorrectly if one needs to be replaced, even though it is clearly marked as to the positive and negative posts. If one of the cells is installed backward, he says, and charged in the reverse direction, then the cell should be taken out of service. "The material in the cells at this point has been disturbed to the point where its future performance becomes unreliable," he says.

He also recommends occasionally rotating the position of the individual cells; moving the cells from the center of the battery to the outside of the battery and vice versa. The reason for this is that heat is concentrated in the center of the battery, and the cells located in the center are subjected to higher heat for longer periods of time. Make sure, however, that the cells are arranged in series and that all hardware is installed as directed by the maintenance manual.

Finally, nickel-cadmium batteries can be stored, says Scardaville, for extended periods of time in either the charged or discharged state. In either case, the ideal temperature for storage is between 32°F and 86°F. These batteries, he says, also lose charge over an extended period of time, so applying a trickle charge at a rate of 1 milliampere per ampere-hour of capacity should keep the battery in a fully charged state.

November/December 1991

Accessory Technology

Recip engine synchronization

By Jeff Rogers

People react in different ways to the physical sensations of flying. Whether they find flying frightening or fun, passengers become concerned at the slightest suggestion that something is wrong with the airplane. A common culprit which may alarm even an experienced flier is the throbbing growl of unsynchronized engines.

Regardless of engine type, multi-engine aircraft are prone to undesirable vibrations generated by engines operating at unequal speeds (rpms). Unsynchronized propellers produce one "beat" per blade per revolution of speed. The effect is most noticeable when the engines are turning within a few rpm of each other.

As engine speeds are brought closer together the throbbing slows, until it diminishes altogether when the engines reach synchronization. Although it poses no significant structural problem, the blending of normal vibrations and noise from each engine can make pilots and passengers very uncomfortable unless the resultant vibration and noise are virtually nil.

Synchronization methods

Early efforts toward keeping engines synchronized relied on pilot technique. Pilots learned to manually limit the duration of the out-of-sync condition by matching propeller speeds with the throttles and prop controls. Eventually, increasingly complex powerplant technologies and expanding pilot workloads made it desirable to develop an automatic means of synchronization.

Many engine synchronization systems used in Beechcraft twins, Gulfstream IVs, Cessna Citations, DC-9s and other aircraft types are produced by the Woodward Governor Company. The theory of synchronization is not complex, and the various Woodward systems provide effective examples of the process.

The Woodward Type I electronic synchronizer designates one engine as the "master." The other engine becomes its "slave." The synchronizing system increases or decreases the speed of the slave engine, over a limited range, until it matches the speed of the master engine. The limited adjustment range is a safeguard against excessive power loss. Without this limit, in the event of master engine rpm drop or complete failure, the slave engine would also lose power as it tried to follow the master engine.

The system consists of a control box, a speed setting actuator, a propeller trimmer assembly, a flexible rotary shaft which connects the actuator and the trimmer, and two synchronizer-type propeller governors. The control box is the brain of the system. It is fully solid state and operates on 28v DC power. Current draw is less than 1 amp. The only required control is the on-off switch, as the system is automatic. The prop governors have built-in electromagnetic speed pickups. The electric signals produced by the rotating propellers are fed into the control box, where they are compared. Any difference in frequency will cause a control command to be sent to the actuator in the slave engine nacelle. The trimmer assembly on the slave engine prop governor will increase propeller pitch to slow the engine or will decrease propeller pitch to speed up the engine. When the propellers are turning at the same speed, the control box removes the control command to the actuator.

The system's secret of success lies in the "step-type" actuator drive motor. The step-type motor turns in small steps, rather than in a continuous rotation. It advances or retracts the propeller trimmer in increments until the speed comparator in the control box removes the signal to the actuator. Then, the actuator remains at that "step" until the next control command is received. This eliminates the need for a constant current motor (servo) to hold the trimmer in position and reduces system weight and electrical load.

The actuator is built with mechanical stops which prevent actuator over-rotation from damaging the flexible shaft or trimmer rod end by trying to force the trimmer beyond its limit of travel.

Limit switches in the actuator housing tell the control box when the actuator is out of center, and in which direction. When system power is turned off, the control box automatically provides the correct rotation command needed to center the actuator. This prevents the trimmer from affecting slave propeller pitch when the system is not on.

Operation

The synchronization system operates independently of the pilot-controlled propeller governors and throttles. For safety reasons, it is not used on takeoff or landing. When power is reduced after takeoff, the engines must be manually synchronized to within the "holding range" of the sync system (approximately plus or minus 50 rpms for the Woodward Type 1).

When the engines are roughly in sync, the system is turned on. Slave engine speed will automatically be adjusted to match the master engine. To change rpm settings for desired flight maneuvers and still maintain synchronization, the pilot should move both prop control

Accessory Technology

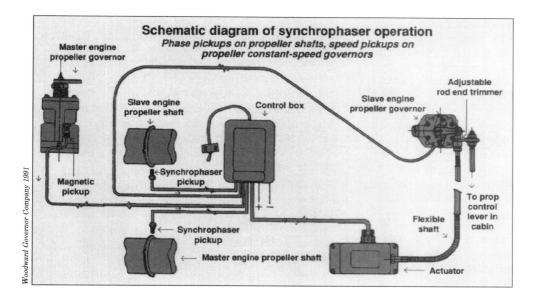

Schematic diagram of synchrophaser operation
Phase pickups on propeller shafts, speed pickups on propeller constant-speed governors

levers together. This is intended to keep any rpm difference which may occur during the speed change from falling outside the system's holding range.

If, after an intentional speed change, the synchronizer can't bring the rpm indications back to equal, the actuator is at either limit of its travel. The system must be turned off, the engines manually resynchronized and the system turned back on. The synchronizer will continuously monitor and adjust slave engine speed as required.

If it becomes necessary to feather either propeller, the synchronizer should be turned off. If the system is left on, and the master prop is feathered, the slave engine will experience a decrease in rpm as the master engine stops. If the slave is feathered, the actuator/trimmer will be uselessly trying to speed it up by adjusting an inert prop governor. The sync system should also be off when unfeathering a propeller. If left on, when unfeathering the master prop, the slave prop rpm will drop until the master comes up to speed. When unfeathering the slave, the system will command reduced pitch until it speeds up. Any of these situations will needlessly "bottom out" the actuator. Unintended propeller response could also complicate an already critical situation.

Propeller blade synchronization

An additional source of vibration and noise is the action of the propeller blades themselves. Although a synchronization system will adjust the props to turn at equal rpms, the phases of propeller rotation may still be out of sync. One revolution of a three-bladed propeller produces three noise "spikes" and three noise "lulls." The desirable phase relationship between the two propellers will allow the lulls of one blade to cancel out the spike of the other. The correct phase offsets (relative propeller blade positions that provide noise cancellation) depend on fuselage shape and engine/propeller combination. Phase offsets are specified by the aircraft manufacturer.

The Woodward Type I synchrophaser performs identically to the synchronizer, with the additional capability of maintaining the desired phase offsets. The synchrophaser uses the same components as the synchronizer, but adds a set of phase sensors to each prop shaft. Prop phase signals are sent to an additional circuit in the control box. If the phases are out of sync, a corrective phasing signal is sent to the actuator. This same signal temporarily overrides the synchronizing function of the system. The actuator pulses the slave propeller to momentarily speed it up. Each successive phasing pulse causes the slave propeller phase to move closer to the master prop phase. When the props are in phase, the phasing signal is removed from the actuator, and the synchronizing function of the system is reactivated.

Flight procedures for synchrophaser operation are identical to the synchronizer, but several requirements must be met to achieve maximum synchrophaser performance. The propellers must be very responsive to prop governor pressure changes. This allows the slave propeller to effectively respond to the subtle phasing pulses. The prop governors must be well-maintained. There can be no "slop" in the squared or splined ends of the flexible shaft linking the actuator to the trimmer. Without a mechanically sound propeller/governor combination and flex shaft, the phasing band (the possible propeller blade positions that register as being in phase) widens. An excessively wide phasing band may give the impression that the synchrophaser is not working properly, when the problem is really mechanical.

Accessory Technology

The Woodward Type II synchrophaser eliminates the actuator, flex shaft and trimmer by using a specially designed, electrically adjusted prop governor. A single pickup in each engine senses both speed and phase. The pickups can be mounted on the prop deice brush blocks, with the phase targets on the aft spinner bulkhead or on the prop deice slip rings.

Testing and troubleshooting

Synchronizer operational checks must be performed in flight. It is not practical to ground-run engines at cruise speeds long enough to complete the tests, and small aircraft will bounce around too much. The flight tests can verify system holding range and automatic actuator centering.

Manually synchronize the props at cruise speed and note engine rpms. Turn the system on. Steadily increase master engine speed. Watch the tachometers to verify that rpms increase together. When the slave engine speed stops increasing, the actuator is at its "speed up" travel limit. Note the slave engine's speed. The rpm increase from cruise speed is the plus follow range. Slow the master engine back to cruise speed. Observe that the slave engine follows. Now, slowly reduce master engine speed below the speed at which the slave engine speed stops decreasing. The actuator is now at its "slow down" travel limit. This rpm change is the minus follow range. Return to cruise speed to perform the actuator centering test.

With the synchronizer on, increase the master engine speed until it is close to the "speed up" limit established in the range test. The slave engine should follow. Next, turn the system off. Slave engine speed should drop back to cruise rpm as the actuator automatically centers. Turn the system back on. Slave engine speed should increase to resynchronize with the master.

The Woodward system manuals describe test procedures using Woodward test equipment to check wiring continuity and resistance values. Troubleshooting flowcharts and symptom/cause tables are provided, but according to Mark Rein, Woodward synchronizer/synchrophaser programs manager, the best troubleshooting tools are a strong working knowledge of the system and sound troubleshooting procedures.

- "Most of the control boxes sent back to us have nothing wrong with them," Rein observes. "Many technicians use a shotgun approach to troubleshooting and overlook the more frequent mechanical problems." Rein suggests a common-sense approach and offers these insights:
- Many of the Woodward systems are old, up to 25 years. Engine-mounted components get dirty. The flex shafts can bind without periodic lubrication. Woodward recommends a mixture of 30W oil and Molykote-G. Hold the flex shaft vertically and introduce the lubricant at the top end. Spin the shaft with your fingers until the oil drips through.
- A word of warning about lubrication: Never oil the trimmer rod end. Oil will attack the baked-on lubricant coating, causing it to break off in chunks. Many technicians have oiled these and thought they were doing the customer a favor, when they were actually dooming the parts to failure.
- The system control boxes generally last the longest, as they are mounted inside the cabin. A Type I system component that is prone to failure is the four-pole on-off switch, due to fatigue buildup over the years. Just because the switch lights up doesn't mean the system is turned on.
- Remember that if the props are swapped side to side, the system will still sense and adjust, but will maintain a phase relationship other than that specified by Beechcraft.

Rein concludes that with conscientious and informed maintenance practices, the synchronizer system can provide trouble-free operation for a long time.

Jeff Rogers is a free-lance writer and avionics technician based at K-C Aviation in Appleton, WI.

Accessory Technology

Turbine engine synchronization

By Jim Sparks

The vibration of turbine engines can be described as a back-and-forth motion in relation to a central point. Anytime an engine or propeller is producing thrust, both aerodynamic and mechanical forces are present and result in blade vibration. Anytime two objects have the same natural vibrational frequency, resonance can occur. This is the transmission of vibration from one molecule to another at a greatly amplified level. By synchronizing propellers or engines, the resulting vibration of one powerplant can be used to offset the other. The concept is really the same as with recip synchronizers, but the technology is slightly different.

For an engine synchronizer to perform, it has to be able to observe the rpm of all powerplants. When engines utilize more than one rotating section (N1, N2, propeller), provisions are generally made so synchronization can work with any group. As with recip engines, a master engine must be designated to supply the standard for the remaining engines. On aircraft incorporating four or more powerplants, an alternate may be designated so if the master engine fails, synchronization between the remaining engines is still possible.

Synchronizer systems typically consist of a control box which receives speed signals from the engines, processes this information, then supplies an output to the slave engine to match the rpm of the master.

The speed signal for the control box is of great importance. In many cases, turbine engines obtain this signal from the engine tachometer generators. These generators usually provide a three-phase alternating current output. This output may also be used for the engine speed indicator in the flight deck. The synchronizer control box observes tach generator output voltage (typically 10 to 21 volts), but is more concerned with the output frequency (cycles per second).

As engine speed increases, so do the CPS. The validity of this signal is very important to obtain proper synchronizer operation. Many conditions exist that can modify the speed signal resulting in synchronizer malfunction.

A very simple, yet effective test is to remove the electrical connector from the tach generator and with the engine shut down, do a resistance check of all three generator windings. These values should all be approximately equal. Probing points can usually be determined by using the applicable wiring diagram. This check might be duplicated while applying a heat gun to the generator. Make sure, however, not to use excessive temperatures. The idea is to duplicate those found in the nacelle under normal operation. A bleed air leak directed onto the generator could cause a weak voltage output at higher power settings.

Signal loss can also be caused by wiring problems. A breakdown in insulation used on a tach generator wire may sufficiently reduce the output of one phase and can make the synchronizer operation erratic.

After performing a resistance check on the generator wiring, it's worthwhile to disconnect the tach indicator and the input wires to the synchronizer and then do a complete continuity check followed by a resistance breakdown test of the entire wiring harness.

High-voltage wiring in the proximity of the synchronizer components may also interfere with synchronizer operation. Stray signals from systems such as strobes or fluorescent lighting may get induced in the sync system. This induction will tend to cause erratic sync operation.

Once it is determined that the control box input is valid from both slave and master tachometer generators, a response is issued to a mechanical actuator. This actuator is best described as a stepping motor installed on the slave engine. Opposing solenoids connected to a rotary device provide the drive. One solenoid is energized repeatedly, or in steps, to running the unit one way or increase rpm. The other solenoid is used for the reverse direction or rpm reduction.

The mechanical response from the actuator is transmitted to the throttle using a flexible cable. Any excessive force required to drive this cable may be sufficient to nullify or desensitize the sync response. This transmission cable is usually maintenance free and any binding usually requires cable replacement.

Lack of synchronizer response at altitude may be an indication of moisture in the cable. When this water freezes, all sync operation stops. In some cases, this moisture may be purged and the cable relubricated.

Synchronizer systems have a limited range. The usual operating range is about 1.5 percent rpm on a turbine engine. These systems also incorporate a self-centering capability. In normal operation, anytime the sync control switch is selected "off," the centering system operates allowing the slave engine throttle linkage to return to its normal position.

During takeoff and landing, synchronization is selected "off." This way if the master engine fails, the slave should not experience a power reduction. Once the aircraft has been stabilized in climb, the pilot makes the initial power reduction and engines are manually matched; then the synchronizer is turned on and will

Accessory Technology

adjust the speed of the slave engine to match the master within its 1.5 percent tolerance.

During ground testing of turbine engines' sync system, it's usually advisable to do so with engines at mid-power, then follow specific system manufacturers' recommendations.

Maintenance on most sync systems is limited to electrical inspection and checking for condition and freedom of movement in the mechanical components.

When diagnosing a synchronizer malfunction, a good pilot debrief is essential. A variable AC frequency generator is a great help when troubleshooting this system. It enables speed simulation to the sync box and allows output checks to be completed with the engine shutdown. Many manufacturers supply plans to build additional test equipment.

Thorough electrical checks are essential when troubleshooting. A shorted output wire to an actuator may cause an internal failure in the control box. Replacing a control box without a complete system check could result in a second failed box.

It is important to refer to a current Airframe Maintenance Manual as well as a specific synchronizer system manual. Most sync manufacturers publish manuals complete with wiring diagrams and troubleshooting flowcharts. *September/October 1991*

Jim Sparks is the director of maintenance training for FlightSafety International and is based out of Houston, TX.

Accessory Technology

Braking tradition
Minnesota shop puts "more" into its brake overhauls

By Greg Napert

The business jet is in a category all by itself when it comes to brakes. Unlike smaller light aircraft, the business aircraft uses braking systems that are much more complex.

Typically, these brake assemblies consist of a number of plates, called stators and rotors, stacked together like pancakes between a pressure plate and a back plate which contain wear pads. The stators are keyed to a component called the torque tube, that is attached to the landing gear, and the rotors are keyed and turn with the wheel. When hydraulic pressure is applied to the pressure plates, the friction created as a result of the layers pressing together stops the wheel from turning.

Stators and rotors aren't made from the same material. This is critical as steel can't rub against steel or the discs would gall and fuse together. To prevent this, the rotors are typically manufactured with a layer of tri-metallic material on their surface which serves as a wear layer.

Overhaul of the assembly at first appears to be quite simple, but trying to overhaul them without any previous knowledge can result, at best, in a short-lived brake, and at worst, disaster.

Lester Sawyer, aircraft technician for Page Avjet Airport Services Inc. at the Minneapolis/St. Paul airport, has been overhauling these types of brakes for over 13 years. In fact, Page's elaborate shop is set up specifically for light corporate jet brake overhaul.

Sawyer says they've seen more than one instance where shops have tried to overhaul the brake assembly themselves, installed the brake in the aircraft without properly testing it and have had the brake overheat the first time the aircraft is taken out. This can be prevented if the overhaul manual is followed and if you're familiar with using the special tools required to overhaul the brake.

In fact, there aren't that many special tools required to overhaul the brake assembly, and most of them can be manufactured very easily. The manufacturer's overhaul manuals contain drawings and instructions for making the tools required. One tool, for instance, for setting the gripper assemblies on the Westwind brake, consists of a 1/4-inch washer welded to a metal block and a bushing to set the gripper height. This tool allows the proper running clearance to be preset into the overhauled brake. Although it's a simple tool, it performs a critical job.

Overhaul

When the brakes are received into the shop, they are broken down into their various components. All parts are stripped of any paint or primer and cleaned up. Any ferrous parts are magnetic particle inspected, and any non-ferrous parts such as aluminum or magnesium are dye-penetrant inspected.

"One of the components," says Sawyer, "that we replace regardless of their condition, is the wear pads that are located on the pressure plate and back plate assemblies." These pads, he says, are the primary wear surface for the brakes and should always be replaced if you want the full life out of the brakes.

Many people remove the pads by pressing the rivets out of the assembly, but Sawyer says that this practice can damage the back plate and pressure plates.

All pads and reused stators are ground to perfect flatness on a surface grinder.

Accessory Technology

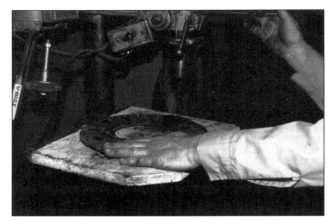

Pads are drilled from plates instead of pressed. This prevents damage to the plate.

"Instead," he says, "we drill the rivets out. This assures us that no additional stress is placed on or damage is done to the plates."

After the pads are removed, the plates are bead blasted, magnetic particle inspected and inspected dimensionally and visually. It's important to remove any burrs, sharp edges or ridges around the rivet holes, he says. Ridges around the rivet holes can prevent the new pads from seating properly and cause warping of the new pads.

Typically, there isn't much damage to these plates, says Sawyer. "They're pretty beefy and hold up to a lot of abuse. As does the brake housing and the torque tube. We usually don't see any damage at all to the brake housing," he says. The torque tube does see some amount of wear at the lugs, where the stators engage it. These lugs should be checked dimensionally. There tends to be more wear toward the outside of the tube because there is more movement, he says, so inspect these areas carefully to ensure that they're within limits.

The pressure return springs are another item that should be inspected carefully. The first thing to keep in mind with them is that they shouldn't be bead blasted. They're specially coated, he says, and blasting them will remove the coating. Instead, use a brass brush to brush any dirt or minor corrosion off of them. They then must be inspected with a calibrated spring tester to assure that they aren't too weak. Weak springs should be discarded and replaced with new.

"The steel stator assemblies often need replacement," says Sawyer, "as do the tri-metallic coated rotor assemblies. We always replace the rotors because all of the wear material is worn off. There are limits that should be observed in the overhaul manual but they usually don't make it for a second run. And if you did reuse them they wouldn't last very long. We have the opportunity, however, to save some of the steel stator assemblies. If they're within limits, we will reflatten them by heating them in an oven between two steel plates, then

All parts are dimensionally checked and non-destructive tested (NDT'd) before returning to service.

regrind them flat. Pressure and backplates are then coated with a primer to prevent rusting."

After all components are inspected and dimensionally checked, new pads are riveted onto the pressure plate and back plate. This procedure requires some experience in setting up the riveting machine. The correct pressure must be used to set the rivets—too much pressure and the pads warp—too little and they won't be held tight enough. The correct setup should result in a pad that can be moved when between 200 and 100 foot-pounds of torque are placed on it. This check can be done, says Sawyer, by simply placing a screwdriver between the pads and twisting. If you can't move the pads by twisting the screwdriver, they're set too tight. And if they move with little or no effort, they're not riveted tight enough.

A check should then be made for warping by trying to slide a .015 feeler gauge between the outer edges of the pad and the plate where the pads make contact. If you're able to force the feeler gauge more than 50 percent of the way across the pad, they're either curled too much or the rivets aren't set tight enough. This should be determined and corrected.

After the pads are riveted, they are ground so that they're perfectly flat. "You don't have to remove much material, just a couple of thousandths, to make them perfectly flat," says Sawyer. "The overhaul manual does say that you don't have to grind them at all if they are all within .010 of each other, but we find that if they're not ground flat, the brakes will chatter and squeal during the initial break-in," he explains. "We've found that we get much better performance out of them if they're ground."

Accessory Technology

After all subassemblies are properly rebuilt and inspected, it's time to recoat them to prevent corrosion. Page has found that protecting all parts can significantly extend the life of the brake components. Some of the OEMs don't place any protection on them at all, and some overhaulers don't use a good enough product to protect them. "For instance, we've found that red-oxide primer is far superior in protecting the steel components to zinc chromate primer," says Sawyer. "We see our brakes come back in good condition, complete with primer still on them, while those parts that aren't properly protected oftentimes have to be replaced."

Also, when repainting the brake housing, it's important that you follow the manufacturer's specifications when selecting the type of paint to use. Brake assemblies that will be exposed to Skydrol™, for instance, must be painted with an epoxy-based paint. Other hydraulic fluids typically can be painted with lacquer-based paints.

After all of the components are reassembled and torqued per overhaul manual instructions, it's time to test the assembly.

The test basically consists of attaching a hydraulic pressure line to the brake assembly and actuating it on the bench. One thing to keep in mind before performing the check is to make sure that you're using the correct hydraulic fluid. Depending on the type of brake, it will either require Skydrol™, Braco™ or MIL-M-5606 fluid.

Once the correct fluid has been determined, the brake needs to be tested according to the manufacturer's instructions. Typically, this means testing it at a predetermined low and high pressure. The low pressure

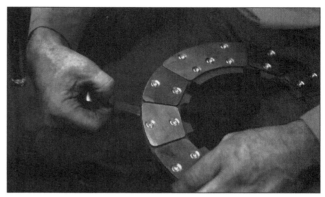

Pads are riveted and checked for security by inserting a screwdriver to check for movement, and a feeler gauge to check for warpage.

assures that there's a significant running clearance. You should be able to turn the rotors in the assembly. If you can't, either the grippers are set incorrectly, are worn or the return springs are too weak and should be replaced. This problem must be investigated before you proceed or the brakes will drag and overheat when they're first used. During the high-pressure check, the brakes should be inspected carefully for leaks.

"Simply following the overhaul manual is usually sufficient when overhauling brakes," says Sawyer, "but we feel that it's important to take the overhaul one step further—adding a few of our own touches so that the brake assembly is sure to make it to its next overhaul without premature failure." *July/August 1991*

Accessory Technology

Rotor head inspection tips

By Greg Napert

Nestled in the small community of West Chester, PA, sits a somewhat unexpected sight. It's easy to forget that all airfields are not located in wide-open flatland areas, taking up a minimum of 8 to 10 square miles. For Keystone Helicopters, space isn't a major requirement; expert technicians are.

Keystone technicians regularly repair, overhaul and maintain a fleet of helicopters for corporate and Emergency Medical Services (EMS) and others around the area. And with helicopter maintenance, it's imperative that you perform maintenance tasks and operations by the book with no shortcuts.

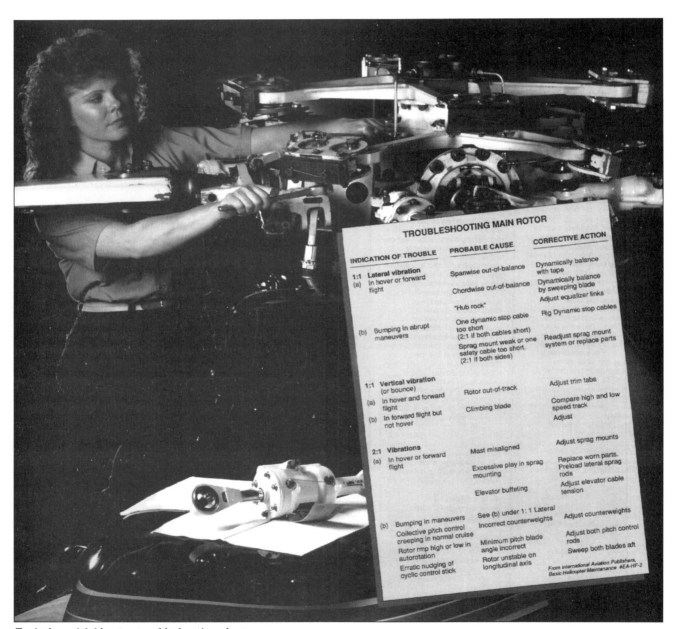

Typical semirigid rotor troubleshooting chart.

Accessory Technology

Like anything else in aviation, it's the small things that get you. Washing the rotor assembly, for example, if not done correctly, can be one of the most destructive things to the mechanism, says Mark Gould, component shop manager for Keystone. Water that seeps into bearings can lead to corrosion and cause premature damage. Also, some soaps may not be compatible with the materials used in the rotor assembly. It's imperative, he says, to make sure that whatever soap is used is approved by the manufacturer of the aircraft.

Jim Malarsie, director of maintenance for another helicopter operation, Rocky Mountain Helicopters in Provo, UT, agrees with this point. Malarsie adds that pressure washers can be very damaging to bearings and rubber seals. If pressure washers are used, he says, it should be done judiciously.

Lubrication and inspection for damage is the crux of the maintenance performed on the helicopter rotor assembly. It's important, says Gould, to maintain a regular program of inspection and lubrication for the helicopter. Most of the damage seen in the overhaul shop is due to insufficient lubrication. The lubrication interval should be adjusted for environmental conditions, he explains. If a helicopter is flown in sandy or dusty conditions, for instance, it's a good idea to grease the bearings more often to force the contamination out of the bearing areas.

There are some helicopters (one example being the Aerospatiale Lama), says Malarsie, that have to be greased with caution. In this model, if the seals between the grip and the rotor head are damaged, it's possible to pump grease into the grip without knowing it. If this is done, the grease will be slung out to the end of the grip and create an imbalance that cannot be corrected. "If they tell you to put 4 ounces of grease," says Malarsie, "put 4 ounces—not 8. Some people think that if some grease is good, more is better. But that isn't necessarily the case."

He adds that it's important to keep the area clean. "Many technicians never clean the grease off of the head and you end up not being able to see anything. Cleaning the rotor head area thoroughly allows us to perform inspections without missing anything."

The technician's job doesn't stop at lubrication. Component inspection, disassembly and replacement are important parts of preventive maintenance. Gould explains that you can never be too careful when inspecting components. If you suspect a crack in a certain area and dye penetrant doesn't show anything, you may want to consider having that part X-rayed.

Cross section of elastomeric bearing used on a Sikorsky S-76 helicopter.

During overhaul of the rotor head assembly, all components are NDT'd (Non-Destructive Tested) for cracks and all components tracked for time limits. There are a number of components, Gould says, that are life limited on the helicopter rotor assembly, and a number of them that are on condition. Keeping track of which components are to be removed from service is a major part of the job.

An abundance of new technology has emerged in the past decade. While rotor head systems are becoming more advanced and contain fewer parts, inspection of these components is still critical. Pam Tressler, lead technician for the component shop at Keystone, explains that elastomeric bearings, a laminated assembly consisting of layers of rubber and metal, must be inspected regularly.

Deterioration of the rubber due to its flexing action results in flaking of the surfaces and separation of the laminations. "It's kind of like residue from an eraser," says Tressler. "The manufacturer usually specifies certain limits that they are allowed to deteriorate. If they exceed those limits, they've got to be replaced." These bearings, she says, are very sensitive to hydraulic fluid and oils. "If any oil or hydraulic oil accidentally comes into contact with them, make sure that it is washed off immediately, or it'll eat away at the rubber."

The same goes for composite components. For example, the composite straps that secure the rotor blades to the rotor head should be very delicately handled. Don't use any solvents to clean this material, she says.

Another material, says Gould, that's becoming more and more common is the Teflon bearing. These bearings don't require any grease or oil, which can damage or adversely affect the operation of these bearings. Surgical gloves should also be worn when handling any rotor components. Many of the components used in the rotor assembly are highly polished metals that are affected by the acid and oils naturally present on your hands.

Accessory Technology

Some rotor heads, instead of being greased, are oil-filled. Leakage on oil-filled heads is not that common; rather, it is more common for them to leak during cold weather operation. Once a seal starts leaking, it usually continues to leak, says Gould. The only option is to replace the offending seal.

An area that commonly requires attention, says Gould, is the lead/lag dampener assembly. The lead/lag dampener on the Sikorsky S-76, for instance, is a shock absorber that dampens the action of the blade as it moves fore and aft. Some dampeners are elastomeric, some are multiple disc, and some are hydraulic. If a dampener is bad, it can cause unusual vibration problems.

The elastomeric dampener is a remove-and-replace item. The multiple disc works on a braking principle: a series of discs dampen the motion of the blade.

The hydraulic dampener is probably the type that most commonly fails and should be checked regularly. Checking them is quite easy, Gould explains. Simply grab the end of the blade and move it back and forth comparing all of the dampeners to each other. You'll be able to tell, by feel, if the dampener is not functioning. Either the dampener will have a spongy feeling, or a dead spot (area of no resistance) will be felt. A bad dampener will also show up during tracking as an out-of-phase blade.

Usually a bad dampener is the result of something worn or damaged internally, such as a shaft or seal. But it may be that the dampener is in need of adjustment, says Gould. There is an internal valve that can be adjusted in the overhaul shop. Also, he says, the hydraulic dampeners usually have a reservoir that feeds hydraulic oil to the assembly. Make sure that this reservoir has a sufficient supply of oil at all times and that the oil is getting to the dampener.

It's also important to inspect flap stops on helicopters which use them. Flap stops work on usually small arms that work with spring force and centrifugal force and move into position as the rotor assembly spools down. Their function is to prevent the rotor blades from flapping around too much when the rotor falls below a certain speed. If the stops are not functioning properly, says Gould, it's possible for the blade to flap enough that it'll damage the helicopter by striking the tail boom, or could endanger passengers while exiting the helicopter.

Gould cautions that the rotor assembly is a balanced assembly, and that whatever components are removed should be put back in the same position.

May/June 1991

Accessory Technology

Recip engine oil
What's the difference?

Oil... is oil... is oil. Or is it? Many in the maintenance industry are concerned with oil only to the extent that it provides customers with a reasonable level of satisfaction. We don't spend much time worrying about getting a few extra hours out of engines with oils that may only provide a marginal level of improvement. We don't have time for it.

But our customers, who have invested heavily into their aircraft, are concerned with squeezing those additional hours. Those hours convert directly to improved margins of safety, decreased wear and reduced costs. And because of this, we should be concerned.

The development of aviation oil over the story of aviation has certainly not been overly impressive. It wasn't until the '50s that the first improvements in oil were seen with ashless dispersant technology (used to prevent the buildup of sludge and varnish) introduced and not until the '70s when multiviscosity oils were introduced. Synthetics came on the scene in the late '70s.

Acceptance of new oils has historically been slow in this industry and continues to be even with the introduction of synthetics.

Ashless dispersants

According to Dennis Boggs, technical director of lubricants for Phillips 66 Company, ashless (meaning non-metallic) dispersant packages were first introduced by Shell Oil Company in 1958.

The automotive industry had already introduced detergent oils (which was an additive package containing metallic compounds) that were successful in protecting automotive engines from sludge and varnish buildups. But when these oils were used in aviation, they created problems. Many aircraft engines, says Boggs, were coated heavily with sludge and varnish, and when detergent oils were added, the sludge and varnish would break loose and clog oil journals and occasionally cause engine seizures.

This prompted Shell Oil to introduce an ashless dispersant package. "The ashless dispersants did not remove already existing engine deposits," says Boggs, "but merely prevented new ones from forming. Residues were dispersed throughout the oil system and held in suspension, preventing buildups. They then could be drained off when the oil was changed. And because these additives were ashless (or non-metallic), they did not form combustion chamber deposits."

Ashless dispersant packages were made available to other manufacturers shortly thereafter and these oils became the standard until the late '70s, says Boggs.

Many of today's most popular oils continue to use ashless dispersant packages as the primary means for holding contaminants in suspension. Phillips uses them in its multiviscosity X/C 20W-50 oil, Shell uses them in its multiviscosity AeroShell® W and single grade oils, and Mobil uses them in its multiviscosity mineral oils.

Viscosity index

Probably the most confusing property of aviation oil and the most important is its viscosity—its resistance to flow. The confusion lies in that the ideal oil should resist flow enough to prevent the oil from easily flowing out from between gear teeth and bearing journals, but should also be fluid enough to flow readily into areas that require constant lubrication and cooling.

According to John Schwaner, president of Sacramento Sky Ranch Inc. and author of the *Sky Ranch Engineering Manual,* "There is no ideal viscosity for oil. Since oil viscosity influences fuel consumption, automakers want lower viscosity oils. Lower viscosity oils mean lower film thicknesses and more wear. The proper oil viscosity for the engine is a compromise. Everything wouldn't be too bad but the oil's viscosity changes with temperature. Ideally, aircraft oil should have the viscosity of a 20-weight oil at low temperatures and the viscosity of a 50-weight oil at high temperatures." To achieve this, manufacturers developed multiviscosity oils.

The first multiviscosity aviation oils were introduced during the '70s, says Boggs. The increasing number of owners of recreational aircraft during the '70s were putting few man-hours on their aircraft in a given year. Because of this, oil that was poured into engines during the summer months remained in the crankcase into the winter months. This further supported the need for multiviscosity oils and resulted in the widespread acceptance of these oils, he says.

To make a multigrade oil, says Ben Visser, engineer for Shell Oil, polymers are added to straight weight oils. These are complex molecules, says Visser, that expand as temperatures increase and contract as temperatures fall. The result is an oil that remains relatively thin at cooler temperatures—preventing damage and undue wear during cold temperature starting and also thickens up slightly at higher temperatures.

Viscosity ratings, according to John Esser, chief engineer at Mobil, are based on the rate of flow of oil at 212°F. At this temperature, oils are measured for

Accessory Technology

viscosity and are classified accordingly; for instance, 10, 20 or 30 weight. Multiviscosity oils have an additional number associated with the rating. The 15W in a 15W-50 oil, for instance, is a winter rating (the W stands for winter) and means that the oil has a viscosity of 15 at a specific temperature as recommended by SAE. The -50 means that it has a specific viscosity, as recommended by SAE, at a temperature of 212°F. This rating system applies to all oils in the industry, including minerals, synthetics and semisynthetics.

Multiviscosity development

Because of the increase in recreational flying during the '70s, engine manufacturers encouraged oil manufacturers to look into the development of new and improved multiviscosity oils. In the late '70s, Phillips 66 Company introduced a multiviscosity oil X/C SAE 20W-50. This oil continues to this day to be the only mineral-based multiviscosity oil available.

This oil was produced by adding a VI (viscosity-index) improver to a base stock of mineral oil.

"Multigrade oil," says Boggs, "is not limited to only cold weather advantages. When operating under high heat conditions, oil consumption rates can be improved in most aircraft."

Shortly after the release of Phillips X/C oil, Shell introduced an oil that was a combination of mineral oil and synthetic oils. According to Ben Visser of Shell, the development of its multigrade oil for aviation began with a look at oils that were based on pure mineral oil. "We added polymers to mineral oils to turn them into a multiviscosity oil and found that the thin base stock used to develop these oils did not give us enough anti-wear protection to the cam and lifters," he says. The polymers could not provide the protection because of the fact that they tended to break down after a period of time.

The breakdown of the polymers in multiviscosity, says Schwaner, is due to two things: The mechanical shear loads that deform the molecular structure and thermal oxidation degradation (due to turbocharging). Because of this, says Schwaner, "Regular oil changes, especially under severe thermo-oxidation conditions, (turbocharging) are important. Any decrease in oil viscosity will increase wear rates and oil consumption. If you are changing your oil at 50 hours and you notice that your oil consumption is less during the first 25 hours than the second 25 hours, you know that this oil is only viscosity stable for 25 hours and you should adjust your oil change interval accordingly, or switch to a different brand of oil," he says.

Visser says because of this viscosity breakdown Shell began experimenting with synthetics. It found that the synthetics would serve well as a multigrade oil without the addition of polymers. This meant that the oil was less likely to break down. However, he adds, it found through testing that pure synthetics would not absorb the lead salts that were produced in the process of combustion. It was found that the lead would form deposits in the engine over a period of time.

"So in order to provide the ability to wash out the lead, we decided to combine the mineral oils with synthetics. This gives us the base oil viscosity of a single grade," says Visser, "yet provides us with enough lead solvency to sufficiently clean the engine."

In 1985, Phillips made an attempt to further the development of multiviscosity oils. It introduced an oil called X/C II. This oil contained an additive package that, according to Boggs, increased the anti-wear characteristics of the oil. As there was no provision for evaluating the performance of anti-wear additives in the Mil-L-22851C (which was the guideline for acceptance by the engine manufacturers) and engine manufacturers did not support this new technology oil development and discontinued testing on it, says Boggs, Phillips 66 removed X/C II from the market making X/C 20W-50 available.

Synthetic oil

Although synthetic oils had been used in aircraft turbine engines for some time (due to their performance characteristics in high temperatures), Mobil Oil didn't begin testing and doing proof-of-performance work on using synthetics on aircraft recips until 1981. And it wasn't until 1988 that it introduced Mobil AV 1 to the market. In 1989, with the flight of the Voyager I (non-stop around-the-world flight), Mobil launched a campaign to market its product.

Specifically, Mobil says that synthetics offer the following advantages over conventional mineral-based oils: Better resistance to oxidation, extended oil drain intervals (up to 200 hours), reduced evaporation and less internal friction, up to 30 percent less oil consumption, fuel savings of up to 5 percent, and a pour point as low as minus 60°F.

According to John Esser, Mobil originally had to get an STC on the oil because Textron Lycoming would not approve it for use in its engines. Textron Lycoming, at the time, required oil manufacturers to meet its already established specifications, and one of the requirements was that the oil be a mineral-based oil, says Esser. However AV 1 did meet all of the physical test requirements.

Eventually, after reviewing a great deal of proof-of-performance data, Textron Lycoming approved the use of Mobil AV-1 in its engines, he says. Mobil has also received an STC from the FAA, says Esser, for a 200-hour oil drain interval.

Engine manufacturers still recommend a 50-hour oil drain interval. If you wish to conform to the STC, you are within your legal rights to do so. Keep in mind, however, that because the engine manufacturers recommend 50-hour oil drain intervals, you may end up voiding the warranty on the engine if something goes wrong. "We

Accessory Technology

are going to try to get the engine manufacturers to raise their recommended oil change intervals," says Esser, "but I don't know if it will happen at this point."

Esser says that synthetics offer the greatest amount of protection in extreme environments. "Whether you are flying in a hot environment where the high temperature capabilities of the oil help protect the engine and form less carbon, or whether you're flying in a cold environment where cold starting and immediate flow of oil to the bearings is important, synthetics provide obvious benefits," he says. "In mild temperature ranges, the benefits that you get are not as dramatic as they are at the extremes."

Dennis Boggs at Phillips says that engines should not be operated at extreme temperatures. In cold temperatures, the engine should be preheated prior to starting. Even if the engine can be easily started when cold, says Boggs, the coefficient of expansion of the different metals in the engine can cause damage to bearing areas. Similarly, the engine should not be run at extremely high temperatures. Crankcase temperatures normally run below 220°F and, at these temperatures, mineral oils are stable offering as much protection as synthetics, he says.

Break-in and the importance of wear

Practically all of the oil manufacturers agree that the break-in process involves wear at the rings to seat them against the cylinder walls. But there is some disagreement over how to go about it.

Both Esser at Mobil and Visser at Shell say that a straight mineral oil with no additives should be used for the break-in process. The reason, says Visser, is that this type of mineral oil allows the rings to wear during this critical period. Additionally, the lack of ashless dispersants allows metal particles and combustion products to build up in the ring area, creating a sort of lapping compound. This allows the rings to seat quickly. As soon as the rings are seated and oil consumption and temperatures drop off, go to an oil with an additive that will keep the engine clean.

Phillips 66 does not agree with this procedure. It claims that its multiviscosity, ashless dispersant X/C should be used even for break-in. The reason, says Bill Coleman of G&A Communications (an agency for Phillips 66), is that the oil does break down in the ring area enough to allow the rings to quickly seat. The ashless dispersants suspend the contaminates and keep the area clean. He adds that because the material produced by the break-in is held in suspension, the dispersants are prevented from getting into other areas of the engine, eliminating unnecessary wear.

Frequent oil changes

John Schwaner, in the *Sky Ranch Engineering Manual*, says that oil must do the following: Lubricate, clean, protect, seal and cool the engine; allow easy starting at cold temperatures; and provide lubrication to parts under extreme pressure.

It's clear that Phillips X/C 20W-50, AeroShell W 15W-50, and Mobil AV 1 successfully accomplish all of these—to varying degrees.

It's also apparent that salts, condensates, combustion products and wear metals accumulate to varying degrees, regardless of which oil is used. Therefore, frequency of use, the environment and the care that the aircraft receives should all be factors in determining type of oil.

"There's them that care and them that don't care," responded one technician who was asked about oils. "Them that don't care," he says, "will dump whatever they can find at the cheapest price in the engine without paying any attention to how the oil performs. Them that care will track oil consumption, perform oil analysis, evaluate the condition of the engine periodically and come up with the best oil to use for their situation."

March/April 1991

Accessory Technology

Hot cooler tips

The engine oil cooler is key to proper oil temperature and pressure. Should it fail during operation, it can cause major damage to the engine and can contribute to unscheduled landings. This can occur from a leak during flight (causing loss of oil pressure) or from excessive oil temperature due to poor airflow or malfunctioning vernatherm valve.

Maintenance of the oil cooler is important. Over time, oil circuits of the cooler become clogged, contributing to excessive pressure to the unit. This, in turn, causes premature failure of the cooler in the form of leaks. Also, particles such as dust, grass, carbon, etc. can build up on the outside surfaces causing poor airflow, which can lead to overheating of both oil and cooler. If these conditions continue, the cooler will eventually fail.

Dave Campbell, a co-owner of Drake Air Inc., an oil cooler repair/overhaul facility in Tulsa, OK, says that oil coolers are often neglected, at times to the point that they fail in flight.

With few exceptions, oil coolers are used on most engines to keep oil temperatures within the limits recommended by the engine manufacturer. The coolers are either engine-mounted (forward and rear) or remote-mounted. Some use outside air to support cooling in addition to fan/blower-generated air. Others use the aircraft's fuel to support the cooling process (i.e., fuel/oil coolers). No matter what type configuration, the coolers all have one thing in common: They are pressure vessels that lead to serious consequences if they fail.

Oil cooler operation is similar to that of a car radiator. Oil passes through the inner circuits of the cooler while outside air flows across the air fin (or tubes, in some cases). Airflow coupled with the fin provides the cooling, or "heat rejection." If the flow of air or oil is restricted, cooling is negatively affected. Again, since the cooler is exposed to the elements, particles can easily restrict airflow, leading to an internal buildup of carbon and subsequent overpressurization.

Campbell of Drake Air recommends removal and cleaning of the cooler—both inside and out—at every 100-hour inspection. Typically, he says, internal buildup is a result of carbon deposits, or metal generated from a failing engine. To properly remove these deposits, thoroughly clean and flush the cooler with a series of chemicals designed to free up the oil passages. Before putting the unit back into service, test to determine if the cooler is free of blockage. This is an added measure which helps ensure proper operation.

Another common airflow problem that leads to overheating is bent fins, usually caused by flying in inclement weather or from foreign object damage (FOD). Straightening fins is a tedious process, according to Campbell, but shape can be restored by carefully using a thin, blunt piece of metal on each side of the fin. Care must be taken to avoid damaging the cooling tubes or tearing the fin material, he says.

Other common causes of oil cooler damage are the result of improper installation or poor maintenance, Campbell adds. There may be damage due to cross-threading. Or, there may be cracks in the inlet/outlet area because of overtightening of hose fittings (since the inlet/outlet porting is brazed or welded onto relatively thin material, it's easy to damage).

Cracking (especially on engine-mounted coolers) can also occur from improper torquing of the mounting bolts. This, in conjunction with engine vibration, can lead to cracking of the mounting flange and a subsequent oil leak.

Campbell also recommends that if an oil cooler is experiencing repeating cracking, even after proper installation, it may be time to balance the propeller. A smooth-running engine not only reduces vibration but can also help extend the service life of the cooler and other components.

According to Campbell, some aircraft owners seem to have the misconception that a leaking oil cooler must be replaced with a new unit. On the contrary, he says, most coolers can be overhauled and returned to service, at a fraction of the cost. Some exceptions: An obviously deformed cooler, due to crash or sudden impact; severe FOD; or excessive fin erosion.

Adhering to a maintenance program pays dividends in extended oil cooler life. Cleaning, checking for loose fittings and mounting bolts, and straightening bent fins are good measures to follow.

Accessory Technology

The basic overhaul process

Dave Campbell of Drake Air Inc. outlines the steps his company uses when overhauling most oil coolers:

- Preinspection, checking for the obvious which might call for scrapping the unit.
- If the unit appears repairable, cleaning and flushing. This includes stripping, high pressure decarbonizing, etching (inside and out), deoxidizing, sonic cleaning for copper units, chromacoating (corrosion inhibitor), and high pressure flushing of vibrating filter.
- Once unit is free of contaminants, it's hydrostatically tested with nitrogen for leaks.
- Any leaks are repaired, valves are tested, fin damage is repaired, flanges are straightened, stripped threads are repaired, mounting surfaces are machined and pressure tested, and the unit is finaled, tagged and certified.

Campbell says the process can be performed indefinitely. He estimates the cost of a typical overhaul at $140. *January/February 1991*

Accessory Technology

The combustion heater
Servicing solutions

By Greg Napert

There have been two companies that have been the primary manufacturers of combustion heaters over the years. And because of this, competition for service and sales in this market has not been exactly fierce.

The two companies, Southwind (a division of Stewart-Warner and now owned by FL Aerospace) and Janitrol (a division of Midland Ross Corporation) have pretty much been able to name their price on heaters and replacement parts and, according to sources in the marketplace, have not been pressured into providing exactly the highest level of customer service.

To make matters worse, because of the decreased production of aircraft over the last 12 or so years, aviation has become a relatively small portion of these companies' business. These companies, which thrived on providing new heaters for production aircraft in the past, no longer make their primary income from the sale of new heaters.

Southwind continues to provide parts for repair and overhaul of its heaters. However, service is provided through a network of distributors that's approved for overhaul of its heaters. Sources at Southwind say that its primary sales of combustion heaters and components are to the military, for ground units such as tanks.

Janitrol has been a bit more aggressive in providing service and support for its heaters. It has continued to provide support and service for its customers, even though its primary source of income comes from supplying heater-related products to other industries. Like Southwind, it has also established a network of distributors for parts and service.

Regardless of the fact that the demand for new heaters is significantly down, service and parts to support the tens of thousands of combustion heaters currently in use are in high demand.

Many technicians choose to repair and overhaul their own heaters. However, knowledge of where to get parts, costs of parts and the convenience of having replacement heaters available are major deterrents in doing so.

According to Norman Gress, technician for Herold Haskins, a heater repair and overhaul facility in Dothan, AL, and an approved Southwind distributor, there are only about a half dozen repair shops in the United States that are actively repairing and servicing combustion heaters. And most of these are two- or three-man shops.

An exception to this rule is a company in Buchanan, MI, called C&D Airmotive Products Inc. The company employs around 30 people who repair, overhaul and manufacture components for heaters. Recently, it has begun to manufacture complete heaters and has positioned itself to compete with the original manufacturers.

Heater ADs

Dennis Sandmann, owner and president of C&D Airmotive, explains that basically two ADs have had a strong impact on maintenance that has been performed over the last 10 years—one on Janitrol heaters and one on Southwind heaters. These ADs required that practically all of the existing heaters on the market be overhauled and inspected on a regular basis. Both ADs still apply.

Specifically, AD 82-07-03 applies to Janitrol Aero Division heater models B1500, B2030, B3040 and B4050. This AD calls out for the mandatory inspection of the heater once the aircraft has accumulated 500 hours since new, or since it was zero-timed. It also requires a pressure decay test and inspection every 100 hours of operation thereafter or two years, whichever comes first. This AD applies to just about all of the general aviation light aircraft, "possibly up to 80 percent of aircraft that have combustion heaters," says Sandmann.

AD 81-09-09, applies to Stewart-Warner (Southwind division) heaters model Series 8240, 8253, 8259 and 8472. This AD is different from the Janitrol AD in that it does not require any pressure decay test or time limit for this test. The AD does require an inspection of all heaters having 250 hours or more time-in-service and at intervals not to exceed 250 hours. For combustion heaters having 1,000 hours or more on them, it requires that they be overhauled.

"I consider this AD to be dangerous in that it exposes the A&P to liability," says Sandmann.

The reason, he explains, is that the AD only requires an overhaul at 1,000 hours, with no way of accurately determining the condition of the combustion tube in between that overhaul interval. Both manufacturers, says Sandmann, claim that the combustion tube is designed to last around 1,300 hours. So, obviously, if you place a serviceable combustion tube back into service with 1,000 hours on it, you do so knowing that it will fail at around 1,300 hours.

"Based on our experience," says Sandmann, "I would recommend that the Southwind heater be pressure decay tested every 100 hours after the initial 500 hours of operation since new." He explains that all heaters should be pressure decay tested at regular intervals and also recommends that the combustion tube in the Southwind heaters be replaced at the 1,000-hour overhaul as they will never make another 1,000 hours. Other overhaulers, such as Herold Haskins, do not recommend replacement if the unit is serviceable; however, they do recommend keeping a close eye on the combustion tubes by performing periodic pressure decay tests.

Keep in mind, however, that the pressure decay test should not be a substitute for the AD. Legally, the A&P is bound to perform the maintenance that is called for in the AD.

Why the ADs?

According to Sandmann, these ADs were the result of accidents that involved combustion heaters. An example of this, says Sandmann, was in 1978 when the National Transportation Safety Board (NTSB) sent out a letter about a Beech 18 that was on a ferry flight to Lansing, MI. The letter described the pilot's conversation to the tower prior to the fatal crash. The pilot stated that he had a fire in the left wing. He then reported that the entire wing was on fire—and that was the last contact they had with him.

Upon investigation, said Sandmann, the NTSB found the Southwind combustion heater thrown from the airplane. It was discovered that the heater had never been removed from the aircraft for maintenance. The shroud of the heater was dark blue (indicating an overtemperature), the combustion air switch had been stuck in the half-way open position, and the combustion tube had a large opened crack in it. These findings indicated that the fire was the result of fuel pouring out of the crack in the heater, into the wing and consequently igniting.

In short, says Sandmann, the ADs were issued as a result of in-fight fires. But in addition, numerous Malfunction and Defect (M&D) reports that were being submitted also contributed to the ADs being issued.

People often assume that the ADs and resulting inspection requirements were designed to prevent carbon monoxide poisoning, according to Sandmann. In fact, such danger is minimal.

He explains, "Because of the way that the combustion heater is designed, the ram air pressure for the cabin air is greater than the pressure inside of the combustion tube. So if a crack would develop in flight, the ram air pressure would force the cabin air to leak into the combustion chamber—not the other way around. It is possible, however, while sitting on the ground, for some of the exhaust gases to leak into the cabin. But the volume of gases would not be enough to harm anyone. The worst that could happen is that someone would get a headache."

The pressure decay test

The most timesaving inspection technique for the combustion heater is the pressure decay check, which consists of blocking off all openings in the combustion tube and pressurizing with compressed air. The rate of leakage determines if the tube has deteriorated to the point that it needs to be replaced. Some leakage is allowed, to account for screws and gaskets not fully seated and to allow slightly damaged or deteriorated combustion tubes to pass the test.

The beauty of this test is that it can be performed while the heater is installed in the aircraft. Without the aid of the pressure decay test, the heater has to be removed and fully disassembled. A thorough explanation of the pressure decay test and operational inspection can be found in the Janitrol Maintenance and Overhaul Manual P/N 24E25-1. This manual is available from C&D Airmotive as P/N 4030.

Common problems

Spark plug problems are common with the combustion heater. Consequently, Sandmann recommends plug replacement every 12 months, even after only 50 hours. "Coatings will form on the electrode of the spark plug during the summer months while the heater is inactive,"

Accessory Technology

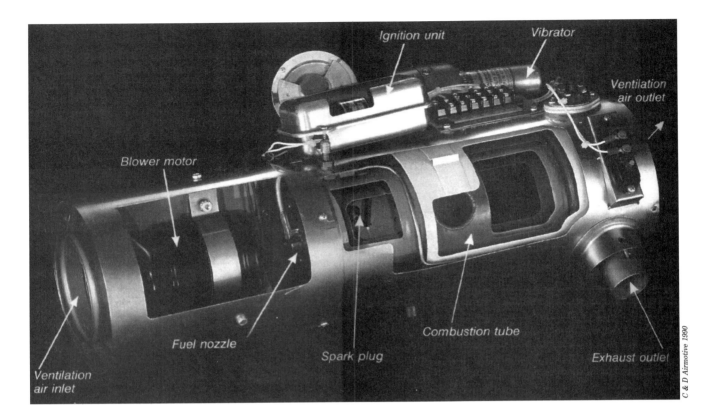

he explains, "and these coatings, along with other deposits remaining on the plug from the previous season can lead to a fouled plug. In addition, spark plugs that appear to be operational on the ground may short out at altitude, due to the difference in barometric pressure.

"I don't recommend cleaning the spark plug with any types of abrasive either. Cleaning it with abrasive will remove the coating from the ceramic and make it more likely to fail. Replacing a $50 to $60 dollar spark plug once a year is an inexpensive way to prevent heater failure."

Spark plugs, of course, don't last forever; electrodes become coated, ceramic wears. Discolorations or brown spots on ceramic indicate arcing.

The vibrator, or voltage regulator, is another potential fault point. However, due to the fact that it is now manufactured with solid-state components, failure is less likely, according to Sandmann. Older vibrators, constructed of mechanical switches and points, were more suspect.

Another component that can fail is the coil. Sandmann explains that an oily film on top of the heater or a weak spark indicates coil failure.

Safety switches can also be troublesome. Two switches control the operation of the heater; another two switches act as safety switches. Failure of any one of these switches can render the heater non-functional, failing to protect the heater from damage. Understanding the operation and relationship between these switches is critical for troubleshooting.

Plugged nozzles can be another problem area, says Sandmann, but usually only on turbine-fueled aircraft since jet fuel is more likely to clog nozzles than gasoline. This means that the nozzles should be cleaned more frequently, usually at 500-hour intervals. "Nozzles do wear out," he says, "because if a nozzle is not operating properly, it may require replacement."

Jet fuel is quite sensitive to operating pressures. The fuel pressure at the nozzle must be between 95 and 105 psi; once outside this range, there can be difficulty in keeping the heater lit at high altitudes. "This is a problem that is difficult to troubleshoot," says Sandmann, "because it only happens at altitude and can't be simulated on the ground."

Troubleshooting

Troubleshooting, in most cases, is quite simple, says Sandmann, and is made easy by understanding that the combustion heater requires three things to operate: fuel, ignition and air. Take away any one of these components, and the heater will not operate.

He adds that much can be discovered by advancing this logic—for instance, placing your hand (using caution not to burn yourself) near the exhaust pipe of the heater and attempting to start the heater signals if there is ignition. If you feel air blowing, the combustion air blower is working. If you smell fuel on your hand, you know that you have fuel and that it's not igniting. You can also tell if the spark plug is operational by simply listening for the snapping of the plug.

Observing heater components for signs such as discoloration on the spark plug or oil leakage on ignition components can be clues to the condition of these components.

At times, however, the logic is not so simple. For instance, it is quite common for the overheat switch to pop after shutting the heater down. This is not due to a defective switch, but from excessive residual heat that results when the heater is shut down abruptly without allowing it to cool. The switch can usually be manually reset.

There are actually two categories of troubleshooting problems related to the ignition heater: ones that occur on the ground that are repeatable, and others that occur only at altitude which are intermittent. "It is the intermittent problems that can become very troublesome and can cause the technician to change parts without actually knowing that they are bad," says Sandmann.

Whatever the problem or the solution, he explains, the technician must assure that the problem is dealt with properly. Trial and error doesn't set well with a pilot who has to operate the aircraft in subzero temperatures. Failure of the heater can mean failure of other systems that are dependent upon operation in temperatures above freezing.

"The heater is not just a luxury," says Sandmann, "it is critical to the operation of the aircraft."

Run your heater on a Hobbs™

Russ Bargo, director of maintenance for Wisconsin Aviation in Watertown, WI, claims that the company has saved considerable expense and time by running its combustion heaters on separate Hobbs meters.

The reason—the ADs require that the heaters are overhauled at specific time intervals. Without any way to document the time on the heater, the actual aircraft time must be used. Heaters, especially during the summer months, accumulate much fewer hours on them than the aircraft. "If we followed the aircraft time, we would be inspecting heaters all summer long," says Bargo.

Placing the Hobbs meter on the heater gives you an accurate indication of how much the heater is being used and is a much more practical way to track time.

Check Mate® expedites troubleshooting

C&D Airmotive Products Inc. has introduced a product to help expedite the troubleshooting of aircraft combustion heaters. The troubleshooting kit, called Check Mate®, is a self-contained unit that includes gauges and instruments needed to check individual components on the heater. The kit also incorporates what C&D says is the only ignition vibrator tester in the industry.

The kit incorporates a harness that connects directly to the heater electrical bus. This allows you to monitor switches and heater controls on the ground, or if necessary, in flight. The unit also includes a clip that allows the operator to sense spark plug operation.

November/December 1990

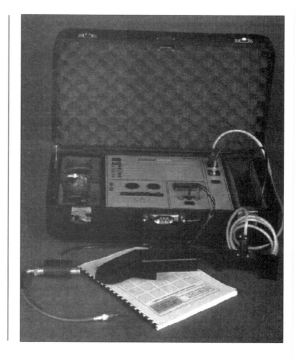

Accessory Technology

Maintenance of the Precision Airmotive RSA fuel system

By Rudy Swider

The Precision Airmotive RSA fuel system? Your aircraft has a Bendix RSA fuel system, you say? For those of you that haven't received word yet, Bendix sold the RS/RSA product line to Precision Airmotive Corporation of Everett, WA, on January 20, 1988.

Whether you refer to your aircraft's fuel injection system as Precision Airmotive or Bendix (unless you're equipped with a Continental system), it does require a certain amount of routine preventive maintenance. And although this system is almost 30 years old very few technicians are familiar with required maintenance items. Proper preventive maintenance is the easiest way to help keep this system operating to its full TBO. It's a simple procedure.

Where to start

Maintenance on the fuel servo unit is required at 50-hour intervals. The inlet filter requires inspection and cleaning after the first 25 hours of operation and at 50-hour intervals thereafter. Surprised? Check the operation and service manual to verify.

The filter should be inspected and cleaned at each annual regardless of accumulated hours since the last inspection. This only makes sense if you realize that contamination will show up here long before it can cause any kind of operational problem. Catch it early, locate and correct the source of contamination, and keep your customers flying.

Before you start tearing into the fuel system, obtain the set of O-rings that will be required. It is best to always replace these packings each time the filter is cleaned.

These can be procured from Precision Airmotive Corporation or any of its designated repair facilities that repair/overhaul the Precision servo units. A word of caution here: Do not substitute standard AN or MS O-rings for the Precision part numbers even though they may appear identical. Using the wrong O-ring on the inlet fitting has been known to exert sufficient force to crack the servo unit housing. A few pennies saved is hardly worth the $1,000-plus price tag of a new housing.

To get at the filter, you need to first remove the inlet fitting. There are technicians that remove the plug oppo-

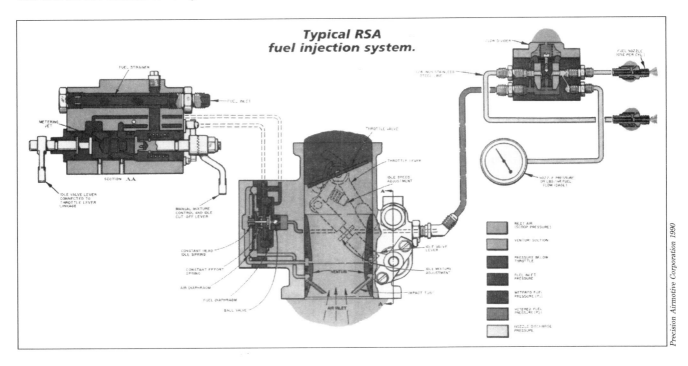

Typical RSA fuel injection system.

Precision Airmotive Corporation 1990

236 The Best of Aircraft Maintenance Technology Magazine

Accessory Technology

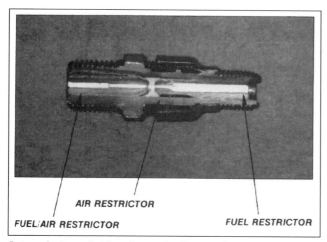

Internal view of old style nozzle. Do not damage this calibrated restrictor with safety wire.

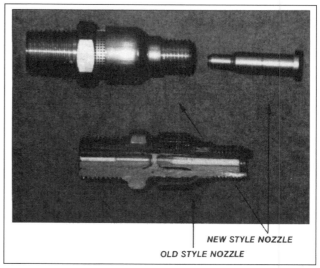

Old style nozzles (bottom cutaway) had restrictors pressed into them. New style (top) is manufactured as a two piece assembly.

site the inlet fitting to get at the filter. This shortcut will introduce any contamination that is on the filter into the servo unit.

Inlet fittings vary from union-type fittings to 90-degree elbows. Be careful with these fittings since they are specially modified for the filter assembly. Remove the fitting using clean wrenches of the appropriate size. Once the fitting is removed, it is a simple task to remove the filter.

If the filter is permanently attached to the inlet fitting, consult and comply with Bendix bulletin RS-48 revision 2 or a later Precision Airmotive revision. This changes the filter to a bypassing type—the theory being that dirty fuel is better than no fuel should the filter ever become plugged.

Inspection of the filter is relatively easy. However, merely looking down the middle or looking at the outside surface and inspecting for particulate matter is not sufficient to tell if it's clean (fuel flow is from the inside out when installed in the servo unit).

The best inspection method is to first dry the filter with air and tap it, open side down, on a clean piece of paper. You will want to examine any contamination and determine the type and source. Next, look into the center of the filter while shining a light through the outside. You should be able to see light through the weave on most of the surface areas. Another method is to breathe through the dried filter. There should be very little restriction to airflow.

Cleaning can be accomplished using Acetone or M.E.K., followed by a rinse in Stoddard solvent and then air drying. If the fitting is badly damaged or corroded, it must be replaced. Corrosion is a good indication that the aircraft fuel system contains a relatively high level of water contamination.

While you're in the area, inspect all fuel hoses for signs of deterioration. The line(s) installed between the servo unit and flow divider, pressurizing valve or splitters should be (in most cases) Teflon lines with a silicone-coated fire sleeve. If it isn't this type, now is a good time to make the change.

Fuel nozzles

The next step in maintaining the system is cleaning the fuel injection nozzles. Despite what has been published elsewhere on the subject, a nozzle is not clean and satisfactory for use when the cleaning solution no longer changes color. Only proper inspection can verify a nozzle has been properly cleaned.

An increase in indicated fuel flow at various power settings is generally the first indication that nozzles need cleaning. If contamination becomes extreme, engine problems will occur. Engine operational problems due to dirty fuel nozzles will be experienced at some later date when contamination has reached an extreme.

Fuel stains around a nozzle also indicate cleaning is necessary. During cleaning, as with the filter, any unusual contaminant should be identified and its source located and corrected.

More information related to cleaning nozzles is contained in Bendix bulletin RS-77 revision 2 (or later Precision Airmotive revision) and Lycoming S.I. 1414.

The best cleaning solution found to date (recommended in bulletin RS-77) is Hoppes No. 9 gun cleaning solvent, available at most local sporting goods stores. Usually a 20- to 30-minute soaking is all that is necessary, followed by a Stoddard solvent rinse and air dry prior to inspection. Don't use lockwire, pins or other metal items to remove contamination from the nozzles. This will affect calibration.

When cleaning two-piece nozzle assemblies, be sure each restrictor is kept with its respective body. This is

Accessory Technology

accomplished by using separate containers for each nozzle assembly. If you're only cleaning the restrictors (permitted between annual inspections), work with each cylinder separately by removing, cleaning, inspecting and reinstalling the restrictor, and reconnecting the fuel line. Keep in mind that if you lose a fuel restrictor, you will have to buy an entire new nozzle body assembly. These restrictors are flow matched to their respective bodies and are only sold as assemblies.

Inspecting nozzles

The nozzles used with the Precision Airmotive (Bendix) fuel injection system have a fuel orifice diameter of approximately 0.028 inch. The only proper method of field inspecting these assemblies is with the use of a 10-power magnifying glass. Both fuel and fuel air restrictions should be "shiny clean" with no evidence of film or particulate contamination.

With older style nozzles, check the top threads (at the fuel line connection) for damaged threads and/or cracks. Damage indicates the fuel line nut has been overtorqued. This can cause a reduction in the size of the air restrictor. Operationally, this only affects engine idle, but the fact that the nozzle has been damaged is grounds for replacement.

New style and old style nozzles for normally aspirated engines are interchangeable with one another and may be used in any combination on an engine.

Not all Precision Airmotive nozzles flow alike anymore. At one time all Precision Airmotive (Bendix) nozzles for turbocharged engines were calibrated the same as normally aspirated nozzles and could be interchanged between cylinders of like engines. For many applications this is still true. There are, however, assemblies referred to as "high flow" nozzles. These nozzles flow 32 pounds per hour at 8 psi. Standard nozzles flow 32 pounds per hour at 12 psi. The inserts of these nozzles are identified with a step on their circumference and have a larger diameter to prevent installation into the wrong body. Always refer to the engine manufacturer's publications prior to ordering replacement nozzle assemblies.

Nozzle fuel lines

Before installing your freshly cleaned nozzles, inspect the nozzle fuel lines. Though these lines are supplied by the engine manufacturer, their condition is critical to the proper operation of the system. Items to check:

- The inside diameter of lines used on most engines should be 0.085 to 0.095 inch (reference Lycoming S.I. 1301). Technicians have been known to substitute other lines such as the smaller I.D. primer lines when a replacement was required. A small line on any one cylinder can cause that cylinder to run leaner than the others. Line length is not critical to the operation of the Precision Airmotive system.

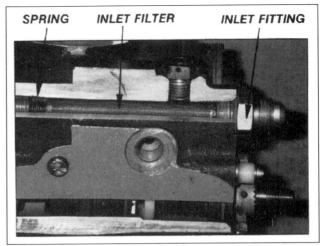

Cutaway of servo body showing inlet filter in its correctly installed position.

- Lines for signs of longitudinal twisting—a sign of overtorqued nuts. Inspect for kinks. The minimum bend radius for a line is 0.062 inch.
- Nuts for cracks.
- Ferrule braze joint and surrounding area for cracking evidenced by fuel dye stains.

Final assembly

Install the nozzles using a clean six-point, deep well socket. In many instances, you will have to install your socket over the nozzle first and then attach your extensions and torque wrench. All too often, nozzles are damaged by trying to force a socket and extension past engine baffling and over a partially installed nozzle (not to mention damage to baffling). This is the primary reason for loose shrouds and screens.

Torque nozzles (or nozzle bodies) to 40 inch-pounds. Sixty inch-pounds is the absolute maximum torque to be applied. If your installation requires alignment of the "A" (very few do), then increase torque from 40 inch-pounds until alignment is obtained. Do not exceed 60 inch-pounds of torque under any circumstances. On nozzles installed horizontally, the "A" should point down +30 degrees.

If you have the newer two-piece assemblies, check to ensure that the fuel restrictors are properly installed. If you are missing one, lock the doors and search everyone. On engines that have nozzles installed horizontally, it is best to leave the shipping cap (tire valve cap) installed until you connect the fuel line.

Improper line connection is a common source of damage. If you have the newer two-piece nozzles, the cost for failing to follow procedures can be a new set of nozzles. When installing nozzle fuel lines, it is first necessary to install the nut finger-tight (provided all threads

Accessory Technology

are clean). At this point you have two options: Try to use a torque wrench with adapters, if room permits, and torque the nut 25 to 50 inch-pounds; or, you can use a standard 7/16-inch open-end wrench and continue to tighten the nut 1/2 to 1 flat (of the nut) from the finger-tight position, then stop. The latter has proven to give you the 25 to 50 inch-pound torque limit. Do not exceed the 50 inch-pound torque limit as nozzle damage may and usually does occur, resulting in rough, rich running at idle.

Once you have completed the fuel system maintenance, recheck that nothing was overlooked. Pressure test the system (with mixture in idle cutoff) for fuel leaks—then you should be ready for ground run, minor adjustment of idle speed and mixture if necessary, and return to service. Don't forget the log book entry.
September/October 1990

Rudy Swider is currently employed by Bendix Engine Products Division and was responsible for technical support for the RS/RSA product line prior to the sale of the line to Precision Airmotive. He has over 19 years' experience in the industry.

Accessory Technology

Starter-generators
Avoid the quick-fix syndrome

By Greg Napert

BURBANK, CA—One piece of equipment performing two separate tasks that are of primary importance to the operation of the aircraft is a lot to demand of any accessory. Consequently, the role that the starter-generator performs demands that it be resilient, powerful and, most of all, reliable.

Tom Hodges, vice president of PacAero, a parts supplier and repair station that specializes in starter-generator electrical and electronic overhaul, says that the starter-generator is taken for granted, not inspected as frequently as it should be, and often frequently abused.

Hodges tells of the time that his company kept overhauling a starter-generator for a customer, only to have the unit repeatedly sent back for overhaul with very low hours on it. Upon investigation, Hodges found the customer was using the unit to perform compressor washes, spinning the engine for long periods of time without paying any attention to the limitations of the starter-generator.

Not built for heat

"I think we're safe to say," says Hodges, "that the most destructive thing to a starter-generator is heat." Eliminate the heat, caused by friction, inadequate cooling and excessive electrical flow, and you should see relatively few problems, he explains.

The interior of a starter-generator, he adds, is susceptible to heat damage. Improperly seated brushes or improper operation can generate enough heat to damage the insulation, varnish and other components used to insulate components from one another. This can cause electrical shorts, and eventually failure of the unit.

In order to reduce the possibility of damage, there are a number of things that can be done.

Inspections

Wayne Root, manager of repair and overhaul and a 15-year veteran of PacAero, has had the opportunity to see every conceivable type of damage related to starter-generators. "Aircraft manufacturers usually provide inspection and overhaul times," says Root, "but it often seems as though the scheduled times are only a few hours before the unit falls."

Root suggests increasing the frequency of inspections to account for operating conditions. Operating environment and conditions have a definite impact on how long the unit lasts. Although inspection frequency is specified

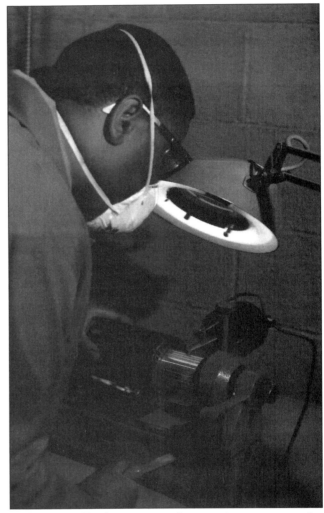

Technician at PacAero ensures that armatures are properly balanced prior to reinstalling. This prevents premature failure of the bearings and unnecessary vibration.

by the manufacturer, adverse operating conditions may dictate more frequent inspections.

New brushes

Paying close attention to break-in instructions and wear limits when installing brushes is a must, says Root. "Its important that the technician goes through the proper steps when doing brush changes, not just take the brushes out and slap a new pair in and off you go. You might get away with it once or twice, but sooner or later it's going to catch up with you and you're going to end up with a piece of junk.

"We see them come in where the brushes are brand new, but the commutator surface is burnt to a crisp with big grooves in it. It's obvious that there was an attempt to get additional time out of it without using proper brush run-in procedures."

According to Hodges, a typical commutator (which is part of the armature assembly) should be able to be turned five or more times before it reaches its minimum limits. If you find yourself buying armature assemblies more frequently than that, it may mean that it's time to take a look at your maintenance program and/or the operating practices of the pilot.

To assure long commutator life, the brushes must be seated according to the maintenance manual. It is important that the brushes are fully seated before operating the starter-generator in the aircraft. Because the diameter of the commutator changes as it is machined, the brushes will require different degrees of seating. Root suggests placing a piece of sandpaper on the commutator and sanding the brushes to the proper curvature when new brushes are installed.

A proper run-in needs to be performed, says Root, to condition the surface of the commutator and the brushes so that, when they are operated at full power and at altitude, the wear will be minimal.

The procedure for running in the brushes differs between makes and models of starter-generators. Some, according to Root, require the starter-generator to be motorized for a period of time. And one model, a Bendix starter used on the Learjet requires that the unit be run as a generator for a specified period of time before flight.

Vibration

Vibration can also have detrimental effects on the life of the unit. The armature assembly must be balanced at overhaul. An unbalanced armature will vibrate the entire starter-generator assembly causing wires to rub, components to loosen, unusual forces on the bearings and premature failure of the unit. Normal wear of the assembly and machining of the commutator are good reasons for balancing the armature.

The armature is normally balanced by removing the required amount of metal from the balance rings — located on either end of the armature windings. These rings are varnished into place and are susceptible to heat damage. If the armature assembly is overheated, the balance rings may loosen and throw off the balance of the entire assembly.

Make sure, that the facility you choose to overhaul your starter-generator, routinely balances the armature during overhaul.

Bearing replacement

Hodges says he has occasionally found bearings installed in the starter-generator that were not approved for use. The bearings, says Hodges, appear to be standard bearings that can be purchased at any supply store. In some cases, these bearings even have the same part numbers. But, in fact, they should not be used. Legitimate bearings are purchased by the manufacturer, inspected and filled with a special grease that has been tested for this application. The use of any other bearings could result in premature bearing failure.

Another problem with bearings, says Root, is that they're not always installed properly. PacAero checks the circumference of the bearing housing to make sure that it's not distorted at all. Pressing a bearing into a distorted housing damages the bearing race. "I have actually seen cases," says Root, "where the bearing housing was staked (punched with a chisel) to hold an otherwise loose bearing in place." Staking the housing will certainly distort the bearing race and lead to premature failure.

Cleaning

Because of the fact that the starter-generator is air-cooled with a fan that is an integral part of the armature assembly, there is ample opportunity for contaminants to enter and plug up the cooling passages.

Also, says Root, cooling may be impeded by carbon residue that is left from the brushes wearing against the commutator which accumulates and packs inside of the armature, fields and cooling passages. "If you don't clean it, then your cooling isn't efficient and the starter-generator stands a chance of overheating," he says.

Therefore, it's very important to blow out these cooling passages at the time when the brushes are changed. The best way to do that, according to Root, is to blow pressurized air through the cooling passages opposite the normal direction of flow. Make sure, however, to contain the "black cloud" of dust that typically results from this cleaning procedure. Otherwise you will spread the contamination to other components.

It's now or never

Many operators, in order to enable the starter to reach TBO, are making it a habit to rebrush and clean the starter at the halfway point to overhaul.

"As the saying goes, you either pay now or pay later," says Hodges. "If the starter-generator is run until it's down to its last leg, the bill to repair it is going to be much larger than if it had been pulled 200 hours before, when it was possible to save some of the components."

Taking a few simple precautions along with regular cleaning and inspection of the starter-generator can save money, headaches and is just good common sense.

July/August 1990

Accessory Technology

COMMUTATOR CHECK CHART

HOW TO GET THE MOST VALUE FROM THIS CHART

This chart will help you to spot undesirable commutator conditions as they develop so you can take corrective action before the condition becomes serious. It will also aid in recognizing satisfactory surfaces.

The box chart below indicates the importance of selecting the correct brush and having the right operating conditions for optimum brush life and commutator wear. CPO offers a complete line of carbon brushes designed to meet all operating conditions and requirements of integral horsepower machines.

CAUSES OF POOR COMMUTATOR CONDITION

Frequent visual inspection of commutator surfaces can warn you when any adverse conditions are developing. The chart below may indicate some possible causes of these conditions, suggesting the proper productive maintenance.

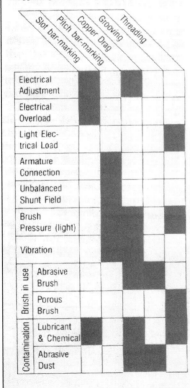

SATISFACTORY COMMUTATOR SURFACES
Commutator Surface Not Damaged

LIGHT TAN FILM over entire commutator surface is one of many normal conditions often seen on a well-functioning machine.

MOTTLED SURFACE with random film pattern is most frequently observed condition of commutators in industry.

SLOT BAR-MARKING, a slightly darker film, appears on bars in a definite pattern related to number of armature conductors per slot.

DARK FILM can appear over entire area of efficient and normal commutator and, if uniform, is quite acceptable.

STREAKING on the commutator surface can occur with long life brushes and is not detrimental if the commutator surface is not damaged.

WATCH FOR THESE DANGER SIGNS
Commutator Surface Damaged

THREADING of commutator with fine lines results when excessive metal transfer occurs. It usually leads to resurfacing of commutator and rapid brush wear. See chart for possible causes.

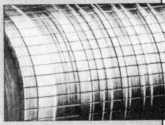

GROOVING is a mechanical condition caused by abrasive material in the brush or atmosphere. If grooves form, start corrective action.

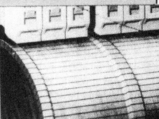

COPPER DRAG, an abnormal buildup of commutator material, forms most often at trailing edge of bar. Condition can cause flashover if not checked.

PITCH BAR BURNING produces low or burned spots on the commutator surface. The number of these markings equals half or all the number of poles on the motor.

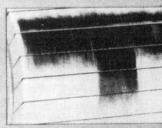

SLOT BAR BURNING can involve etching of trailing edge or commutator bar. Pattern is related to number of conductors per slot.

Carbon Products Operation Inc., 1990 (800) 627-6661

Accessory Technology

Wheel and brake servicing

By John Bakos

The service life of most wheels and brakes has been extended considerably in the last few years. Modern materials see up to three times the life, when compared to older braking materials. But good and frequent maintenance practices are still in order. This frequently overlooked but most important part of the aircraft must be in proper working order at all times.

General

Disc brakes come in different styles: internal, external, multiple disc and single disc. Internal brakes are identified by the brake and discs located inside the wheel with the rotating discs driven by the wheel. Internal brakes can be single or multiple discs. External disc brakes are single disc with the disc bolted to the wheel and the brake or caliper positioned over the disc.

Wheels and brakes are designed based on the weight and speed of the aircraft. When it returns to land, the aircraft's energy of motion must be dissipated to stop the aircraft. This kinetic energy is transformed into heat through friction generated by the brakes.

In order to dissipate the heat generated during the stop, a material with good thermal conductivity, such as high carbon steel, is required. Newer brake designs used on military and commercial aircraft use carbon brakes. Carbon discs are lighter than steel with the ability to withstand high temperatures. However, because this technology is relatively new and expensive, its application is limited.

Brakes are classified as organic or metallic. Organic linings are a resin bonded homogeneous composition while metallic linings have a sintered, powdered metal composition bonded to a steel support plate. Organic linings have traditionally been asbestos based; however, non-asbestos materials are now being used with greater frequency. The new non-asbestos organic brake lining has improved service life by two to three times that of old asbestos based materials.

An important point to obtaining the designed lining service life is to perform the proper lining conditioning (breaking-in) procedure. Consult each wheel and brake manufacturer for the correct procedure for their brakes.

Larger aircraft use internal brakes which are comprised of alternating rotating and stationary discs housed between a cylinder and backplate, supported by a torque tube. The assembly is mounted to the landing gear and fitted inside the wheel. Similar in concept to a bicycle coaster brake, the rotating and stationary sides are squeezed together resulting in the stopping of the aircraft. Lining material used on internal brakes is

Technician installing brake caliper. Be sure not to overtorque bolts.

metallic composition and takes on several designs, from friction material bonded (sintered) to a disc, cups holding friction material which are riveted to a disc, to bare steel discs. Again proper conditioning and use are the keys to obtaining the designed service life of the brake.

Brake linings

Worn linings are the easiest to replace on external brakes, since the aircraft does not have to be jacked up. Backplate attachment bolts are removed, backplates slide out from between disc and wheel, brake cylinder is slide out of the torque plate and pressure plate is removed from the brake. The two types of lining are attached differently.

Organic linings are riveted to the pressure and backplates; metallic linings are held in place by solid head rivets (pins) which fit into holes in the back of the metal support plate. Metallic linings can be pried off with a screwdriver and new linings snapped into place.

Accessory Technology

Nicks or grooves in wheel develop stress concentrations which cause 95 percent of bead seat failures.

A shot of spray adhesive can be used to hold the lining in place until installation on the aircraft. The pins need not be removed unless they're damaged or worn down due to excessive use of worn-out linings.

Rivets attaching organic linings may be drilled out with a 5/32 diameter drill, being careful not to enlarge the attachment holes in backplates and pressure plates. Attachment of new linings is accomplished with new rivets and rivet setting tool, after cleaning of any debris or dirt on the surface of the pressure and backplates. It is important that linings fit flush and tight against the support surfaces.

Because internal brakes must be removed from the aircraft for lining replacement it is generally accomplished with an overhauled/exchange brake which has had new rotating and stationary discs, wear pads, etc. replaced.

Upon installation of new linings, discs or overhauled brakes, lining conditioning must be accomplished prior to flight of the aircraft. Failure to perform a conditioning procedure could result in premature wear of the linings and discs.

In conjunction with lining and disc replacement, other maintenance should be performed to keep the wheels and brakes in top operating condition. New O-rings should be installed if the existing ones show signs of wear or deterioration. O-rings of the proper compound should be used for the type of fluid being used. Red oil and Skydrol-type fluids are incompatible and require different O-ring compounds. Nicks or scratches on pistons or in piston bores should be polished out if possible; if not, the parts should he replaced.

In the case of external brakes, the anchor bolts (pins) that slide in the torque plate bushings, and the bushings themselves, should be cleaned and polished to remove dirt and corrosion. A dry graphite or silicon spray is the recommended lubricant, as wet oil or grease tends to collect dirt, sand, etc. that will interfere with the free sliding motion of the pins in the bushings. Binding of the anchor pins leads to dragging brakes, uneven wear or short brake lining life. Dragging brakes are also caused by overtorquing backplate attachment bolts which results in a crushing of the brake cylinder by the backplate. The backplate follows the depression resulting in a tapered opening for the brake disc. The backplate lining then rubs on the disc causing drag. A depression more than .005 inch deep in the cylinder face (around the bolt holes) requires replacement of the brake cylinder.

The use of a solid brake line can also lead to dragging external brakes and a minimum of 12 inches of flexible brake line is recommended to maintain the free-floating characteristic of the external brake design.

Upon installation of the brake on the aircraft, bleed the brake using a pressure pot to bleed from the bottom up. This is the recommended bleeding procedure. Since air travels up, carefully inspect the installation for proper actuation, full brake release, tire pressure leakage, brake fluid leakage or anything that would cause improper operation.

Wheels

Wheel maintenance is also an important part of the landing equation. Supporting the weight of the aircraft through the tires, the wheels transmit the stopping forces to the tire and ground. At any given time approximately 60 degrees of the wheel circumference is supporting the aircraft. As the aircraft moves along the runway or taxiway, this 60-degree arc moves around the wheel, setting up alternating stress and relaxation cycles in the wheel halves and bolts. Materials used for wheels, as well as brake cylinders, include aluminum and magnesium, both cast and forged.

The bead seat, the area where the tire rests against the wheel, is the most highly stressed area of the wheel. Great care must be used to ensure that no nicks, scratches or gouges are inflicted. These become stress concentrations which soon lead to failure.

Small nicks, scratches or gouges up to a maximum depth of .015 inch in the bead seat may be blended out and polished using a fine (600 grit) sandpaper. Wheel half registers, bearing hubs and bolt bosses may be blended to a maximum depth of .010 inch. In all other areas, the maximum depth that can be blended is .030 inch. Specific repair and refinishing procedures are contained in the applicable maintenance or overhaul manuals.

Do not pry tires away from the wheel flange with screwdrivers, pry bars or other sharp tools. Use a tire bead breaker and press evenly around the wheel to separate the bead adhesion. Cracks in the wheel, particularly around the bolt holes and bead seats, are cause for replacement.

Axle nut torque is critical to wheel half and wheel bearing life. Too tight and the wheel won't rotate freely, too loose and the wheel wobbles. Either case will ruin the bearings and wheel. The bearing manufacturers don't specify axle nut torque values, so the best thing is to consult the aircraft manufacturer's manuals.

Bolts, nuts and washers used on wheels and brakes should be inspected prior to using again. Inspect bolts for thread wear, thread damage, corrosion or cracks that may have developed at the shank where it intersects the bolt head. Self-locking nuts should be checked for thread wear, damage and integrity of the locking feature. You want to be sure the nut doesn't back off in service. Check washers for cracks, damage or other items that would render them unserviceable.

Bolts should be installed using the crisscross method, first bringing torque up to about half of the required value. Repeat the procedure to torque them to the final value. Never use an impact wrench on any item of hardware where torque is critical. These wrenches produce load spikes that cause stress greater than the bolts and nuts were designed for. Use a recently calibrated hand torque wrench.

Wheel bearings

Wheel bearings, when properly maintained, will provide many trouble-free flights. Suggested maintenance intervals vary depending on the manufacturers. The recommended repacking frequency, according to Cleveland Wheels & Brakes, is 500 roll miles. It is also a good idea to inspect them at every tire change or annual inspection.

Bearings may be cleaned with solvent to remove grease, dirt, etc. Soaking them will soften the grease and any hardened deposits. A soft bristle brush may be used to remove the residue. Bearings may be dried with low pressure-dry compressed air by blowing parallel to the rollers. Never blow across the rollers, causing them to spin. Also, bearings should never be cleaned with steam, as the heat and excess oxygen will damage the surfaces.

Inspection of bearings is essentially a visual inspection. Discoloration of the metal due to overheating, lubrication breakdown, moisture contamination or incorrect grease will show up in different colors (black, blue, yellow, brown), Brinelling or spalling, or a chipping or indentation of the bearing cup or cone. If any of the above conditions exist, the bearings should be replaced.

Repacking bearings is a simple, if somewhat messy task. An approved grease should be used. Place a golf ball-sized amount in the palm end of the cone through the grease thus forcing it between the rollers. Continue this process working around the bearing until the grease is forced all the way through the rollers.

The bearing cone is now ready for installation in the wheel. If you are not immediately installing the bearing, wrap it in wax paper or store it in a sanitary location so the grease does not become contaminated.

Tires

Tire pressure is also critical for wheel as well as tire life. Follow the tire manufacturer's recommendations for inflation of tires particular to your aircraft. Variation as small as 10 percent of inflation pressure can have significant impact on wheel life. Wheels are designed to the Tire and Rim Association specifications for location and amount of load applied. One of the factors in wheel design is tire inflation which, if out of tolerance, changes the location of the applied load, thus altering the stresses the wheel must endure. *May/June 1990*

John Bakos is aftermarket sales manager for Parker Hannifin, Cleveland Wheels & Brakes.

Accessory Technology

Aging props prompt closer attention to maintenance

By Greg Napert

EAST HADDAM, CT—Pride. That's the best way to describe the feeling that technicians at New England Propeller have for their work. There is a sense that the quality of life in this small New England community transcends into the work that is performed in this full-service propeller overhaul shop.

Attention to detail and communications with customers about improvements in maintenance practices, based on what the technicians see in the shop, are the result of this pride.

But there is also a feeling of concern that is sensed as Arthur D'Onofrio Jr., president and owner of N.E.P., explains the recent onslaught of propeller failures. Many problems, D'Onofrio explains, are simply the result of propellers that have been in service too long. D'Onofrio also puts some of the blame on the manufacturers, stating that propeller designs have not significantly changed over the past 20 or so years, resulting in old technology still being used in propeller design. "We have learned a lot over the years that should be applied to new propeller design," says D'Onofrio.

D'Onofrio also points to the fact that maintenance practices can have a direct impact on the condition of propellers. Aging propellers, says D'Onofrio, combined with designs that are not quite as good as they could be, are making it even more important than in the past to pay strict attention to maintenance practices.

For example, explains Courtney Clark, vice president of operations and product support, practices that were once accepted, such as using blade paddles to move blades in and out of feather, should be discontinued. Also, says Clark, pulling or pushing the aircraft around by the propeller should not be allowed. Doing so places unusual loads on the propeller. This could damage the blade sockets and bearings, and could affect blade tracking. It is important to remember that when a prop is at rest, it is much weaker than when in motion. Centrifugal forces of 70,000 pounds and greater that seat the blades and add stiffness give the propeller its strength while in motion, says Clark.

According to Clark, blade filing, in many cases, is not done properly. "AC 43.13-1A gives good instructions on how to file blades," says Clark, but many times they are not followed. "You have to file all of the damage out of the blade in order for it to be effective." If any portion of the nick is left in the blade, the stressed area has not been removed, allowing for the possibility of failure in that area. Clark claims that he has seen cases where the nicks have either been burnished or lightly filed and this is not sufficient to remove the damage.

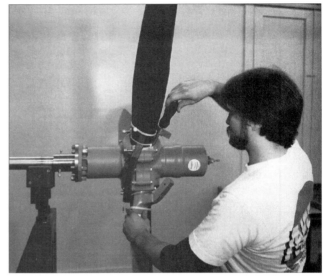

Propeller is statically balanced at the overhaul facility and should not be changed during the life of the prop.

Restoring the proper contour to the blade is also important, says Clark. Improperly contoured blade tips and blunt leading edges are common errors. Also, says Clark, "If you remove metal from one blade, the other blade should have an equal amount removed so that the assembly is kept in balance."

Greasing propeller hubs is another area that is often abused. Mixing incompatible greases can destroy O-rings and seals, and not thoroughly purging the grease from the hub can allow moisture to remain, resulting in imbalance. Many technicians simply pump one or two pumps of grease into each prop. This can result in excessive pressure, damaged seals, cause imbalance and doesn't purge the old grease from the prop. To properly grease the hub, says Clark, remove one zerk fitting from the hub and position the hole toward the ground. Be sure that the proper grease is selected, and pump grease into the top zerk fitting until the grease that is running from the bottom hole is clean. Use caution, Clark says, when reinserting the zerk fitting not to use excessive torque. The threads, after repeated use, can be easily stripped out.

"Corrosion, if left unattended, can often scrap a prop," explains Clark. Removing any corrosion as soon as it begins by sanding the affected area and anodizing or treating as recommended can extend the life of the propeller indefinitely.

Dynamic balancing has come a long way in recent years, and affordable equipment is available from a host of manufacturers. "There is no question that dynamic balancing is a good thing. Eliminating vibration can definitely add to the safety of an airplane. Propeller manufacturers are currently working on approving procedures for dynamic balancing," says Clark. Because of the many different types of installations, it is impossible to write one generic procedure that would be applicable for all. The important thing is that it be done correctly.

A big mistake, says Clark, when balancing a propeller is to remove weights that were used to statically balance the propeller in the factory. Removing the static weights to attain dynamic balance may appear to work when the propeller is not running at cruise speed, but severe imbalance can result when aerodynamic forces at different speeds change the characteristics of the propeller.

Also, by removing static weights, you have destroyed the balance if the prop is to be used on another engine or airplane.

Where to add weight to achieve dynamic balance poses the biggest problem, says Clark. Without knowing whether or not it is OK to add weights to bulkheads, prop spinners or other areas of the prop and what effect these weights will have on structural integrity or performance characteristics, it is imperative that the technician be properly trained before attempting to dynamically balance the prop. (Any approval questions should be directed to your local FAA office.)

Clark stresses the importance of overhaul at TBO. Many Part 91 operators, says Clark, tend to let the prop go beyond TBO. "This is a horrendous injustice. They are not forced by law to overhaul their props.

"The A&P is the one who must recommend and push the owner to overhaul when the prop comes due." Even though the prop may appear to be in good condition, corrosion, wear and damage such as cracks may be present inside the hub. Disassembling the prop and inspecting it internally is the only way to verify the integrity of the propeller.

Accessory Technology

McCauley meeting produces common questions and good answers

The following are excerpts from a McCauley memorandum dated November 14, 1989 and distributed to all McCauley service stations from Chuck Rocco, manager of marketing and product support. The questions were generated at the McCauley 1989 Worldwide Service Station meeting which was held in September 1989.

Q. *Is there any McCauley approved dynamic balance information for reciprocating engine equipped aircraft?*
A. As the bulkheads of many recip aircraft are not made by McCauley, specific balance weight limitations, mass placement, etc., cannot be determined until each individual bulkhead is analyzed. This analysis is currently underway and at its completion appropriate service information will be released.

Q. *How do you feel about changing static balance of the propeller?*
A. It is important to maintain the static balance of the propeller as supplied from the factory or from propeller shops for several reasons. It provides a good initial propeller balance to initiate dynamic balance of the powerplant; therefore, the propeller is not matched to the installation and can be changed, i.e., on a twin engine from the left side to the right side, without major impact to the balance. Furthermore, it is difficult, if not impossible, to obtain the full-blade angle range during static run-ups when performing dynamic balance. Few people actually conduct a dynamic balance with flight tests. By adjusting the static balance weights on the propeller, you may inadvertently make it acceptable at low-blade angle ranges and detrimental at high angles. We therefore feel it is in the best interest to keep the process as simple and uniform as possible and to not change the static balance of the propeller.

Q. *Does McCauley intend to allow the reuse of propeller blades showing signs of direct lightning strike?*
A. In the past there appeared to be few propellers affected by lightning strike. Operators are now requesting that we look into this. McCauley would like to cautiously explore its feasibility... at present, any McCauley blade showing signs of direct lightning strike is to be scrapped. It is difficult to locate entry and exit points from lightning strike or potential electrical shortages.

Q. *Why not drill and tap fixed-pitch propellers?*
A. Balancing of fixed-pitch propellers without hub balance weights by grinding along the blade hub is not recommended. We are evaluating the possibility of adding balance weights to these models.

Q. *Is McCauley going to establish a TBO period for fixed-pitch propellers?*
A. The issue of fixed-pitch TBO limitations is currently under consideration. However, no conclusion has been reached as yet.

Q. *Does McCauley endorse the use of shot to clean out corrosion?*
A. We do not allow removal of corrosion on hubs.

Q. *What is your feeling on removing paint by plastic media?*
A. Although still under consideration, McCauley's initial reaction to the use of plastic is a positive one. From all evidence we have seen thus far, plastic media is acceptable for paint removal on McCauley parts.

Q. *I have a lot of oil leaks that I can't find. Everyone tries to blame the propellers. How do I find where the leaks are coming from?*
A. If propeller leakage is suspected, but the source of leakage is not readily apparent, before removing the propeller from the aircraft:

1. Remove the spinner.
2. Wipe clean all propeller, flange and spinner bulkhead parts.
3. Use white "Dye-check" or equivalent and coat the hub and blade shank areas. Do not attempt engine run-up without spinner shell installed unless spinner bulkhead fillets and deice leads have been removed.
4. After solution dries, reinstall spinner and run engine for at least five minutes.
5. Shut down engine and examine coated surfaces. The sources of any leakage will show as a stain on the coated surfaces.
6. If it is definitely established that propeller is leaking, remove it and mark so that the proper inspection can be made during disassembly.

March/April 1990

Accessory Technology

Starter systems
Simple truths

By Greg Napert

The starter circuit

It's a common error to assume that when a starter doesn't turn over, there's a problem with the starter itself, explains Kevin Scully, owner of SaFlite Accessories in Oshkosh, WI. Scully says that it's important to think of the starter in terms of a "starting circuit."

The starter circuit includes the battery, solenoid starting switch, manual starting switch, starting motor, the connecting wires and the frame of the aircraft. "Actually, there is potential for trouble at any one of those points," Scully says.

The first step in trouble-shooting should be to check all of the wiring, including the battery cables, to make sure that all connections are good and that the terminals are not corroded.

Scully also suggests checking the specific gravity of the electrolyte in the battery. He has occasionally found instances where technicians have serviced the aircraft battery with automotive electrolyte.

"They have a specific gravity difference," Scully says. "The fact is that the aircraft battery design is slightly different than automotive because they have to get more power at less weight. If you service an aircraft battery with automotive electrolyte, you've reduced the capacity." This can cause the starter to turn slowly, or not at all, while everything else appears to be working properly.

Scully cautions that a number of things can be wrong with the system at one time. He points out one instance where a customer came to him with a bad starter, voltage drop across the solenoid and automotive electrolyte in the battery. He says the customer took the aircraft to other shops but that the problem was never completely solved. After correcting all of these problems, Scully said that the starting system was like new and the customer was happy.

Problem areas

One of the more prevalent problems, Scully says, can appear with the starter solenoid. After many cycles of operation, the contacts within the solenoid have a tendency to burn. This burning causes a voltage drop that is significant enough to prevent the starter from turning. An easy way to check this is to disconnect the starter from the circuit, activate the relay and check the voltage on each side of the relay. According to Scully, if there is "more than a couple of tenths of a voltage drop," then there is a problem with the relay and it

Kevin Scully, owner of SaFlite Accessories in Oshkosh, WI, removes the Bendix assembly from a starter.

should be replaced. Burned contacts can also cause the solenoid to stick, which will lead to a burned-out starter motor (see Starter warning light).

Bad key switches are another problem area. Because of the regular wear and tear, they have a tendency to either stop working or begin sticking. A sticky switch can quickly lead to a burned-out starter motor, and there is usually no way to tell that the motor is still activated

Accessory Technology

during the flight. But when it's time to start the aircraft again, it's found that the starter motor is burned out and needs to be replaced.

Installation pointers

One of the most common starter installation errors is overtightening the field terminal stud. According to Scully, "The field terminal on most starters is a copper stud, and frequently I've seen this stud stripped out or damaged." On many older Delco-Remy starters, Scully explains, the field terminal stud doesn't have any provision to keep the stud from turning in the housing.

If the stud is allowed to rotate, internal damage of the field coils can result. "I recommend applying Loctite 242 (or equivalent) and using a split washer," Scully says. "Tighten the connection until the washer just becomes flat and no further."

Newer starters, however, incorporate a machined stud that fits into the housing so that it cannot rotate.

Continental uses a starter that is quite different from the others. The Bendix assembly does not exist. Instead, Continental incorporates a clutch assembly in the accessory section of the engine which serves the same purpose as the Bendix. As a result, the starter is mounted directly on the accessory case. When installing this type of starter, it is important, for the purpose of preventing oil leaks, to keep the mounting flanges clean and to assure that there are no burrs or old gasket material left on the flange.

Bendix blues

Due to the positioning of the Bendix on many starter installations, the arrangement is very susceptible to dirt, water and contamination. A sticky Bendix can result in failure of the starter to engage, or worse, failure of the starter to disengage after it has been activated. Scully claims that some repair shops recommend cleaning and lubricating the Bendix every 50 hours with silicone spray (Lycoming service instruction 1278 recommends cleaning with Varsol or equivalent and spraying with silicone at every 100-hour inspection). He suggests that it need not be done that frequently.

"I use the recommended lubricant when I assemble the Bendix onto the shaft during overhaul," Scully says, "so I don't recommend all that washing and spraying with silicone. If the airplane is utilized, the Bendix will work. I've not been able to get a sticky one to work with any reliability by just spraying it with silicone. I've tried that route and it worked for a short time, then started acting up again."

Scully suggests that the Bendix is relatively simple to remove, clean and replace. And he also suggests that if it's time for the starter to be overhauled, the Bendix should be replaced.

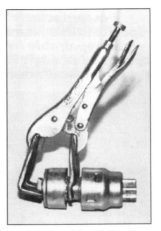

Modified welding pliers that Scully uses to compress the Bendix assembly.

There are really no special tools required to replace the Bendix assembly explains Scully. However, Prestolite models EBB 131 and EBB124 incorporate a steel roll pin that holds the Bendix onto the shaft. This requires compression of the Bendix in order to remove the pin. Scully has modified a pair of welder's vice-grips by filing down the inside of the fingers that slide over the center of the Bendix (see photo) to the point where they slide without resistance over the Bendix. He can then compress the Bendix, which frees him to extend the pin collar (using a screwdriver) and remove the pin. Scully cautions, however, to use just enough force to overcome the pressure of the spring. Excessive force with this tool can damage the Bendix.

After the shaft is removed from the Bendix assembly, it is important that it's cleaned and properly inspected. Scully says that prior to reassembly, crocus cloth should be used to polish the shaft as smooth as possible so that "the Bendix doesn't encounter any resistance." He suggests using a thin film of Lubriplate 777 or equivalent to lubricate the shaft upon reassembly. Scully also recommends that the appropriate overhaul manual be followed when removing and installing the Bendix assembly.

For liability's sake

Scully suggests that more and more shops are turning their starters into an approved accessory overhaul or repair facility due to "technician's liability and customer satisfaction." He explains that because these accessory shops specialize in starter repair and overhaul, they are more familiar with the starter and are more likely to perform a dependable repair.

"If he (the technician) is dealing with someone who's reputable, and most of them are, he's got a good product and can put it on with confidence," Scully says. "He's installed something that's yellow tagged, and his liability is limited to making sure the installation is correct. And his customer is happy because he's got a reliable product that's right the first time."

Cold weather operation

With winter promising to stick around for a few more months (in a large portion of the country), it's wise to keep a few cold weather operating tips in your back pocket.

Lycoming brings you a few pointers that are sure to keep you in good standing with the boss.

It is important to remember that these tips apply to temperatures above -25°F.

Preheat

At ground temperatures below 10°F, the engine and oil system should be preheated. This does not mean that the engine will not start below 10°F, but due to the lack of circulation of the oil, it is possible that engine damage will occur. Scored cylinders, scuffed piston skirts and broken piston rings are typical damage from very cold weather starts without preheating. Keep in mind that the application of heat to the cylinder area only does not guarantee that the entire oil system is properly heated. *On engines that are equipped with oil coolers, be sure to heat the oil cooler as well as the oil sump.*

Pull the prop through

Rotation of the propeller by hand provides an indication that the engine is free to turn and can add lubrication to areas that are otherwise dry. Make sure that the mag switches are in the OFF position and pull the propeller through several times. Always assume the mags are hot and stand clear of the prop.

Don't overwork the starter

If an engine requires an excessive amount of cranking, be cautious not to run the starter for more than 30 seconds at a time, with one-minute cooling periods between attempts. Pushing the starter beyond this could result in a burned-out motor.

Watch for oil pressure

Immediately upon starting the engine, check for an indication of oil pressure. Remember that different model aircraft exhibit different characteristics. On most single engine aircraft an almost immediate response is noted. On twin-engine aircraft the engines are located considerably farther from the oil gauges and as a result, the response may be much slower. On some twins, the oil pressure may go up and during warm-up may drop again for a short period of time, then rise back to normal. Check with someone who is familiar with the operation of the aircraft prior to starting the engines.

Keep the RPM up

After start, do not idle engine below 1,000 rpm. It's not good practice to idle engines below 1,000 rpm at any time (there are some exceptions to this rule, such as the Piper Pressurized Navajo). This is particularly true during cold weather operation to prevent lead fouling of spark plugs. *January/February 1990*

Accessory Technology

Vacuum pumps
New technology has made them a very delicate device

HARTLAND, WI—Vacuum pumps are simple devices. But simple devices can often get you into trouble if taken half-heartedly.

"We have torn down a lot of pumps and we see what makes them fail," says Michael White, president of RAPCO Inc.

In fact, RAPCO has torn down thousands of Airborne's 200/400 Series vacuum pumps and claims that it continuously sees pumps that have been mishandled, mainly, because of the unfamiliarity with the new technology.

White explains that much of the misunderstanding stems from the fact that the old style vacuum pumps were built like tanks. Most technicians never get to see the inside of the pump and don't understand the changes that have taken place over the years.

Notice that the old style pump (see photo) had a heavy-duty housing that could be placed in a vice when fittings were removed or installed. The new pumps (by contrast) incorporate a housing that has a much thinner and more delicate wall. Not to mention the fact that the rotor is closely fitted to the housing (.002 to .003 clearance). Placing the new style pump in a vice incorrectly could crack the housing or damage the rotor.

The principal materials that are used on the new style pumps are aluminum (the housing), carbon (the rotor) and graphite (the vanes and bushings).

The aluminum and carbon components are unaffected by the oil, water and solvent that may enter the system. But the graphite components are susceptible to damage. Solvent and oil can cause the graphite to swell. Oil-soaked bushings can cause the drive shaft to freeze up and shear. And oil-soaked vanes can swell so they become lodged in the rotor, rendering the pump useless.

Dave Schaefer, plant supervisor at RAPCO, explains that the phenomenon known as "intermittent pump" is a result of oil-soaked vanes. When the engine is operated at low speed, the vanes are stuck in the rotor and won't pump air. At high rpms, however, the vanes break loose and begin to work properly.

Old style pumps had a much longer life due to the fact that they incorporated a lubrication system that bathed the components in oil at all times. Also, the vanes were considerably thicker and wore much more gradually.

Today's pumps have much thinner vanes. "It's like lead in a pencil," says White. The vanes wear down to the point where they break off and cause the pump to fail.

Airborne's new style vacuum pump (left) vs. old style vacuum pump (right). Notice the differences in the thickness of vanes and housings between old and new models.

According to White, "All pumps are destined to fail." But knowing the inner workings of the vacuum pump (as with any other component) can lead to smarter maintenance and extended pump life.

Tips from Airborne for installation of vacuum pumps

- Never install a pump that has been dropped.
- Clamp pump on mounting flange surface only. Clamping the housing will cause damage to the rotor or housing.
- Change all filters in the system. Failing to do so could void the warranty on the pump.
- Spray the fittings with silicone spray. Never use Teflon tape, pipe dope or thread lube.
- Never tighten fittings on the housing more than one and one-half turns beyond hand tight.
- Always replace all lockwashers when installing a new pump.
- Clean the pump inlet line thoroughly.
- Blow out all hoses using high pressure air before installing.
- Replace all old, hard, cracked or brittle hoses. Sections of the inner liners may separate causing pump failure.
- Use silicone spray on the hoses to slide them straight onto the fittings. Wiggling the hose from side to side can cause particles of hose to be cut from the inside diameter of the hose and damage the pump.

- Double check the routing of the hoses. Improper routing could cause damage to the gyro system.

To overhaul or not to overhaul—that is the question

The differing opinions regarding the effectiveness of vacuum pump overhaul exist. Parker Hannifin Corp., Airborne claims that its vacuum pumps cannot be overhauled. In fact, it has been placing a warning not to overhaul the pump directly on the pump body.

Syd Reames, aftermarket accessories manager for Airborne, explains that they do not believe that the overhauled pumps will see half of the life of a new Airborne vacuum pump.

Reames claims that the standards for overhaul of its pumps do not meet its stringent manufacturing guidelines. He recommends that, for the safety of everyone involved, the cores be returned directly to Airborne as exchange for a new pump.

On the contrary, Michael White, president of RAPCO Inc., claims that his company's standards for overhaul meet or beat the standards that were adhered to during original manufacture.

White says the RAPCO uses a new Airborne pump as the reference point when testing each one of its overhauled pumps. He also claims that because of the improved coating that is used on the liner, the overhauls will actually outlast Airborne's originals based on warranty claim information that has been compiled.

He points to the fact that RAPCO operates under FAR Part 145.33(c) as a limited repair station and that its overhaul process is approved by the administrator.

What's a technician to do? The roughly 15 to 25 percent savings realized by having the pumps overhauled is certainly enticing. And the safety concerns presented by Airborne are definitely valid. But in the end the decision to overhaul or not to overhaul is up to you.

Slick's new magneto timing procedure is truly "slick"

Forget about conventional internal timing procedures for Slick magnetos. Slick has introduced a couple of tools that greatly simplify the procedure.

Its new "E" gap gauge (tool No. T-150) is a precisely machined .062-inch feeler gauge. The two ends of the gauge are machined differently to allow use with the older style rotors (no slots on the magnet head) and with the new CNC style rotors (slots on the magnet head). The base (tool No. T-100) holds the magneto in position during the timing procedure.

To set the timing, place the magneto on the T-100 base and position the magneto so that the coil is at 12 o'clock.

Old style rotors... insert the flat end of the "E" gap gauge between the raised laminations on the rotor magneto head and the laminations in the frame. Place the gauge against the right lamination for right-hand rotation magnetos and against the left lamination for left-hand rotation magnetos. Turn the rotor in the direction of the magnetos rotation until the laminations on the rotor head contact and hold the "E" gap gauge in place.

New style rotors... insert the notched end of the "E" gap gauge into the appropriate "L" or "R" timing slot in the rotor magneto head. Use the "R" slot for right-hand rotation magnetos and the "L" slot for left-hand rotation magnetos. Turn the rotor in the direction of magneto rotation to hold the "E" gap gauge against the laminations in the frame.

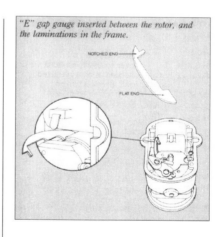

"E" gap gauge inserted between the rotor, and the laminations in the frame.

To proceed, connect a timing light to the points and adjust them until they are just opening. Tighten the adjusting screws (torque the pivot screw to 15 to 18 inch-pounds and the adjusting screw to 18 to 20 inch-pounds).

For Retard Breaker Magnetos, the secondary points will now have to be set. For all others, the timing is now complete.

A very important point to consider when working on magnetos, says Harry L. Fenton, technical representative at Slick, is that there are continuous design changes that take place with new magnetos as well as existing magnetos. "It is nearly impossible for the customers to keep up to date unless they subscribe to Slick's Master Service Manual," Fenton says.

Slick's F-1100 Master Service Manual contains maintenance and application information that pertains to most of its magnetos. The cost to subscribe is $30 for the initial subscription and $20 a year for revisions.

Accessory Technology

Hot brakes or cool?

The controversy continues! Whether or not to use steel, chrome or stainless steel for your brake discs.

What is the purpose of a brake disc? To dissipate the heat generated during braking of a vehicle. How is this accomplished? Through radiation and convection. During the braking process, friction between the brake linings and disc creates heat, which must be removed. Initially, the disc surface gets hot and during the stop, heat (or temperature increase) travels through the disc, thus equalizing material temperature.

By accomplishing this heat transfer quickly, high surface temperatures, which cause brake fade, are minimized and the coefficient of friction required to produce required stopping performance is realized. High surface temperatures reduce coefficient of friction. That results in brake fade, which requires higher brake pedal effort to stop. Also, brake linings exposed to high temperature will experience rapid wear and degradation of the bonding material (resins).

Thermal conductivity (k) measured in Btu/hour foot °F, is the ability of a material to transfer heat. Higher thermal conductivity values mean a faster heat transfer, which in the case of brake discs, results in lower surface temperatures. Lower thermal conductivity values mean higher surface temperatures due to a slower heat transfer. Higher temperature at the lining/disc interface means a reduced coefficient of friction, brake fade and degradation of linings.

Reviewing the various materials used for brake discs shows cast iron, steel and stainless steel to be the primary choices. Cast iron and steel show thermal conductivity values from 32 to 24 in the temperature range of 32° to 750°F, while stainless ranges from 13 to 10 over the same temperature range—clearly a distinct difference in the ability to transfer or dissipate heat.

Another factor that needs consideration in the discussion of brakes is the service life of brake linings. Just as high temperatures reduce lining life, rough brake discs do as well. We all know that bare steel subjected to the atmosphere corrodes. If you were to look at the discs on your automobile after it sat a few days, corro-

Brake disc comparison. From left to right—standard, chrome and stainless.

sion would appear. However, since it takes some effort to do this, we tend to ignore them. On light aircraft the brake discs are visible and tend to attract attention when turning brown. Left in this condition, the light surface corrosion soon turns to pits, which due to their roughness, causes rapid lining wear. Frequent use of the brakes cleans the surface corrosion and keeps the discs clean. Not every aircraft is fortunate to be flown frequently enough.

Bare steel offers little protection against the elements. Alternatives include chrome plating the steel disc or using stainless.

Stainless, as we know, takes a long time to corrode. However, its lower thermal conductivity means that higher surface temperatures, brake fade and reduced lining wear result.

Chrome plating offers corrosion resistance when applied to bare steel with the advantage of steel's thermal conductivity. Eventually the plating does wear off, and you then have a bare steel disc, which can be worn to its limits.

Which is best? Depends on what you want a brake disc to do! Look pretty or work.

November/December 1989

Reader Response

Dear Aviation Maintenance Technician:

Thank you for choosing this book from *The Best of Aircraft Maintenance Technology Magazine Series*. We hope you are pleased with your selection. Your input is invaluable to us. Please take a moment to provide us with your comments and suggestions.

— *Aircraft Maintenance Technology Magazine*

Please print clearly.

Name _____ Date _____

Title *(Student, Instructor, other)*_____

Business/School _____

Address _____

City_____ State _____ Zip code/Postal code _____

Country _____ Telephone *(optional)* _____

Where did you purchase this Best of AMT **Airframe/Accessory Techonology** text? _____

Was the text recommended to you? _____ By whom?_____

Do you intend to purchase the two additional texts in the series? Professional/Legal _____

Recip/Turbine Technology _____

Comments and Suggestions:
Please tell us what you liked or disliked about the text: content, subject matter, ease-of-use, illustrations and figures, etc.

_____
_____
_____
_____
_____
_____
_____

Mail this form to:

AMT Magazine
Attn: Greg Napert
1233 Janesville Avenue
Fort Atkinson, Wisconsin 53538

Please photocopy or remove this page.